Electron Microscopy of Interfaces in Metals and Alloys

ELECTRON MICROSCOPY IN MATERIALS SCIENCE SERIES

Series Editors

B Cantor and **M J Goringe**

Forthcoming books in the series

The Measurement of Grain Boundary Geometry
V Randle

Electron Microscopy of Interfaces in Epitaxial Layers and Artificially Layered Materials
A Petford-Long and J Hutchison

Electron Microscopy of Quasicrystals
K Chattopadhyay and S Ranganathan

Electron Crystallography of Organic Compounds
J R Fryer and D L Dorset

ELECTRON MICROSCOPY IN MATERIALS SCIENCE SERIES

Electron Microscopy of Interfaces in Metals and Alloys

C T Forwood and L M Clarebrough

Division of Materials Science and Technology, CSIRO, Australia

Adam Hilger, Bristol, Philadelphia and New York

British Library Cataloguing-in-Publication Data

Forwood, C. T.
Electron microscopy of interfaces in metals and alloys.—(Electron microscopy in materials science)
I. Title II. Clarebrough, L. M. III. Series
502.8

ISBN 0-7503-0116-3

Library of Congress Cataloging-in-Publication Data

Forwood, C. T.
Electron microscopy of interfaces in metals and alloys / C. T. Forwood and L. M. Clarebrough.
p. cm.—(Electron microscopy in materials science series)
Includes bibliographical references and index.
ISBN 0-7503-0116-3
1. Surfaces (Physics)—Technique. 2. Electron microscopy. 3. Crystallography. 4. Transmission electron microscopes.
I. Clarebrough, L. M. II. Title. III. Series.
QC173.4.S94F67 1991
620.1′699—dc20 90-29110
CIP

Published under the Adam Hilger imprint by IOP Publishing Ltd
Techno House, Redcliffe Way, Bristol BS1 6NX, England
335 East 45th Street, New York, NY 10017-3483, USA

US Editorial Office: 1411 Walnut Street, Philadelphia, PA 19102

Typeset by P&R Typesetters Ltd, Salisbury, England
Printed in Great Britain by J W Arrowsmith Ltd, Bristol

Contents

Foreword

Materials science has evolved as a crucial engineering discipline during the last 20 years. The basic approach of the materials scientist is to investigate the microstructure of a material, so as to optimise its manufacturing process and resulting properties for subsequent engineering use. Electron microscopy has proved to be by far the most powerful technique for examining and understanding material microstructures, and electron microscopy methods are essential for developing new engineering materials of all types. The objective of this series of monographs is to provide overviews of the impact of electron microscopy in different branches of materials science. The series is designed to be broadly based across the materials spectrum (including metals, ceramics, polymers and semiconductors) and to deal with the full range of available electron microscope techniques, such as electron diffraction, lattice imaging, scanning electron microscopy and microprobe analysis. In general, individual monographs will concentrate on a particular type of material or a particular problem in materials science, and will review the use of electron microscope techniques to characterise and understand the relevant material microstructures. The series is intended to be of interest to a wide variety of academic and industrial research scientists and engineers.

In this book, the first of the series, Drs Forwood and Clarebrough have achieved these aims in an authoritative and comprehensive manner. They cover all aspects of the use of the transmission electron microscope in the study of grain boundaries and interphase interfaces in metals and alloys although, of course, the rationale of the approach may be extended to all material systems. We hope that the book will act as a handbook for practising electron microscopists studying grain boundaries and interphase interfaces and will help all who read it to appreciate the power of the transmission electron microscope in that field.

B Cantor and **M J Goringe**

Acknowledgments

We thank all those who have helped in the preparation of this book. We are grateful to R W Balluffi, R Bonnet, F Cosandey, P J Goodhew, J M Howe, K M Knowles, H Mykura, J M Penisson and A P Sutton for permission to reproduce electron micrographs, diagrams and tables. We thank all our colleagues in the CSIRO Division of Materials Science and Technology for their encouragement and in particular G Roderick for line drawings, R Lamb and M Fergus for photographic work and R Wijewardena for typing the manuscript.

C T Forwood and **L M Clarebrough**

Introduction

Advances in electron microscopy are now enabling the defect structure of interfaces in metals and alloys to be studied experimentally in a quantitative manner. Quantitative experiments are being carried out as a result of various factors. One of these is the ability to identify positively the Burgers vectors of interfacial dislocations using their contrast in two-beam images. A second factor is the use of geometric analysis of arrays of interfacial dislocations to determine the Burgers vectors of individual dislocations in these arrays in situations where the Burgers vectors cannot be identified from two-beam image contrast. Another factor is that the advances in the design of electron microscopes have not only facilitated quantitative two-beam microscopy, but have also recently allowed atomic resolution in *n*-beam images of metals and alloys. In addition, the electron microscopy of specially prepared interfaces (particularly grain boundaries) has played a very significant role in the identification of the nature of interfacial defect structures. The description of quantitative experiments which make use of these advances and the interpretation of these experiments form the major part of this book.

The technique used for positive identification of the Burgers vectors of interfacial dislocations from their double two-beam images is a development of the 'image matching' procedure of Head *et al* (1973) which involves comparing experimental and computed electron micrographs. This development using computed theoretical electron micrographs of interfacial dislocations is described, but the listing of the computer program PCGBD (Personal Computer program for Grain Boundary Dislocations) which is used for the computations in this book is not included because the program is readily available on floppy disc, for use in personal computers, from the CSIRO Division of Materials Science and Technology.

The positive identification of the defect structure of interfaces requires that electron microscopy is done in such a way that all the information necessary to determine the crystallography of the interface and the defects in it is collected, and the methods used to collect the data and determine the crystallography are described. Worked examples of the analysis of the interfacial structure of various grain boundaries and interphase interfaces, using image contrast and/or the geometry of the interfacial structure, are presented in detail so as to demonstrate each step in the procedure.

The approach adopted in writing this book reflects the directions of the authors' own research and does not cover all aspects of work dealing with the electron microscopy of interfaces in metals and alloys, but rather concentrates on the dislocation structure of interfaces. The subject matter of the book is confined to metals and alloys so that work on silicon and germanium, although referred to, is not reviewed in any detail. Furthermore, although computer modelling of the atomic structure of interfaces has advanced in parallel with advances in electron microscopy, and is discussed in cases where it relates to experimental work, all such computations are not reviewed.

A common theme that emerges in the book is the applicability of the coincident site lattice model to the structure of high-angle grain boundaries and the applicability of its extension in the form of a constrained coincidence model to interphase interfaces. The experimental results presented suggest that these models provide a good description of the structure of high-angle grain boundaries and some interphase interfaces. However, these models cannot be used to predict the details of the interfacial dislocation structure for a particular orientation relationship and interface plane, as they are based solely on geometric considerations and take no account of the energy of the system. Thus, experimental investigation is essential to the determination of interface structure, for any given case of interest, and the emphasis in this book is on experimental, rather than theoretical, investigations.

1

Dislocation Theory of Interfaces

1.1 INTRODUCTION

The internal interfaces in metals and alloys are those between crystals of different orientation (grain boundaries) and those between crystals of different structure (interphase interfaces). The structure of such interfaces, particularly grain boundaries, has been an exciting subject of research for over a century. Briefly, this research has shown that an interface has a localised structure of its own, which is only a few atomic layers thick, and in which the atoms occupy sites that are displaced from normal lattice sites in the two adjoining crystals. These atomic displacements accommodate the different orientations of the neighbouring crystals and, in general, give rise to regions in the interface where there is 'good' atomic fit separated by localised regions of 'bad' atomic fit. In most cases, the atomic displacements are associated with interfacial dislocations, the cores of which form the localised regions of bad atomic fit.

The earliest ideas concerning the structure of grain boundaries were that the grains in a polycrystal were held together by some sort of amorphous grain boundary cement, and the formation and development of these early ideas are described in a very interesting review by King and Chalmers (1949). The amorphous cement model of grain boundaries lasted, with minor modifications, until the international conference at Bristol in 1939 on Internal Strains in Solids. At that meeting Bragg (1940) put forward a major argument against the existence of an amorphous phase by pointing out that the high atomic mobility, which had been demonstrated in metals at relatively low temperatures, would preclude the formation of such a phase. At the same meeting, Burgers (1940) described a model for the structure of boundaries between crystallites formed by plastic deformation, involving a periodic array of dislocations. This dislocation model, which is illustrated schematically in figure 1.1, was the first demonstration that the elastic displacement fields associated with dislocations could accommodate the misorientation between two grains. The boundary in figure 1.1 is a symmetrical low-angle tilt boundary in a simple cubic lattice with lattice parameter b and consists of edge dislocations spaced equally at a distance d apart. The Burgers vector of each dislocation has a magnitude b and the misorientation between the grains is given by

$$\theta = b/d$$

for small values of θ. The accommodation of a misorientation by dislocations was particularly appealing at that time as, in the model, the atomic displacements are localised at the boundary, and distant from the boundary the grains are tilted with respect to one another and unstrained. Such a dislocation model would lose its physical meaning for high-angle grain boundaries, say θ equal to 15° or more, because then the dislocations would become so closely spaced as to lose their characteristic elastic displacement fields. For example, for $\theta = 15°$ the spacing of the dislocations in figure 1.1 would be just less than $4b$. However, it will be seen later in Chapter 5 that the concepts of a dislocation model can still be applied, but in a more sophisticated way, to the structure of high-angle grain boundaries.

The first quantitative evidence supporting a dislocation model for the

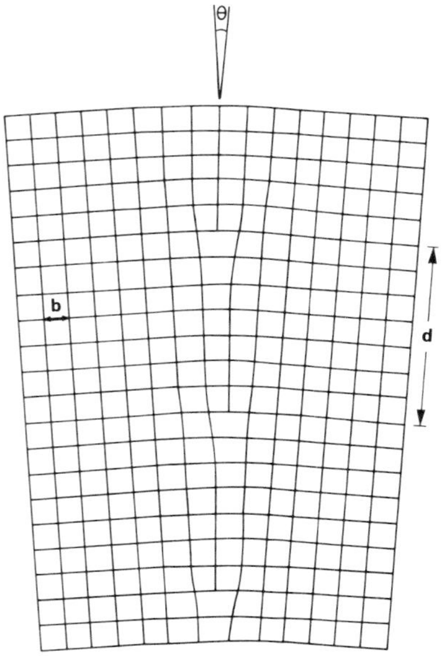

Figure 1.1 Schematic illustration of the dislocation model due to Burgers (1940) of a symmetrical low-angle tilt boundary in a simple cubic lattice with lattice parameter b. The projection is along a cube direction parallel to the tilt axis and the boundary consists of a periodic array of edge dislocations, with spacing d, parallel to the tilt axis.

structure of grain boundaries was obtained by Read and Shockley (1950). They used elasticity theory to calculate the energy of a boundary composed of dislocations, as a function of the misorientation between the grains, and obtained good agreement with experimental measurements of boundary energy in a silicon–iron alloy and in pure tin and lead (see Read and Shockley 1952). A more direct indication that a grain boundary consists of an array of dislocations, of the type envisaged by Burgers, was demonstrated in a spectacular way by the experiments of Washburn and Parker (1952). They showed that a low-angle symmetrical-tilt boundary in a zinc bicrystal moved in one direction under the action of an applied shear stress and in the opposite direction when the stress was reversed. This result was exactly that predicted by Shockley (1949) for a boundary consisting of an array of dislocations as envisaged by Burgers. More direct evidence for the dislocation structure of grain boundaries was given by the etching experiments of Vogel *et al* (1953) and Vogel (1955). They were able to correlate the spacings of etch pits in low-angle boundaries in germanium with the spacings of dislocations predicted by the Burgers model from the measured misorientations between the grains. However, the most direct evidence for the presence of dislocations in grain boundaries came with the first observations by transmission electron microscopy of dislocations in aluminium by Hirsch *et al* (1956). They found that many grain boundaries in their aluminium specimens were composed of arrays of dislocations, and obtained a correlation between the spacings of the dislocations and the misorientations across the boundaries.

Advances in the technique of transmission electron microscopy and accompanying developments in diffraction theory since 1956 have enabled the structure of interfaces to be described quantitatively in terms of displacements in and between the neighbouring grains. One type of displacement is that associated with interfacial dislocations and is specified by their Burgers vectors, and the other type is that associated with rigid-body displacements between the grains. The identification of both types of displacement and the way in which they contribute to the atomic structure of interfaces will form the major part of this book.

In this chapter an introduction to the formal dislocation theory for interfacial dislocations will be presented, first for a grain boundary and then for the more general case of an interphase interface. In addition, anisotropic elasticity theory will be introduced so as to give a derivation of the elastic displacement fields and stresses associated with interfacial dislocations.

1.2 DISLOCATIONS IN A GENERAL GRAIN BOUNDARY

The misorientation between two grains at a grain boundary can be specified by a rotation θ around a common axis $\boldsymbol{u}$ with the same crystallographic indices in both grains†. The limiting cases are a pure tilt boundary where $\boldsymbol{u}$ lies in the boundary plane and a pure twist boundary where $\boldsymbol{u}$ is normal to the boundary plane. A general grain boundary has mixed tilt and twist character so that $\boldsymbol{u}$ and the boundary plane have no special relative orientation. Frank (1950) was the first to develop a theory which provided a dislocation description for this case and the original paper should be consulted for his generalised treatment.

Frank's geometric theory is illustrated in figure 1.2(*a*)–(*e*). In figure 1.2(*a*), a reference crystal lattice has been cut along AA′ by a plane, with a normal specified by the unit vector $\boldsymbol{v}$, so as to divide the lattice into two crystals, represented by '+' and '−'. In figure 1.2(*b*), crystal + is rotated by an angle of $+\theta/2$ and crystal − by one of $-\theta/2$ in a right-handed sense about an axis defined by a unit vector $\boldsymbol{u}$ passing through the lattice point O and directed into the page. In figure 1.2(*c*) these two misoriented lattices are extended until they juxtapose at the original cut, thus forming a grain boundary. A vector $\boldsymbol{x}$, which can have any direction in this boundary plane, is then chosen to extend from the origin O over several unit cells, as illustrated in figure 1.2(*c*). The net Burgers vector of the dislocations in the grain boundary which intersect $\boldsymbol{x}$ can be determined by comparing a Burgers circuit containing the vector $\boldsymbol{x}$ in the bicrystal with an equivalent circuit in the reference lattice. This is done using the FS/RH convention with the closure failure being made in good crystal (see, for example, Hirth and Lothe 1968). The circuit in the bicrystal (figure 1.2(*c*)) is a closed circuit made in a right-handed sense around an axis parallel to $(\boldsymbol{x} \wedge \boldsymbol{v})$, where $\boldsymbol{v}$ is defined as pointing from crystal − into crystal +. This circuit is also made in a right-handed sense with respect to the rotation axis $\boldsymbol{u}$. The circuit starts at S, the end point of the vector $\boldsymbol{x}$, extends through crystal + to the origin O and then returns through crystal − to the point F which is coincident with S. In the reference lattice (redrawn in figure 1.2(*d*)) the first part of the circuit, SO in crystal +, is represented by S_+O and the second part of the circuit, OF in crystal −, is represented by OF_-. Clearly there is a closure failure $F_-S_+ = \boldsymbol{B}$ in the reference lattice and this defines the net Burgers vector of those dislocations contained in the boundary which are intersected by the vector $\boldsymbol{x}$. In general the vector $\boldsymbol{x}$ will make an angle α with $\boldsymbol{u}$ as shown in figure 1.2(*e*), thus the vector $\boldsymbol{B}$ will have the magnitude

$$|\boldsymbol{B}| = |\boldsymbol{x}|\, 2\sin(\theta/2)\sin\alpha$$

with a direction along $(\boldsymbol{x} \wedge \boldsymbol{u})$, and since

$$|\boldsymbol{x} \wedge \boldsymbol{u}| = |\boldsymbol{x}| \sin\alpha$$

† The parameters $\boldsymbol{u}$ and θ will depend on the choice made for indexing the grains and the implications that this choice has for the geometric analysis of general grain boundaries will be discussed in sections 1.4 and 5.2.1.

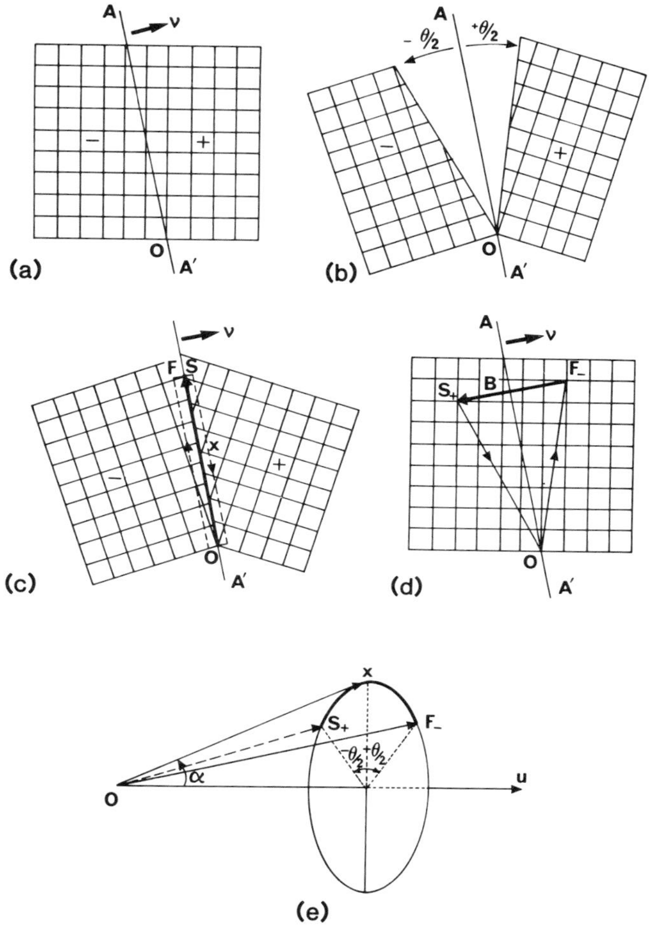

Figure 1.2 Schematic illustration of the derivation of the net Burgers vector $\boldsymbol{B}$ crossing a vector $\boldsymbol{x}$ in a planar grain boundary AA′ with unit normal $\boldsymbol{v}$, where lattice + is rotated with respect to lattice − by an angle θ in a right-handed sense about an axis $\boldsymbol{u}$ directed into the plane of the page through the point O.

then

$$\boldsymbol{B} = 2\sin(\theta/2)(\boldsymbol{x} \wedge \boldsymbol{u}). \tag{1.1}$$

Equation (1.1) is the general formula, originally derived by Frank, for the net Burgers vector of the dislocations required geometrically to accommodate the misorientation at a general grain boundary. Although it has been derived with respect to the reference lattice of figure 1.2(d), it applies to any axis system so that $\boldsymbol{B}$ could be expressed with respect to the lattice of crystal + or crystal − of figure 1.2(c). When applying equation (1.1) to boundaries in real polycrystals, the Burgers vectors of the individual grain boundary dislocations that make up the net Burgers vector content $\boldsymbol{B}$ are defined by the lattices of the neighbouring grains and will be lattice vectors, or linear combinations of lattice vectors, of these grains. Equation (1.1) specifies $\boldsymbol{B}$ as lying in the plane that is normal to the rotation axis $\boldsymbol{u}$ and for a general grain boundary this will not be a simple low-index crystallographic plane. Therefore, in this situation, the net Burgers vector content $\boldsymbol{B}$ must be the resultant of at least three non-coplanar Burgers vectors associated with at least three independent families of grain boundary dislocations. Another property of equation (1.1) is that $\boldsymbol{B}$ is linearly dependent on $\boldsymbol{x}$, and this can only be satisfied, for any vector $\boldsymbol{x}$ in the plane of the boundary, if the grain boundary dislocations comprising each family are parallel and equally spaced. In other words, each family of grain boundary dislocations must take the form of a periodic array lying in the plane of the interface. Frank's theory for the net Burgers vector $\boldsymbol{B}$ applies to the physical situation in which there is no long-range stress field associated with the boundary. Thus, in applying Frank's theory to a particular case, the arrays of dislocations chosen to comprise $\boldsymbol{B}$ must not generate any long-range stress. Saada (1979) has derived the conditions that must be satisfied for this to be the case.

Frank's formulation will now be developed to examine in more detail the spacings and directions required in arrays of grain boundary dislocations in order to accommodate a given misorientation and boundary plane. The analysis is similar to that given by Amelinckx and Dekeyser (1959) and Hirth and Lothe (1968). The general case involving three independent arrays of grain boundary dislocations will be treated first, followed by the special case which only requires two arrays.

1.2.1 Three Independent Arrays of Grain Boundary Dislocations

When the net Burgers vector $\boldsymbol{B}$ of equation (1.1) arises from three independent arrays of grain boundary dislocations with non-coplanar Burgers vectors $\boldsymbol{b}_1$, $\boldsymbol{b}_2$, $\boldsymbol{b}_3$, with line directions parallel to unit vectors $\boldsymbol{r}_1$, $\boldsymbol{r}_2$, $\boldsymbol{r}_3$, and with spacings d_1, d_2, d_3, then equation (1.1) can be written as

$$n_1\boldsymbol{b}_1 + n_2\boldsymbol{b}_2 + n_3\boldsymbol{b}_3 = 2\sin(\theta/2)(\boldsymbol{x} \wedge \boldsymbol{u}). \qquad (1.2)$$

In this equation n_i is the number of dislocations in the ith array intersected by the vector $\boldsymbol{x}$ and, in accordance with the FS/RH convention, will be positive or negative according to the sign of $\boldsymbol{r}_i \cdot (\boldsymbol{x} \wedge \boldsymbol{v})$. The values of n_i, with their appropriate signs, are given by $n_i = \boldsymbol{N}_i \cdot \boldsymbol{x}$, where $\boldsymbol{N}_i$ is a vector in the plane of the boundary and normal to the line direction of the dislocations in the ith array, with a magnitude equal to the reciprocal of their spacing.

It is defined as

$$N_i = (\boldsymbol{v} \wedge \boldsymbol{r}_i)/d_i. \quad (1.3)$$

Thus equation (1.2) can be rewritten as

$$(\boldsymbol{N}_1 \cdot \boldsymbol{x})\boldsymbol{b}_1 + (\boldsymbol{N}_2 \cdot \boldsymbol{x})\boldsymbol{b}_2 + (\boldsymbol{N}_3 \cdot \boldsymbol{x})\boldsymbol{b}_3 = 2\sin(\theta/2)(\boldsymbol{x} \wedge \boldsymbol{u}). \quad (1.4)$$

After dot multiplication of both sides by $(\boldsymbol{b}_2 \wedge \boldsymbol{b}_3)$ equation (1.4) becomes,

$$(\boldsymbol{N}_1 \cdot \boldsymbol{x})\boldsymbol{b}_1 \cdot (\boldsymbol{b}_2 \wedge \boldsymbol{b}_3) = 2\sin(\theta/2)(\boldsymbol{x} \wedge \boldsymbol{u}) \cdot (\boldsymbol{b}_2 \wedge \boldsymbol{b}_3).$$

That is

$$(\boldsymbol{N}_1 \cdot \boldsymbol{x}) = \frac{2\sin(\theta/2)}{\boldsymbol{b}_1 \cdot (\boldsymbol{b}_2 \wedge \boldsymbol{b}_3)} [\boldsymbol{u} \wedge (\boldsymbol{b}_2 \wedge \boldsymbol{b}_3)] \cdot \boldsymbol{x}. \quad (1.5)$$

When $\boldsymbol{x}$ is positioned parallel to $\boldsymbol{r}_1$, $\boldsymbol{N}_1 \cdot \boldsymbol{x} = 0$, so that from equation (1.5)

$$[\boldsymbol{u} \wedge (\boldsymbol{b}_2 \wedge \boldsymbol{b}_3)] \cdot \boldsymbol{r}_1 = 0. \quad (1.6)$$

Thus $\boldsymbol{r}_1$ is perpendicular to $\boldsymbol{u} \wedge (\boldsymbol{b}_2 \wedge \boldsymbol{b}_3)$ and since $\boldsymbol{r}_1$ is also perpendicular to $\boldsymbol{N}_1$ (equation (1.3)) and the boundary normal $\boldsymbol{v}$, it follows from the vector diagram of figure 1.3 that $\boldsymbol{r}_1$ is a unit vector parallel to $[\boldsymbol{u} \wedge (\boldsymbol{b}_2 \wedge \boldsymbol{b}_3)] \wedge \boldsymbol{v}$ and that

$$\boldsymbol{N}_1 = \lambda(\boldsymbol{u} \wedge (\boldsymbol{b}_2 \wedge \boldsymbol{b}_3) - \boldsymbol{v}\{\boldsymbol{v} \cdot [\boldsymbol{u} \wedge (\boldsymbol{b}_2 \wedge \boldsymbol{b}_3)]\}) \quad (1.7)$$

where λ is an unknown coefficient. Dot multiplication of both sides of equation (1.7) by $\boldsymbol{x}$ gives

$$(\boldsymbol{N}_1 \cdot \boldsymbol{x}) = \lambda[\boldsymbol{u} \wedge (\boldsymbol{b}_2 \wedge \boldsymbol{b}_3)] \cdot \boldsymbol{x} \quad (1.8)$$

since $\boldsymbol{v} \cdot \boldsymbol{x} = 0$.

Comparing equation (1.8) with equation (1.5) gives,

$$\lambda = \frac{2\sin(\theta/2)}{\boldsymbol{b}_1 \cdot (\boldsymbol{b}_2 \wedge \boldsymbol{b}_3)}.$$

It follows from figure 1.3 that

$$|\boldsymbol{N}_1| = 1/d_1 = \frac{2\sin(\theta/2)}{\boldsymbol{b}_1 \cdot (\boldsymbol{b}_2 \wedge \boldsymbol{b}_3)} |\boldsymbol{u} \wedge (\boldsymbol{b}_2 \wedge \boldsymbol{b}_3)| \sin\alpha$$

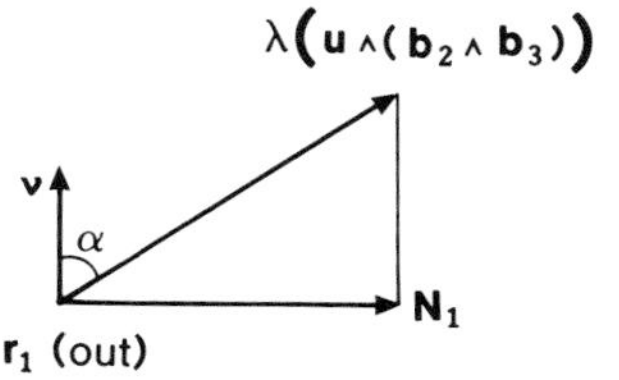

Figure 1.3 Vector representation of equation (1.7) leading to the spacing of dislocations in array 1.

that is,

$$1/d_1 = 2\sin(\theta/2)\left|\frac{[\boldsymbol{u}\wedge(\boldsymbol{b}_2\wedge\boldsymbol{b}_3)]\wedge\boldsymbol{v}}{\boldsymbol{b}_1\cdot(\boldsymbol{b}_2\wedge\boldsymbol{b}_3)}\right|.$$

In summary, the directions and spacings of the dislocations in the three arrays of grain boundary dislocations are given by

$$\boldsymbol{r}_1 \parallel [\boldsymbol{u}\wedge(\boldsymbol{b}_2\wedge\boldsymbol{b}_3)]\wedge\boldsymbol{v} \tag{1.9}$$

and

$$d_1 = \left(2\sin(\theta/2)\left|\frac{[\boldsymbol{u}\wedge(\boldsymbol{b}_2\wedge\boldsymbol{b}_3)]\wedge\boldsymbol{v}}{\boldsymbol{b}_1\cdot(\boldsymbol{b}_2\wedge\boldsymbol{b}_3)}\right|\right)^{-1} \tag{1.10}$$

with similar expressions for $\boldsymbol{r}_2$, d_2 and $\boldsymbol{r}_3$, d_3. Expressions similar to these have been derived by Amelinckx and Dekeyser (1959) and in a different form by Read (1953).

The relationships (1.9) and (1.10) indicate that if a grain boundary has a variation in grain boundary normal $\boldsymbol{v}$, then the change in grain boundary dislocation structure necessary to accommodate the change in boundary plane, need merely involve changes in the line directions and spacings of the dislocations in each of the three grain boundary dislocation arrays, without any alteration in their Burgers vectors. The form of this geometry can be pictured from the schematic illustration in figure 1.4. This figure shows a faceted grain boundary ABCDHGFE consisting of three boundary planes ABFE, BCGF and CDHG with boundary normals $\boldsymbol{v}'$, $\boldsymbol{v}''$, and $\boldsymbol{v}'''$ respectively. The relationships (1.9) and (1.10) for the array of grain boundary dislocations with Burgers vector $\boldsymbol{b}_1$ are represented by intersecting the boundary with a set of equally spaced parallel planes (... a, b, c, d ...) with plane normal parallel to $\boldsymbol{u}\wedge(\boldsymbol{b}_2\wedge\boldsymbol{b}_3)$ and spacing

$$D_1 = \left(2\sin(\theta/2)\left|\frac{\boldsymbol{u}\wedge(\boldsymbol{b}_2\wedge\boldsymbol{b}_3)}{\boldsymbol{b}_1\cdot(\boldsymbol{b}_2\wedge\boldsymbol{b}_3)}\right|\right)^{-1}.$$

The lines of intersection of these planes with the different facet planes of the boundary can then be taken to represent the array of grain boundary dislocations with Burgers vector $\boldsymbol{b}_1$, which have line directions $\boldsymbol{r}'_1$, $\boldsymbol{r}''_1$, $\boldsymbol{r}'''_1$ and spacings d'_1, d''_1, d'''_1 that satisfy relationships (1.9) and (1.10). It can be seen from figure 1.4 that the grain boundary dislocations form parts of parallel planar loops with plane normal parallel to $\boldsymbol{u}\wedge(\boldsymbol{b}_2\wedge\boldsymbol{b}_3)$ and, if one grain surrounded the other, then the grain boundary dislocations would form closed planar loops in the interface.

The geometry of the arrays of grain boundary dislocations with Burgers vectors $\boldsymbol{b}_2$ and $\boldsymbol{b}_3$ can be described similarly by the intersection of the facet planes by two additional sets of parallel planes; one set with plane normal

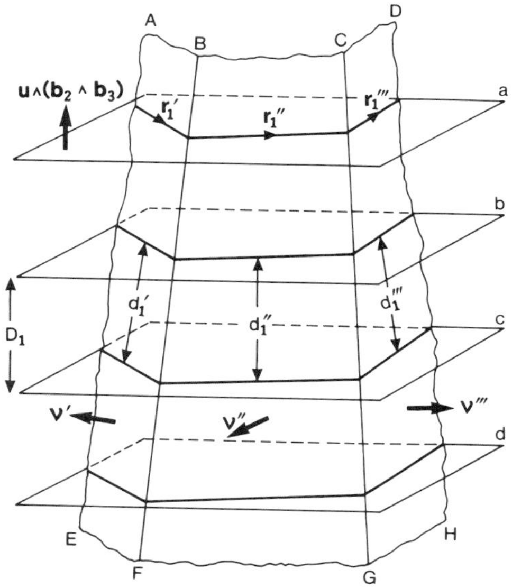

Figure 1.4 Schematic illustration of the geometry of an array of grain boundary dislocations with change in boundary plane (see text).

parallel to $\boldsymbol{u} \wedge (\boldsymbol{b}_3 \wedge \boldsymbol{b}_1)$ and spacing

$$D_2 = \left(2\sin(\theta/2)\left|\frac{\boldsymbol{u} \wedge (\boldsymbol{b}_3 \wedge \boldsymbol{b}_1)}{\boldsymbol{b}_2 \cdot (\boldsymbol{b}_3 \wedge \boldsymbol{b}_1)}\right|\right)^{-1}$$

and the other set with plane normal parallel to $\boldsymbol{u} \wedge (\boldsymbol{b}_1 \wedge \boldsymbol{b}_2)$ and spacing

$$D_3 = \left(2\sin(\theta/2)\left|\frac{\boldsymbol{u} \wedge (\boldsymbol{b}_1 \wedge \boldsymbol{b}_2)}{\boldsymbol{b}_3 \cdot (\boldsymbol{b}_1 \wedge \boldsymbol{b}_2)}\right|\right)^{-1}.$$

The three sets of planes are related in that they have a common zone axis parallel to the axis of rotation $\boldsymbol{u}$.

In later chapters it will be shown that relationships (1.9) and (1.10) can be applied to the results of electron microscopy to determine directly the dislocation structure of low-angle grain boundaries (Chapter 4) and, indirectly, that of high-angle grain boundaries (Chapter 5).

1.2.2 Two Independent Arrays of Grain Boundary Dislocations

For some grain boundaries the plane containing the net Burgers vector $\boldsymbol{B}$ of equation (1.1), i.e. the plane normal to the rotation axis $\boldsymbol{u}$, will be a plane containing simple lattice vectors from the grains. In this situation $\boldsymbol{B}$ can result from two independent arrays of grain boundary dislocations with

Burgers vectors which lie in the plane normal to $\boldsymbol{u}$. If two such arrays of grain boundary dislocations with Burgers vectors $\boldsymbol{b}_1$, $\boldsymbol{b}_2$, line directions $\boldsymbol{r}_1$, $\boldsymbol{r}_2$ and spacings d_1, d_2 are designated 1 and 2 in such a way that

$$\boldsymbol{u} = \frac{\boldsymbol{b}_1 \wedge \boldsymbol{b}_2}{|\boldsymbol{b}_1 \wedge \boldsymbol{b}_2|}$$

then equation (1.1) can be written as

$$(\boldsymbol{N}_1 \cdot \boldsymbol{x})\boldsymbol{b}_1 + (\boldsymbol{N}_2 \cdot \boldsymbol{x})\boldsymbol{b}_2 = \frac{2\sin(\theta/2)}{|\boldsymbol{b}_1 \wedge \boldsymbol{b}_2|}[\boldsymbol{x} \wedge (\boldsymbol{b}_1 \wedge \boldsymbol{b}_2)]$$

that is

$$(\boldsymbol{N}_1 \cdot \boldsymbol{x})\boldsymbol{b}_1 + (\boldsymbol{N}_2 \cdot \boldsymbol{x})\boldsymbol{b}_2 = \frac{2\sin(\theta/2)}{|\boldsymbol{b}_1 \wedge \boldsymbol{b}_2|}[(\boldsymbol{b}_2 \cdot \boldsymbol{x})\boldsymbol{b}_1 - (\boldsymbol{b}_1 \cdot \boldsymbol{x})\boldsymbol{b}_2] \qquad (1.11)$$

where $\boldsymbol{N}_i$ is defined by equation (1.3). Since equation (1.11) holds for any vector $\boldsymbol{x}$ in the boundary, it follows that $\boldsymbol{N}_1$ and $\boldsymbol{N}_2$ must be equal to the components in the boundary plane of $2\sin(\theta/2)(\boldsymbol{b}_2/|\boldsymbol{b}_1 \wedge \boldsymbol{b}_2|)$ and $-2\sin(\theta/2)(\boldsymbol{b}_1/|\boldsymbol{b}_1 \wedge \boldsymbol{b}_2|)$ respectively. Thus

$$\boldsymbol{N}_1 = \frac{2\sin(\theta/2)}{|\boldsymbol{b}_1 \wedge \boldsymbol{b}_2|}[\boldsymbol{b}_2 - \boldsymbol{v}(\boldsymbol{v} \cdot \boldsymbol{b}_2)] \qquad (1.12)$$

and

$$\boldsymbol{N}_2 = \frac{-2\sin(\theta/2)}{|\boldsymbol{b}_1 \wedge \boldsymbol{b}_2|}[\boldsymbol{b}_1 - \boldsymbol{v}(\boldsymbol{v} \cdot \boldsymbol{b}_1)]. \qquad (1.13)$$

Since $\boldsymbol{N}_i \wedge \boldsymbol{v} = \boldsymbol{r}_i/d_i$, cross multiplication by $\boldsymbol{v}$ of equations (1.12) and (1.13) gives

$$\boldsymbol{r}_1/d_1 = \frac{2\sin(\theta/2)}{|\boldsymbol{b}_1 \wedge \boldsymbol{b}_2|}(\boldsymbol{b}_2 \wedge \boldsymbol{v}) \qquad (1.14)$$

and

$$\boldsymbol{r}_2/d_2 = \frac{2\sin(\theta/2)}{|\boldsymbol{b}_1 \wedge \boldsymbol{b}_2|}(\boldsymbol{v} \wedge \boldsymbol{b}_1). \qquad (1.15)$$

Thus

$$\begin{aligned} \boldsymbol{r}_1 &\parallel (\boldsymbol{b}_2 \wedge \boldsymbol{v}) \\ \boldsymbol{r}_2 &\parallel (\boldsymbol{v} \wedge \boldsymbol{b}_1) \end{aligned} \qquad (1.16)$$

with

$$d_1 = \left(2\sin(\theta/2)\frac{|\boldsymbol{b}_2 \wedge \boldsymbol{v}|}{|\boldsymbol{b}_1 \wedge \boldsymbol{b}_2|}\right)^{-1} \qquad (1.17a)$$

and

$$d_2 = \left(2\sin(\theta/2)\,\frac{|\boldsymbol{b}_1 \wedge \boldsymbol{v}|}{|\boldsymbol{b}_1 \wedge \boldsymbol{b}_2|}\right)^{-1}. \qquad (1.17b)$$

For the special case of a pure tilt boundary the two arrays of edge dislocations have line directions parallel to $\boldsymbol{u}$ and equation (1.14) becomes

$$(\boldsymbol{b}_1 \wedge \boldsymbol{b}_2)/d_1 = 2\sin(\theta/2)(\boldsymbol{b}_2 \wedge \boldsymbol{v}).$$

Dot multiplication by $(\boldsymbol{b}_1 \wedge \boldsymbol{b}_2)$ gives

$$|\boldsymbol{b}_1 \wedge \boldsymbol{b}_2|^2/d_1 = 2\sin(\theta/2)(\boldsymbol{b}_2 \wedge \boldsymbol{v})\cdot(\boldsymbol{b}_1 \wedge \boldsymbol{b}_2)$$

so that the spacings of the two arrays of dislocations are given by

$$d_1 = \left(\frac{2\sin(\theta/2)}{|\boldsymbol{b}_1 \wedge \boldsymbol{b}_2|^2}[(\boldsymbol{b}_1 \wedge \boldsymbol{b}_2) \wedge \boldsymbol{b}_2]\cdot\boldsymbol{v}\right)^{-1} \qquad (1.18)$$

and

$$d_2 = \left(\frac{2\sin(\theta/2)}{|\boldsymbol{b}_1 \wedge \boldsymbol{b}_2|^2}[\boldsymbol{b}_1 \wedge (\boldsymbol{b}_1 \wedge \boldsymbol{b}_2)]\cdot\boldsymbol{v}\right)^{-1}. \qquad (1.19)$$

When analysing grain boundaries containing two arrays of dislocations, it is sometimes convenient to express the spacing of the dislocations in one array along the line direction of the other rather than as the spacing normal to its line direction. If the spacing of the dislocations in array 1 is d'_1, measured along the line direction $\boldsymbol{r}_2$ of the dislocations in array 2, and the spacing of the dislocations in array 2 is d'_2, measured along the line direction $\boldsymbol{r}_1$ of the dislocations in array 1, then

$$1/d'_1 = |\boldsymbol{r}_1 \wedge \boldsymbol{r}_2|/d_1$$

and

$$1/d'_2 = |\boldsymbol{r}_1 \wedge \boldsymbol{r}_2|/d_2.$$

Thus from equation (1.14)

$$d'_1 = \left(2\sin(\theta/2)\,\frac{|(\boldsymbol{b}_2 \wedge \boldsymbol{v}) \wedge (\boldsymbol{v} \wedge \boldsymbol{b}_1)|}{|\boldsymbol{b}_1 \wedge \boldsymbol{b}_2||\boldsymbol{v} \wedge \boldsymbol{b}_1|}\right)^{-1}$$

that is

$$d'_1 = \left(2\sin(\theta/2)\,\frac{|\boldsymbol{v}\cdot(\boldsymbol{b}_1 \wedge \boldsymbol{b}_2)|}{|\boldsymbol{b}_1 \wedge \boldsymbol{b}_2||\boldsymbol{v} \wedge \boldsymbol{b}_1|}\right)^{-1}$$

or

$$d'_1 = \left(2\sin(\theta/2)\,\frac{|\boldsymbol{v}\cdot\boldsymbol{u}|}{|\boldsymbol{v} \wedge \boldsymbol{b}_1|}\right)^{-1} \qquad (1.20)$$

and similarly

$$d'_2 = \left(2\sin(\theta/2)\frac{|\boldsymbol{v}\cdot\boldsymbol{u}|}{|\boldsymbol{v}\wedge\boldsymbol{b}_2|}\right)^{-1}. \qquad (1.21)$$

For the special case of a pure twist boundary

$$\boldsymbol{v} = \boldsymbol{u} = \frac{\boldsymbol{b}_1 \wedge \boldsymbol{b}_2}{|\boldsymbol{b}_1 \wedge \boldsymbol{b}_2|}$$

so that equations (1.20) and (1.21) reduce to

$$d'_1 = |\boldsymbol{b}_1|/2\sin(\theta/2)$$

and

$$d'_2 = |\boldsymbol{b}_2|/2\sin(\theta/2)$$

which for small values of θ can be written as

$$d'_1 = |\boldsymbol{b}_1|/\theta$$

and

$$d'_2 = |\boldsymbol{b}_2|/\theta.$$

Thus, for small values of θ, the expressions for the spacings of the dislocations in a pure twist boundary are the same as that for the spacing of the edge dislocations in a symmetric-tilt boundary of the type illustrated in figure 1.1.

An application involving the two-dislocation model of a grain boundary will be described in section 5.4.2.

1.3 DISLOCATIONS IN A GENERAL INTERPHASE INTERFACE

So far discussion has been confined to grain boundaries which are a particular type of interface in that, for any given boundary, the two misoriented grains are related by a simple rotation around a common axis. In this section the concept of a dislocation content associated with a grain boundary, introduced in section 1.2, will be extended to the case of a general interphase interface between crystals of different structure, where the relationship between the two crystal lattices has to be described by a transformation which is a general deformation rather than a simple rotation. This extension will be introduced by first reconsidering the grain boundary of figure 1.2.

With reference to figure 1.2(d), the net Burgers vector $\boldsymbol{B}$ of the dislocations in the boundary which are intersected by the vector $\boldsymbol{x}$ is given by

$$\boldsymbol{B} = \overrightarrow{\mathrm{F_-S_+}} = \overrightarrow{\mathrm{OS_+}} - \overrightarrow{\mathrm{OF_-}}.$$

If $\mathbf{R}_+$ is defined as the rotation $(\boldsymbol{u}, +\theta/2)$ which transforms the reference lattice into lattice +, and $\mathbf{R}_-$ as the rotation $(\boldsymbol{u}, -\theta/2)$ which transforms the reference lattice into lattice −, then

$$\overrightarrow{OS_+} = \mathbf{R}_+^{-1}\boldsymbol{x}$$

and

$$\overrightarrow{OF_-} = \mathbf{R}_-^{-1}\boldsymbol{x}$$

so that

$$\boldsymbol{B} = (\mathbf{R}_+^{-1} - \mathbf{R}_-^{-1})\boldsymbol{x}. \qquad (1.22)$$

In equation (1.22) $\boldsymbol{B}$ and $\boldsymbol{x}$ are expressed with respect to the reference lattice of figures 1.2(a) and (d). However, if one of the crystal lattices is chosen as the reference lattice, say crystal +, and $\boldsymbol{B}$ and $\boldsymbol{x}$ are expressed in this lattice, then equation (1.22) becomes

$$\boldsymbol{B} = (\mathbf{I} - \mathbf{R}^{-1})\boldsymbol{x}$$

where $\mathbf{R}$ is the rotation which transforms lattice + into lattice − and $\mathbf{I}$ is the identity.

As pointed out by Christian (1965), the advantage of this representation (due originally to Bilby (1955)) is that it can be generalised to apply to interphase interfaces by replacing the rotation $\mathbf{R}$ with a general deformation, $\mathbf{S}$, which transforms the lattice of crystal + to the lattice of crystal − and involves strain as well as rotation. Thus, with respect to crystal + the net Burgers vector of those dislocations contained in the interphase interface which are intersected by a vector $\boldsymbol{x}$ is given by

$$\boldsymbol{B} = (\mathbf{I} - \mathbf{S}^{-1})\boldsymbol{x}$$

which is rewritten as

$$\boldsymbol{B} = \mathbf{T}\boldsymbol{x} \qquad (1.23)$$

where

$$\mathbf{T} = (\mathbf{I} - \mathbf{S}^{-1}).$$

The properties of equation (1.23) are similar to those already described for a grain boundary in equation (1.1), in that, for a general interphase interface, $\boldsymbol{B}$ must be the resultant Burgers vector of at least three independent periodic arrays of interfacial dislocations with non-coplanar Burgers vectors. Following Knowles (1982) (see also Sargent and Purdy 1975), this general case will now be developed in a similar way to that in section 1.2.1 for three arrays of dislocations with non-coplanar Burgers vectors $\boldsymbol{b}_1$, $\boldsymbol{b}_2$, $\boldsymbol{b}_3$, line directions r_1, r_2, r_3 and spacings d_1, d_2, d_3. Thus equation (1.23) can be rewritten as

$$(\boldsymbol{N}_1 \cdot \boldsymbol{x})\boldsymbol{b}_1 + (\boldsymbol{N}_2 \cdot \boldsymbol{x})\boldsymbol{b}_2 + (\boldsymbol{N}_3 \cdot \boldsymbol{x})\boldsymbol{b}_3 = \mathbf{T}\boldsymbol{x} \qquad (1.24)$$

where, as before,

$$\boldsymbol{N}_i = (\boldsymbol{v} \wedge \boldsymbol{r}_i)/d_i$$

but in this case $\boldsymbol{v}$ is the unit vector normal to the interphase interface and is directed from crystal − into crystal +. After dot multiplication of both sides by $(\boldsymbol{b}_2 \wedge \boldsymbol{b}_3)$, equation (1.24) becomes

$$(\boldsymbol{N}_1 \cdot \boldsymbol{x})\boldsymbol{b}_1 \cdot (\boldsymbol{b}_2 \wedge \boldsymbol{b}_3) = (\mathbf{T}\boldsymbol{x}) \cdot (\boldsymbol{b}_2 \wedge \boldsymbol{b}_3). \qquad (1.25)$$

As pointed out by Knowles (1982),

$$\mathbf{T}\boldsymbol{x} \cdot (\boldsymbol{b}_2 \wedge \boldsymbol{b}_3) = \boldsymbol{x} \cdot \tilde{\mathbf{T}}(\boldsymbol{b}_2 \wedge \boldsymbol{b}_3)$$

where $\tilde{\mathbf{T}}$ is the transpose of $\mathbf{T}$, so that equation (1.25) can be rewritten as

$$\boldsymbol{N}_1 \cdot \boldsymbol{x} = \boldsymbol{x} \cdot \tilde{\mathbf{T}}\left(\frac{\boldsymbol{b}_2 \wedge \boldsymbol{b}_3}{\boldsymbol{b}_1 \cdot (\boldsymbol{b}_2 \wedge \boldsymbol{b}_3)}\right). \qquad (1.26)$$

When $\boldsymbol{x}$ is positioned parallel to $\boldsymbol{r}_1$, $\boldsymbol{N}_1 \cdot \boldsymbol{x} = 0$, so that from equation (1.26)

$$\boldsymbol{r}_1 \cdot \tilde{\mathbf{T}}\left(\frac{\boldsymbol{b}_2 \wedge \boldsymbol{b}_3}{\boldsymbol{b}_1 \cdot (\boldsymbol{b}_2 \wedge \boldsymbol{b}_3)}\right) = 0. \qquad (1.27)$$

Thus $\boldsymbol{r}_1$ is perpendicular to

$$\tilde{\mathbf{T}}\left(\frac{\boldsymbol{b}_2 \wedge \boldsymbol{b}_3}{\boldsymbol{b}_1 \cdot (\boldsymbol{b}_2 \wedge \boldsymbol{b}_3)}\right)$$

and since $\boldsymbol{r}_1$ is also perpendicular to $\boldsymbol{N}_1$ and the interface normal $\boldsymbol{v}$, it follows from a vector diagram similar to that of figure 1.3 that

$$\boldsymbol{r}_1 \parallel \tilde{\mathbf{T}}\left(\frac{\boldsymbol{b}_2 \wedge \boldsymbol{b}_3}{\boldsymbol{b}_1 \cdot (\boldsymbol{b}_2 \wedge \boldsymbol{b}_3)}\right) \wedge \boldsymbol{v} \qquad (1.28)$$

and

$$d_1 = \left(\left|\tilde{\mathbf{T}}\left(\frac{\boldsymbol{b}_2 \wedge \boldsymbol{b}_3}{\boldsymbol{b}_1 \cdot (\boldsymbol{b}_2 \wedge \boldsymbol{b}_3)}\right) \wedge \boldsymbol{v}\right|\right)^{-1} \qquad (1.29)$$

with similar expressions for $\boldsymbol{r}_2$, d_2 and $\boldsymbol{r}_3$, d_3.

The similarity between relationships (1.28) and (1.29), and (1.9) and (1.10), indicates that the arrays of dislocations in interphase interfaces can be considered in the same way as has been discussed for grain boundaries. For example, a change in interface boundary plane can be accommodated by a change in the spacings and directions of each of the three arrays of interfacial dislocation without any change in their Burgers vectors. The form of the geometry of the interfacial dislocations can again be appreciated from a schematic illustration of the type shown in figure 1.4, where the interface is now an interphase interface with boundary normals $\boldsymbol{\nu}'$, $\boldsymbol{\nu}''$, $\boldsymbol{\nu}'''$. For this situation the constructed set of equally spaced parallel planes (... a, b, c, d ...) for the interfacial dislocations in array 1 has a plane normal parallel to

$$\tilde{\mathbf{T}}\left(\frac{\boldsymbol{b}_2 \wedge \boldsymbol{b}_3}{\boldsymbol{b}_1 \cdot (\boldsymbol{b}_2 \wedge \boldsymbol{b}_3)}\right)$$

and a spacing

$$D_1 = \left(\left|\tilde{\mathbf{T}}\left(\frac{\boldsymbol{b}_2 \wedge \boldsymbol{b}_3}{\boldsymbol{b}_1 \cdot (\boldsymbol{b}_2 \wedge \boldsymbol{b}_3)}\right)\right|\right)^{-1}.$$

The interfacial dislocations in array 1 are then represented by the intersections of these planes with the facet planes of the interphase interface. There will be similar intersections from two other sets of planes corresponding to arrays 2 and 3. Clearly if one phase is completely enclosed within another then all of the interfacial dislocations in each of the three arrays will form closed loops.

Applications of this analysis to the electron microscopy of interphase interfaces will be described in Chapter 7.

1.4 DISCUSSION OF GEOMETRIC ANALYSIS

In principle, the geometric analysis of sections 1.2 and 1.3 for arrays of interfacial dislocations can be applied to interfaces for any degree of misorientation. However, for misorientations where the interfacial dislocations are very closely spaced, e.g. when spacings are comparable with the interatomic spacing, then difficulties arise in attributing physical significance to the results of such an analysis.

When the geometric analysis is applied to the total misorientation, it does not necessarily provide a unique value for the net Burgers vector $\boldsymbol{B}$ associated with the dislocation structure of a given interface (see, for example, discussions by Sutton (1984) and Christian (1985)). This is because the value obtained for $\boldsymbol{B}$ will depend on the choice made for the way in which the two lattices meeting at the interface are indexed. For example, in the case represented by equation (1.22), if lattice + is re-indexed into a lattice +′ by the unitary matrix $\mathbf{U}_+$ operating on lattice +, and lattice − re-indexed into a lattice −′ by a unitary matrix $\mathbf{U}_-$ operating on lattice −, then, with respect to the reference lattice, the net Burgers vector becomes $\boldsymbol{B}'$ where

$$\boldsymbol{B}' = (\mathbf{U}_+^{-1}\mathbf{R}_+^{-1} - \mathbf{U}_-^{-1}\mathbf{R}_-^{-1})\boldsymbol{x}. \qquad (1.30)$$

This is different from the net Burgers vector $\boldsymbol{B}$ given by equation (1.22). At first sight, there appears to be a serious problem in applying the theory to a real physical situation for which there must be a unique solution. However, in practice this problem can be overcome by choosing an indexing for crystals + and − which gives a dislocation content corresponding to minimum energy. For example, from equation (1.30) there will be a value of $\boldsymbol{x}$ for which $\boldsymbol{B}'$ has a maximum, $\boldsymbol{B}'_{\text{max}}$, and the minimum energy situation is given, in general, by the re-indexing matrices $\mathbf{U}_+$ and $\mathbf{U}_-$ for which $\boldsymbol{B}'_{\text{max}}$ is a minimum. In the case of low-angle grain boundaries, the appropriate indexing of crystals + and − would be that corresponding to the minimum value of the misorientation θ. In addition, it will be seen in section 5.2.1 that the geometric analysis does provide a unique solution for the net Burgers vector $\boldsymbol{B}$ of secondary grain boundary dislocations in high-angle boundaries.

1.5 THE 0-LATTICE DESCRIPTION OF INTERFACE STRUCTURE

Bollmann (1967a, b, 1970 and 1982) has developed an alternative method of analysing the structure of general grain boundaries and interphase interfaces. His method is based on the concept of the 0-lattice†, which he introduced to describe the matching and mismatching of misoriented lattices at an interface. If two misoriented crystal lattices, 1 and 2, are allowed to interpenetrate, there will be a periodic set of points in space (generally not lattice points of either lattice) where, for each point, the internal coordinates in a cell of lattice 1 are identical with the internal coordinates in a cell of lattice 2. It is this set of points which defines Bollmann's 0-lattice. In terms of a general deformation $\mathbf{S}$ which transforms lattice 1 into lattice 2, a point defined by a vector $\boldsymbol{y}$ in lattice 2 is generated from a point defined by a vector $\boldsymbol{x}$ in lattice 1 according to

$$\boldsymbol{y} = \mathbf{S}\boldsymbol{x}.$$

A point on the 0-lattice is therefore defined by a vector $\boldsymbol{x}^{(0)}$ when $\boldsymbol{y}$ differs from $\boldsymbol{x}$ by a translation vector $\boldsymbol{b}_1$ of lattice 1, i.e. when

$$\boldsymbol{y} = \boldsymbol{x}^{(0)} = \boldsymbol{x} + \boldsymbol{b}_1 = \mathbf{S}\boldsymbol{x}.$$

From this set of equations,

$$\boldsymbol{b}_1 = (\mathbf{I} - \mathbf{S}^{-1})\boldsymbol{x}^{(0)}$$

† The 0-lattice concept derives from the earlier concept of a coincident site lattice (CSL). Coincident site lattices are formed by those lattice points which are coincident when the two misoriented lattices of neighbouring grains are allowed to interpenetrate. A particular CSL is specified by a parameter Σ where $1/\Sigma$ of the lattice points are common to both lattices (see section 3.4.2).

that is

$$\boldsymbol{b}_1 = \mathbf{T}\boldsymbol{x}^{(0)} \tag{1.31}$$

where

$$\mathbf{T} = (\mathbf{I} - \mathbf{S}^{-1}).$$

The 0-lattice is defined by three basis vectors which are given by

$$\boldsymbol{x}^{(0)} = \mathbf{T}^{-1}\boldsymbol{b}_1$$

when $\boldsymbol{b}_1$ takes the values of three primitive translation vectors of lattice 1. The resulting 0-lattice may be a point, line or plane lattice depending on whether $\mathbf{T}$ has the rank 3, 2 or 1. For a general interphase interface the 0-lattice is a point lattice. For the case of a grain boundary, where $\mathbf{S}$ is a simple rotation, the 0-lattice is a line lattice in which the lines are parallel to the axis of rotation, and when $\mathbf{S}$ corresponds to a simple shear the 0-lattice is a lattice of planes.

The 0-elements (points, lines or planes) represent regions of 'good' fit between lattices 1 and 2 where the two interpenetrating lattices are in register. In-between the 0-elements, the lattices are out of register and the change in registration on going from one 0-element to another is the appropriate translation vector $\boldsymbol{b}_1$. The 0-lattice is related to the dislocation description of the structure of interfaces by considering that the lack of registration between the 0-elements is localised into cell walls midway between the 0-elements, i.e. the walls of the Wigner–Seitz cells of the 0-lattice. The intersections of the interface with these cell walls are considered to define the lines of the interfacial dislocations, and the Burgers vectors of these dislocations are then taken as the appropriate translation vectors $\boldsymbol{b}_1$.

It has been pointed out by Christian (1976) that equation (1.31), which specifies the Burgers vectors of the interfacial dislocations by the 0-lattice method, is essentially the same as equation (1.23), which is derived from Frank's treatment as generalised by Bilby (1955) and Christian (1965). The difference between the two treatments is that the 0-lattice treatment quantises the Burgers vectors of the interfacial dislocations to the crystal lattices from the outset, whereas the generalised Frank treatment considers an average net Burgers vector content associated with the interface, which is subsequently quantised to Burgers vectors of interfacial dislocations defined by the lattices of the neighbouring grains.

1.6 ELASTICITY AND INTERFACIAL DISLOCATIONS

In an experimental analysis of the structure of interfaces in terms of arrays of interfacial dislocations it is necessary to determine the Burgers vectors, spacings and directions of the dislocations in the arrays. The determination of the Burgers vectors of interfacial dislocations is essential in the analysis, as it is the Burgers vectors which specify the atomic displacements at interfaces. In practice, the Burgers vector of an interfacial dislocation can be determined quantitatively using transmission electron microscopy by making a comparison between the contrast of the dislocation in experimental images with that predicted by theoretical calculations. The image contrast associated with a dislocation arises from changes in the scattering of electrons by the perturbed potential of the crystal caused by the displacement field of the dislocation. Therefore, in order to carry out contrast calculations, it is first necessary to be able to specify the displacement field of an interfacial dislocation. Interfacial dislocations generally occur in arrays which accommodate the misorientation between two neighbouring grains, and solutions for the displacement fields of such arrays have been derived by Rey and Saada (1976) and Bonnet (1981a) in isotropic elasticity and by Saada (1976, 1979) and Bonnet (1981b) in anisotropic elasticity. However, in this book, for computations of the bright-field image contrast of an interfacial dislocation in an array, the displacement field of an isolated dislocation is used, and this has proved to be adequate when the dislocations in the array are coarsely spaced compared with their image widths which are typically 100–200 Å.

In this section an outline will be given of how the displacement field of an interfacial dislocation can be derived using anisotropic elasticity theory for a single dislocation in an interface dividing two different anisotropic elastic media. Anisotropic elasticity theory is required not only because all metals and alloys depart from idealised elastic isotropy, but also because, for two misoriented crystals meeting at an interface, the effect of anisotropy is that, with respect to a single-axis system for the bicrystal, the elastic constants in general will be different either side of the interface. The treatment given here follows that of Tucker (1969) and uses the notation of Stroh (1958). It is based on the general anisotropic elasticity theory for a dislocation in a single elastic medium given by Eshelby *et al* (1953) (see also Hirth and Lothe 1968).

For a set of right-handed cartesian axes, Ox_1, Ox_2, Ox_3, in a single elastic medium, Hooke's law defines the relationship between a stress σ_{ij} (acting in the direction Ox_i across the plane normal to Ox_j) and the strain e_{ij} as

$$\sigma_{ij} = c_{ijkl}e_{kl} \tag{1.32}$$

where i, j, k and l may take values from 1 to 3, with the usual summation convention applying to repeated dummy subscripts, and where c_{ijkl} are the 81 elastic constants. These elastic constants are not all independent since

$$c_{ijkl} = c_{jikl} = c_{ijlk} = c_{klij} \tag{1.33}$$

and they are usually defined using a two-subscript notation such as c_{mn} where $ij \rightarrow m$ and $kl \rightarrow n$ as:

$$11 \rightarrow 1,\ 22 \rightarrow 2,\ 33 \rightarrow 3,\ 23 \text{ or } 32 \rightarrow 4,\ 31 \text{ or } 13 \rightarrow 5 \text{ and } 12 \text{ or } 21 \rightarrow 6.$$

In this notation $c_{mn} = c_{nm}$, so there are only 21 independent elastic constants, and this number may reduce further when account is taken of crystal symmetry, e.g. in the case of cubic symmetry they reduce to three independent constants c_{11}, c_{12}, c_{44}. The elastic strain e_{kl} is defined in terms of the elastic displacements u_i as

$$e_{kl} = \tfrac{1}{2}[(\partial u_k/\partial x_l) + (\partial u_l/\partial x_k)].$$

Thus, using the symmetry relationships of equation (1.33), the stresses in equation (1.32) can be rewritten as

$$\sigma_{ij} = c_{ijkl}(\partial u_k/\partial x_l). \tag{1.34}$$

If there are no body forces acting then the equations of equilibrium are

$$\partial \sigma_{ij}/\partial x_j = 0$$

so that from equation (1.34) it follows that

$$c_{ijkl}(\partial^2 u_k/\partial x_j \partial x_l) = 0. \tag{1.35}$$

For a dislocation which is infinitely long and lies along the axis Ox_3 in a single elastic medium the elastic stresses and displacements will be independent of the variable x_3, so that a solution to equation (1.35) can be sought in the form

$$u_k = A_k f(z) \tag{1.36}$$

where A_k are components of some vector and f is some analytic function of

$$z = (x_1 + px_2)$$

with p being a constant. Substituting equation (1.36) into (1.35) gives, with $i = 1, 2, 3$,

$$(c_{i1k1} + pc_{i1k2} + pc_{i2k1} + p^2 c_{i2k2})A_k = 0 \tag{1.37}$$

which with $k = 1, 2, 3$ gives three independent linear equations in A_k and these have a non-zero solution when the determinant of the coefficients of A_k equals zero, i.e. when

$$c_{i1k1} + pc_{i1k2} + pc_{i2k1} + p^2 c_{i2k2} = 0. \tag{1.38}$$

Expression (1.38) is a sextic equation in p which, in general, has to be solved numerically. However, it has been shown by Eshelby *et al* (1953) that the six roots of p are complex and occur in pairs which are complex conjugates, so the roots can be expressed as p_α and $\bar{p}_\alpha$ where $\alpha = 1, 2, 3$ and the bar denotes complex conjugate. When these roots are substituted in equation (1.37) solutions for the components A_k can be found which are also complex and have the form $A_{k\alpha}$ and $\bar{A}_{k\alpha}$. Thus, in equation (1.36) only the real component has any physical meaning, so equation (1.36) can be rewritten as

$$u_k = \mathscr{R}\left[\sum_\alpha A_{k\alpha} f_\alpha(z_\alpha)\right] \tag{1.39}$$

where $\mathscr{R}$ denotes 'real part of'. From equation (1.34) the two components of stress, σ_{i1} and σ_{i2}, are given by

$$\begin{aligned}\sigma_{i1} &= \mathscr{R}\left[\sum_\alpha \left(c_{i1k1} A_{k\alpha} \frac{\partial}{\partial x_1}(f_\alpha(z_\alpha)) + c_{i1k2} \frac{\partial}{\partial x_2}(f_\alpha(z_\alpha))\right)\right] \\ &= \mathscr{R}\left[\sum_\alpha \left([(p_\alpha)^{-1} c_{i1k1} + c_{i1k2}] A_{k\alpha} \frac{\partial}{\partial x_1}(f_\alpha(z_\alpha))\right)\right]\end{aligned}$$

and

$$\begin{aligned}\sigma_{i2} &= \mathscr{R}\left[\sum_\alpha \left(c_{i2k1} A_{k\alpha} \frac{\partial}{\partial x_1}(f_\alpha(z_\alpha)) + c_{i2k2} \frac{\partial}{\partial x_2}(f_\alpha(z_\alpha))\right)\right] \\ &= \mathscr{R}\left[\sum_\alpha \left((c_{i2k1} + p_\alpha c_{i2k2}) A_{k\alpha} \frac{\partial}{\partial x_1}(f_\alpha(z_\alpha))\right)\right].\end{aligned}$$

Stroh defines a matrix with elements $L_{i\alpha}$ such that

$$L_{i\alpha} = (c_{i2k1} + p_\alpha c_{i2k2}) A_{k\alpha} = -[(p_\alpha)^{-1} c_{i1k1} + c_{i1k2}] A_{k\alpha}$$

so that

$$\sigma_{i1} = -\mathscr{R}\left(\sum_\alpha L_{i\alpha} \frac{\partial}{\partial x_2}(f_\alpha(z_\alpha))\right) \tag{1.40}$$

and

$$\sigma_{i2} = \mathscr{R}\left(\sum_\alpha L_{i\alpha} \frac{\partial}{\partial x_1}(f_\alpha(z_\alpha))\right). \tag{1.41}$$

The appropriate form of $f_\alpha(z_\alpha)$ for a dislocation has been shown by Eshelby *et al* (1953) to be

$$(2\pi \mathrm{i})^{-1} D_\alpha \ln(z_\alpha)$$

which in the notation of Stroh becomes

$$f_\alpha(z_\alpha) = (2\pi)^{-1} M_{\alpha j} G_{ji} b_i \ln(z_\alpha) \tag{1.42}$$

where the matrix with elements $M_{\alpha j}$ is reciprocal to the matrix with elements $L_{j\alpha}$, the matrix with elements G_{ji} is reciprocal to

$$(\mathrm{i}/2) \sum_\alpha (A_{j\alpha} M_{\alpha i} - \bar{A}_{j\alpha} \bar{M}_{\alpha i})$$

and b_i are the three components of the Burgers vector of the dislocation. Thus from equations (1.39) and (1.42) the elastic displacements associated with the dislocation can be written as

$$u_k = \mathscr{R}[(2\pi)^{-1} \sum_{\alpha} A_{k\alpha} M_{\alpha j} G_{ji} b_i \ln(z_\alpha)]. \tag{1.43}$$

In order to consider the case of an interfacial dislocation, a dislocation of infinite length is located along the axis Ox_3 in an interface defined by $Ox_2 = 0$ (i.e. in the Ox_1x_3 plane) which separates two different semi-infinite anisotropic elastic media labelled I and II in figure 1.5. This is a special case of a general problem in bicrystal elasticity originally considered by Tucker (1969) for a dislocation parallel to Ox_3. He treated in detail the case of a dislocation in one of the elastic media, but his general result applies equally well to a dislocation lying in the interface. Following Tucker, the medium occupying the half-space given by $x_2 > 0$ is labelled I and that occupying the half-space $x_2 < 0$ is labelled II and, for convenience, two superscripts g and h are defined such that, for any point in space, g refers to the medium in which the point lies and h to the other medium. That is

$$g = \text{I},\ h = \text{II} \text{ when } x_2 > 0$$

and

$$g = \text{II},\ h = \text{I} \text{ when } x_2 < 0.$$

Tucker showed that, although two elastic media are involved, the displacements associated with a dislocation have the same general form as that given by equation (1.39) for a dislocation in a single elastic medium. Thus for two elastic media he showed that the displacements are given by

$$u_k^{(g)} = \mathscr{R}\left[\sum_{\alpha} A_{k\alpha}^{(g)} f_{\alpha}^{(g)}(z_{\alpha}^{(g)})\right] \tag{1.44}$$

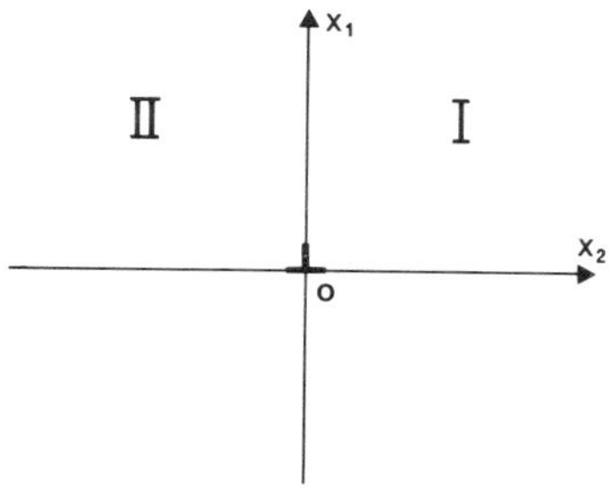

Figure 1.5 System of cartesian axes used for a straight interfacial dislocation of infinite length along Ox_3 lying in a planar interface at $x_2 = 0$ separating two anisotropic elastic media I and II.

but in this case

$$f_\alpha^{(g)}(z_\alpha^{(g)}) = f_{0\alpha}^{(g)}(z_\alpha^{(g)}) + f_{1\alpha}^{(g)}(z_\alpha^{(g)})$$

where $f_{0\alpha}^{(g)}$ is the value of $f_\alpha^{(g)}$ for a dislocation in a single elastic medium as given by equation (1.42) and $f_{1\alpha}^{(g)}$ is an additional term taking account of the other elastic medium h.

The form of $f_{1\alpha}^{(g)}$ depends on the nature of the elastic boundary conditions imposed at the interface between the elastic media g and h. Head (1953) considered the elastic boundary conditions for three types of interface, namely the following: a free surface, a fully slipping boundary and a welded boundary. For stable grain boundaries and interphase interfaces in metals and alloys the appropriate boundary conditions are those for the welded boundary, i.e. where the components of stress acting on the interface, σ_{i2}, are continuous across the interface and where all the components of displacement u_k are also continuous across the interface.

Tucker has shown for a dislocation lying parallel to a welded boundary that†

$$\begin{aligned} f_\alpha^{(g)}(z_\alpha^{(g)}) &= f_{0\alpha}^{(g)}(z_\alpha^{(g)}) \\ &\quad + \sum_\beta M_{\alpha i}^{(g)} G_{il}^{(g,h)} \bar{J}_{lj}^{(g,h)} \bar{L}_{j\beta}^{(g)} \bar{f}_{0\beta}^{(g)}(z_\alpha^{(g)}) \\ &\quad + \sum_\beta M_{\alpha i}^{(g)} (\delta_{ij} + \bar{G}_{il}^{(h,g)} J_{lj}^{(h,g)}) L_{j\beta}^{(h)} f_{0\beta}^{(h)}(z_\alpha^{(g)}) \end{aligned} \tag{1.45}$$

where the matrix with elements $G_{il}^{(g,h)}$ is the matrix reciprocal to the matrix with elements

$$\frac{\mathrm{i}}{2} \sum_\alpha (A_{i\alpha}^{(g)} M_{\alpha l}^{(g)} - \bar{A}_{i\alpha}^{(h)} \bar{M}_{\alpha l}^{(h)})$$

and

$$J_{lj}^{(g,h)} = \frac{\mathrm{i}}{2} \sum_\alpha (A_{l\alpha}^{(g)} M_{\alpha j}^{(g)} - A_{l\alpha}^{(h)} M_{\alpha j}^{(h)}).$$

δ_{ij} is the Kronecker delta. To treat an interfacial dislocation lying along Ox_3 the Burgers vector with components b_i can be considered as being apportioned equally between the two elastic media, i.e. $\frac{1}{2}b_i$ in g and $\frac{1}{2}b_i$ in h. Then from equation (1.42) it follows that

$$f_{0\alpha}^{(h)}(z_\alpha^{(g)}) = (2\pi)^{-1} M_{\alpha j}^{(h)} G_{ji}^{(h,h)} \tfrac{1}{2} b_i \ln(z_\alpha^{(g)}) \tag{1.46}$$

and

$$f_{0\alpha}^{(g)}(z_\alpha^{(g)}) = (2\pi)^{-1} M_{\alpha j}^{(g)} G_{ji}^{(g,g)} \tfrac{1}{2} b_i \ln(z_\alpha^{(g)}). \tag{1.47}$$

† It should be noted that this equation differs from Tucker's equation (27) in which two minus signs are typographical errors.

From equations (1.44), (1.45), (1.46) and (1.47) the elastic displacements associated with the interfacial dislocation are obtained as

$$u_k^{(g)} = \mathscr{R}\left[(2\pi)^{-1} \sum_\alpha A_{k\alpha}^{(g)} M_{\alpha j}^{(g)} G_{ji}^{(g,h)} b_i \ln(z_\alpha^{(g)})\right]. \tag{1.48}$$

Similarly, when equations of the type (1.40) and (1.41) are taken with equations (1.45), (1.46) and (1.47) the elastic stresses associated with the interfacial dislocation are obtained as

$$\sigma_{i1}^{(g)} = -\mathscr{R}\left[(2\pi)^{-1} \sum_\alpha L_{i\alpha}^{(g)} p_\alpha^{(g)} M_{\alpha j}^{(g)} G_{jk}^{(g,h)} b_k (z_\alpha^{(g)})^{-1}\right] \tag{1.49}$$

and

$$\sigma_{i2}^{(g)} = \mathscr{R}\left[(2\pi)^{-1} \sum_\alpha L_{i\alpha}^{(g)} M_{\alpha j}^{(g)} G_{jk}^{(g,h)} b_k (z_\alpha^{(g)})^{-1}\right]. \tag{1.50}$$

Equation (1.48) forms the basis for the calculation of the contrast associated with an interfacial dislocation in electron microscope images, since contrast depends on the strain field in the bicrystal which, for an isolated interfacial dislocation, is given by the derivative of u_k (see section 2.4.1). However, interfacial dislocations generally occur in arrays and the strain field associated with an array cannot be described by a simple linear summation of the strain fields of isolated dislocations obtained from equation (1.48) since, in general, such a summation gives finite stresses and strains at large distances from the boundary (see, for example, Hirth and Lothe 1968). Nevertheless, the strain field associated with a particular dislocation in an array of interfacial dislocations can still be adequately described by the derivative of equation (1.48), provided the strains being considered are at distances from the dislocation that are small compared with the spacings of the dislocations in the array (see, for example, van der Merwe 1950).

In Chapter 2, the elastic strain fields derived from equation (1.48) will be used in the development of the necessary diffraction contrast theory for the determination of the Burgers vectors of interfacial dislocations from electron microscope images.

2

Image Formation and Diffraction Effects in the Electron Microscopy of Interfaces

2.1 INTRODUCTION

The use of transmission electron microscopy to identify defects in metals and alloys relies on theories of electron diffraction contrast for quantitative interpretation of the images of the defects. For dislocations, and more complex defect configurations involving both dislocations and stacking faults within individual crystal grains, the two-beam dynamical theory of Howie and Whelan (1961) has been the basis of methods used for quantitative analysis of two-beam images (see, for example, Hirsch *et al* 1965 and Head *et al*, 1973). The weak-beam method for imaging defects (Cockayne *et al* 1969) has also been most valuable in determining the geometry of various dislocation configurations. Additionally, in recent years, the theory of n-beam diffraction has been finding increased application in the identification of defects from lattice images of metals (see, for example, Spence 1981).

As for the case of defects in single grains, the quantitative identification of defects in grain boundaries and interphase interfaces has also involved the use of two-beam, weak-beam and n-beam imaging. In this chapter the application of these methods to the study of the defect structure of interfaces will be discussed. However, the main emphasis will be on two-beam methods, as these are the most commonly used. In addition, diffraction phenomena associated with interfaces, some of which provide an important means of identifying interface structure, will also be discussed.

2.2 ELECTRON DIFFRACTION—GENERAL INTRODUCTION

When an electron beam of wavelength λ, in the form of a plane wave with wave vector $\boldsymbol{k}$, where $|\boldsymbol{k}| = 1/\lambda$, is incident on a crystalline specimen in the electron microscope, the specimen behaves as a three-dimensional diffraction grating. As a result, the elastically scattered electrons emerge from the exit surface of the specimen not only as a plane wave with wave vector $\boldsymbol{k}$, but also as a set of plane waves with wave vectors $\boldsymbol{k}'_n$. The condition for strong diffraction occurs when

$$\boldsymbol{k} - \boldsymbol{k}'_n = \boldsymbol{g}_n \tag{2.1}$$

where $\boldsymbol{g}_n$ are a set of vectors (diffracting vectors) which are normal to crystallographic planes in the specimen and have magnitudes equal to the reciprocals of the interplanar spacings d_n. They therefore define a reciprocal point lattice for the crystal structure as though the crystal were of infinite extent.

Equation (2.1) is an expression of Bragg's law and, for a particular set of crystal planes defined by the reciprocal lattice vector $\boldsymbol{g}$ ($|\boldsymbol{g}| = 1/d$), can be represented by the Ewald construction of figure 2.1. Figure 2.1(a) shows a set of reciprocal lattice points G_{-1}, O, G_1, G_2, where $\overrightarrow{OG_1}$ is the reciprocal lattice vector $\boldsymbol{g}$. The Ewald sphere, with radius $1/\lambda$, is centred at the point C so that it passes through the origin O and $\overrightarrow{CO}$ represents the incident wave vector $\boldsymbol{k}$. Equation (2.1) is satisfied when the sphere also passes through another reciprocal lattice point such as G_1, so that CG_1 represents the diffracted wave vector $\boldsymbol{k}'$. Clearly the geometry of figure 2.1(a) satisfies the Bragg condition

$$2d \sin \theta_B = \lambda$$

where $2\theta_B$ is the angle between $\boldsymbol{k}$ and $\boldsymbol{k}'$.

In general, the crystal will not be oriented to satisfy the exact Bragg condition for a particular diffracting vector $\boldsymbol{g}$ and in this case the diffraction condition is given by

$$\boldsymbol{k} - \boldsymbol{k}' = \boldsymbol{g} + \boldsymbol{s} \tag{2.2}$$

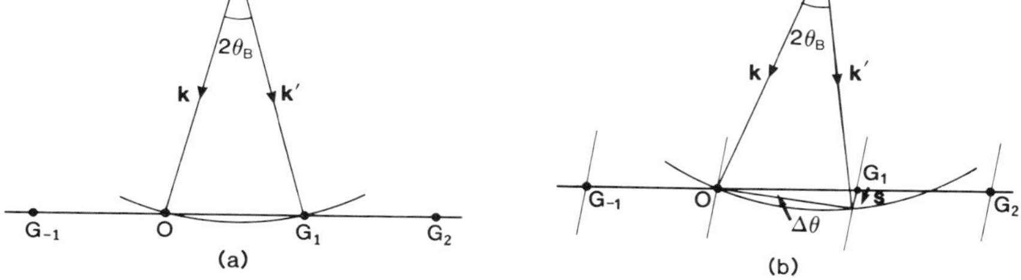

Figure 2.1 Illustration in reciprocal space of the Ewald construction for a crystal at the exact Bragg condition (a), and for a crystal oriented so that the deviation $\boldsymbol{s}$ from the Bragg condition is positive (b).

where $\boldsymbol{s}$ is a vector in reciprocal space defining the deviation from the exact Bragg condition. For an incident electron wave of unit amplitude, the amplitude $\varphi_{\boldsymbol{g}}$ of the diffracted wave with wave vector $\boldsymbol{k}'$ is given in the kinematic approximation by

$$\varphi_{\boldsymbol{g}} = (F_{\boldsymbol{g}}/V_{\mathrm{c}}) \int_{\mathrm{crystal}} \exp(-2\pi \mathrm{i} \boldsymbol{s} \cdot \boldsymbol{r})\, \mathrm{d}\tau \qquad (2.3)$$

where $F_{\boldsymbol{g}}$ is the structure factor for the diffracting vector $\boldsymbol{g}$, V_{c} is the volume of the unit cell and $\mathrm{d}\tau$ is a volume element of the crystal at a position given by $\boldsymbol{r}$ (see Hirsch *et al* 1965). Thus expression (2.2) can be represented in an Ewald construction by spreading out each reciprocal lattice point with a distribution function given by $\varphi_{\boldsymbol{g}}$ of equation (2.3). The specimens used in the electron microscopy of metals and alloys are in the form of thin foils so that each reciprocal lattice point is spiked normal to the plane of the thin foil. The length of the spike and the amplitude distribution in the spike depend on the foil thickness t, with the length of the spike of the order of $1/t$ and the amplitude distribution along the spike given by

$$(F_{\boldsymbol{g}}/V_{\mathrm{c}})[\sin(\pi t s_t)/\pi s_t]$$

where s_t is the component of $\boldsymbol{s}$ in the direction normal to the plane of the foil. The diffracting condition given by equation (2.2) for the case of a crystalline specimen in the form of a thin foil is represented by the Ewald construction in figure 2.1(*b*), where the reciprocal lattice points G_{-1}, O, G_1, G_2 are spiked and the vector $\boldsymbol{s}$ (the deviation from the Bragg condition) is represented by the vector connecting the reciprocal lattice point G_1 with the end point of the wave vector $\boldsymbol{k}'$. The corresponding angular departure from the exact Bragg condition is the tilt, $\Delta\theta$, about an axis normal to the plane of the figure. For the accelerating voltages used in electron microscopy, the radius of the Ewald sphere is very large compared with the magnitude of the low-order reciprocal lattice vector $\boldsymbol{g}$, and as a consequence the portion of the Ewald sphere shown in figure 2.1 can be regarded as approximately planar. The vector $\boldsymbol{s}$ is approximately parallel to the incident wave vector $\boldsymbol{k}$ and in theories of diffraction contrast, $\boldsymbol{s}$ is often taken as being parallel to $\boldsymbol{k}$ and defined by the scalar parameter s. The latter is the component of $\boldsymbol{s}$ in the $\boldsymbol{k}$ direction which, for small angles $\Delta\theta$, is given by

$$s = |\boldsymbol{g}|\Delta\theta.$$

In the electron microscope, the emerging plane waves are focused by the objective lens to form a pattern of diffraction spots in the back focal plane of the lens. When the specimen is tilted with respect to the incident beam, the relative intensity in the diffraction spots alters. By suitably tilting the specimen, a diffraction pattern can be obtained in which most of the intensity is contained in only two strongly excited spots: one, the central beam, corresponding to transmission of the incident beam with wave vector $\boldsymbol{k}$, and the other to one diffracted beam with wave vector $\boldsymbol{k}'$. It is this condition which is known as two-beam diffraction. A two-beam image of the specimen is formed by positioning an objective aperture in the back focal plane of the objective lens so as to select one of these beams which, after magnification, gives either a bright-field image when the central beam is selected or a dark-field image when the diffracted beam is selected. The relative intensities in the central and diffracted beams depend on the deviation from the Bragg condition s, and any strain in the crystal lattice of the specimen associated with the presence of a defect will cause a local change in s. It is this local change in s which gives rise to contrast from the defect when the specimen is imaged.

In weak-beam imaging the specimen is tilted so that, in addition to the central spot, a high-order diffraction spot (such as $2\boldsymbol{g}$) is excited and then a dark-field image is taken with a weakly excited diffraction spot (such as $-\boldsymbol{g}$) for which s is very large. The method of weak-beam imaging is less sensitive to the presence of small strains because s is very large, but it has the consequent advantage that the image width of the defect is narrow, which can lead to improved resolution.

A third imaging condition is when the specimen is tilted so that the incident electron beam is close to a low-index crystallographic direction in the specimen. In this condition several diffraction spots around the central beam are strongly excited. When the objective aperture is placed so as to select the central beam and a chosen number of these other excited beams, subsequent magnification gives an n-beam image of the specimen. The contrast in such n-beam images is in the form of interference fringe patterns (lattice images) which can be related to the crystal structure of the specimen (see section 2.5).

In addition to elastic scattering of electrons, which gives rise to the spots in the diffraction patterns, inelastic scattering also occurs. Inelastically scattered electrons contribute to diffraction patterns in two main ways; as a near-uniform diffuse background spread over a wide angular range, and as pairs of Kikuchi lines superimposed on this background. When an objective aperture is used for image formation, only a small fraction of the inelastically scattered electrons passes through the aperture. The contribution of these electrons to the image is generally neglected in theories of image contrast. However, the majority of inelastically scattered electrons, which are those excluded by the objective aperture, are taken into account in theories of image contrast by considering this apparent loss of electrons in terms of absorption effects.

The inelastically scattered electrons give rise to the formation of pairs of Kikuchi lines and these are used for the detailed analysis of diffraction patterns. Pairs of Kikuchi lines in diffraction patterns can be considered as arising from a secondary source of inelastically scattered electrons which is distributed throughout the specimen and radiates in all directions with a strong peak in the direction of the incident beam. The energy loss in inelastic scattering is small relative to the accelerating voltage, so that the wavelength of the inelastically scattered electrons is very close to that of the incident electrons. Bragg diffraction of these inelastically scattered electrons from a given set of crystallographic planes results in the formation in the diffraction pattern of a pair of Kikuchi lines with an angular separation very close to $2\theta_B$ (see Hirsch *et al* 1965, Chapter 5). Thus, in the diffraction pattern the spacing between the Kikuchi lines can be taken as the same as that between the central spot and the spot arising from diffraction from the same set of planes. The directions of the inelastically scattered electrons will be closer to the direction of the incident beam for one of the Kikuchi lines in the pair than for the other and this results in a greater intensity of electrons in one line (the excess line) than in the other (the deficient line). These properties of Kikuchi lines enable them to be identified with particular sets of crystallographic planes in the specimen so that they can be appropriately indexed.

The importance of Kikuchi lines in diffraction patterns is that they enable the accurate determination of the orientation of a crystal grain relative to the direction of the incident electron beam. Further, in practical electron microscopy, it is necessary to tilt the specimen through large angles so that the direction of the incident electron beam can be aligned with a series of different crystallographic directions, and the Kikuchi lines enable this tilting to be monitored and carried out in a controlled way. This is possible because the Kikuchi lines, unlike the spots in the diffraction pattern, move when the specimen is tilted as though they are rigidly fixed to crystallographic planes in the specimen.

The relationship between Kikuchi lines and diffraction spots in the diffraction pattern is illustrated schematically in figure 2.2. The lines D and E represent the deficient and excess Kikuchi lines associated with the diffracting vector $\boldsymbol{g}$ which lies in the systematic row of diffraction spots $-\boldsymbol{g}$, o, $\boldsymbol{g}$, $2\boldsymbol{g}$. Figure 2.2(*a*) illustrates the position of the Kikuchi lines for the exact Bragg condition when $s = 0$ (corresponding to figure 2.1(*a*)) and figure 2.2(*b*) illustrates the condition $s > 0$ (corresponding to figure 2.1(*b*)). From a comparison of figures 2.1 and 2.2 it can be seen that the separation, x, of the Kikuchi lines corresponds to an angular separation $2\theta_B$ and the distance, Δx, between the Kikuchi line E and the diffraction spot $\boldsymbol{g}$, which represents the departure from the Bragg condition, corresponds to the small angular tilt $\Delta\theta$. Thus, small angular tilts of the specimen are related to corresponding displacements of the pairs of Kikuchi lines by

$$\Delta\theta/2\theta_B = \Delta x/x.$$

It will be seen in Chapter 3 how diffraction patterns and their accompanying Kikuchi lines can be used for the determination of the diffraction parameters

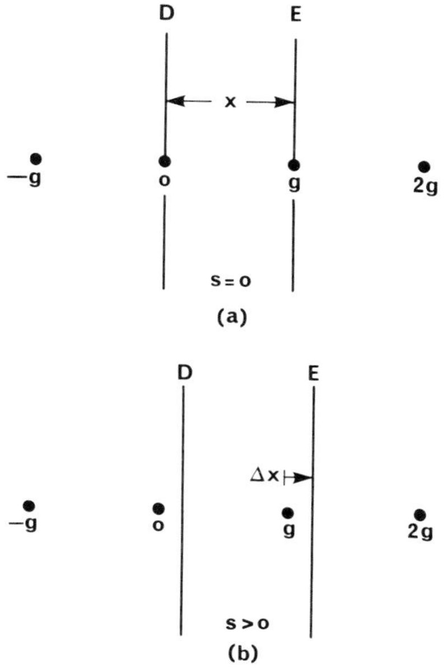

Figure 2.2 Schematic illustration of the relationship between diffraction spots and Kikuchi lines for the exact Bragg condition (*a*) and for a positive deviation from the Bragg condition (*b*).

in equation (2.2) and for the accurate determination of the orientation relationship between two neighbouring crystals meeting at an interface.

2.3 THE TWO-BEAM THEORY OF IMAGE CONTRAST

The two-beam theory of Howie and Whelan (1961) for diffraction from an imperfect crystal, in which atoms are displaced from their normal lattice sites, is introduced in section 2.3.1, and in section 2.3.2 the application of the theory to the computation of theoretical images will be discussed.

2.3.1 The Howie–Whelan Equations

When a crystal is oriented relative to the direction of the incident electron beam so that the two-beam diffraction condition (as described in section 2.2) is satisfied for a diffracting vector $\boldsymbol{g}$, with a small departure, s, from the Bragg condition, then within the crystal only two electron waves need be considered.

One of these waves corresponds to the transmitted beam, with wave amplitude T, and the other to the diffracted beam, with wave amplitude S. As these transmitted and diffracted beams propagate through the crystal, they will be diffracted one into the other and this dynamic exchange between the two beams occurs throughout the thickness of the crystal. The Howie–Whelan theory uses a column approximation in which it is assumed that this exchange takes place within a narrow column taken through the crystal parallel to the incident beam, and that there is no dynamic exchange between the beams in neighbouring columns. In practice, this approximation works well because the Bragg angles for electron diffraction are small (of the order of $\frac{1}{2}°$) and therefore the separation between the transmitted and diffracted beams is always small. For example, in a specimen 1000 Å thick the maximum separation of a transmitted and a diffracted beam at the bottom surface of the specimen would be only approximately 17 Å. For each column the exchange between the wave amplitudes T and S at any point down the column is described by a pair of first-order differential equations which, for centrosymmetric crystals, can be written as

$$
\begin{aligned}
\mathrm{d}T/\mathrm{d}z &= \pi\mathrm{i}(1/\xi_0 + \mathrm{i}/\xi_0')T + \pi\mathrm{i}(1/\xi_{\boldsymbol{g}} + \mathrm{i}/\xi_{\boldsymbol{g}}')S\exp(2\pi\mathrm{i}sz + 2\pi\mathrm{i}\boldsymbol{g}\cdot\boldsymbol{R}) \\
\mathrm{d}S/\mathrm{d}z &= \pi\mathrm{i}(1/\xi_0 + \mathrm{i}/\xi_0')S + \pi\mathrm{i}(1/\xi_{\boldsymbol{g}} + \mathrm{i}/\xi_{\boldsymbol{g}}')T\exp(-2\pi\mathrm{i}sz - 2\pi\mathrm{i}\boldsymbol{g}\cdot\boldsymbol{R})
\end{aligned}
\tag{2.4}
$$

where $\boldsymbol{g}$ is the diffracting vector, z is the coordinate down the column measured in the direction of the incident beam, $\boldsymbol{R}$ is the displacement field in the crystal at the coordinate z in the column and s is a measure of the deviation from the Bragg condition and is the magnitude of $\boldsymbol{s}$ in the z direction; ξ_0 has the dimension of length and is a measure of the mean refractive index for the electron waves in the crystal; $\xi_{\boldsymbol{g}}$ is the extinction distance for the diffracting vector $\boldsymbol{g}$ and is a measure of the distance down the column between consecutive intensity maxima for either of the electron waves; ξ_0' and $\xi_{\boldsymbol{g}}'$ are parameters which are included to take account of absorption effects due to loss of electrons by inelastic scattering and are expressed as the imaginary part of the refractive index and the imaginary part of the extinction distance respectively.

Before contrast calculations are made, equations (2.4) can be reduced to a more convenient form by multiplying the complex wave amplitudes T and S by a suitable phase factor. This is possible because in image contrast the concern is with electron intensities, i.e. the values of $|T|^2$ or $|S|^2$ at the exit surface of the specimen, and these values are independent of the phases chosen for T and S. Equations (2.4) are simplified in this way by taking

$$
\begin{aligned}
T' &= T\exp(-\pi\mathrm{i}z/\xi_0) \\
S' &= S\exp(2\pi\mathrm{i}sz - \pi\mathrm{i}z/\xi_0 + 2\pi\mathrm{i}\boldsymbol{g}\cdot\boldsymbol{R})
\end{aligned}
$$

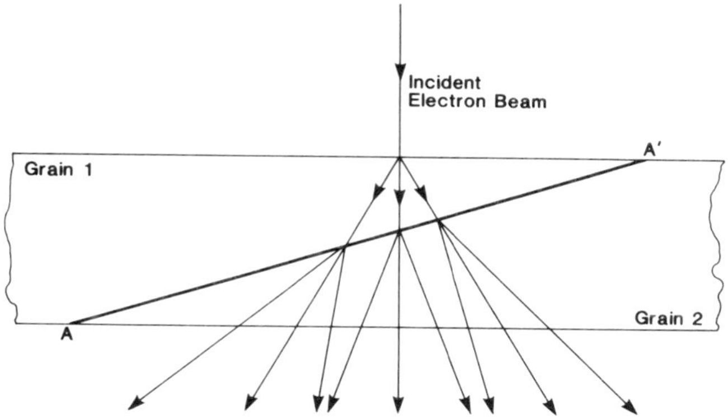

Figure 2.3 Schematic illustration of the electron beams excited by diffraction in a bicrystal.

a bicrystal is more complicated than diffraction from a single crystal because, on crossing the interface AA′, there is a change in orientation and/or structure, so that each of the diffracted beams from grain 1 will generate a new set of diffracted beams in grain 2 which will emerge at the exit surface of the bicrystal. Clearly, in order to apply the two-beam theory of Howie and Whelan to the case of a bicrystal, it is necessary to set up simple well-defined diffraction conditions in both grains. Three such conditions are:

(i) two-beam diffraction operating simultaneously in both grains with *different* diffracting vectors in each grain;
(ii) two-beam diffraction operating simultaneously in both grains, but with the *same* diffracting vector in each grain;
(iii) two-beam diffraction operating in one grain with *no* diffracted beams strongly excited in the other grain.

The Howie–Whelan theory of two-beam diffraction can be applied to each of these diffraction conditions (Humble and Forwood 1975, Forwood and Humble 1975), but since conditions (ii) and (iii) can be regarded as special cases of (i), the application of the theory will first be discussed in detail for the simultaneous double two-beam diffraction condition (i) before discussing the other cases.

2.4.1 Simultaneous Double Two-beam Imaging Condition (i)

In order to set up the simultaneous double two-beam condition (i), one of the grains of the bicrystal, say grain 1 of figure 2.4(a), is tilted with respect to the incident beam so as to obtain a two-beam diffraction condition. This condition for a diffracting vector $\boldsymbol{g}_1$, with its corresponding Kikuchi lines, is illustrated by the schematic diffraction pattern of figure 2.4(b). The bicrystal

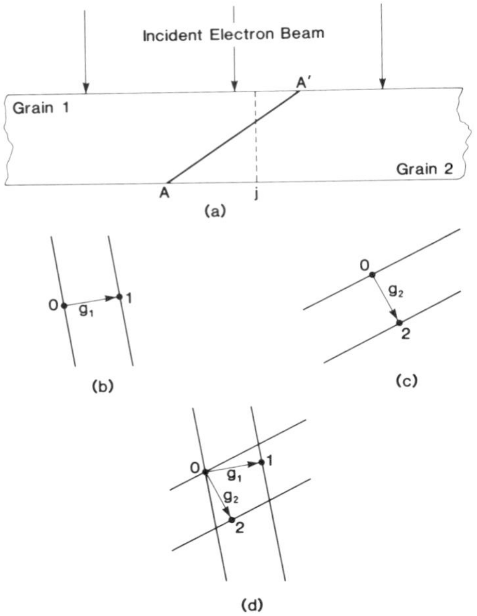

Figure 2.4 Schematic illustration of simultaneous double two-beam diffraction in a bicrystal.

is then tilted around an axis parallel to the diffracting vector $\boldsymbol{g}_1$ until a two-beam condition for a diffracting vector $\boldsymbol{g}_2$ in grain 2 is satisfied (figure 2.4(*c*)). In this way, a simultaneous double two-beam condition is obtained in the region of the interface AA′ as illustrated in figure 2.4(*d*).

For a column through the bicrystal such as that labelled *j* in figure 2.4(*a*), the Howie–Whelan equations will apply, in grain 1, to the central beam 0 and the diffracted beam 1, with diffracting vector $\boldsymbol{g}_1$, from the entrance surface down to the interface. At the interface the two strong beams from grain 1 enter grain 2. The central beam excites the two-beam condition with diffracting vector $\boldsymbol{g}_2$ in grain 2. The diffracted beam 1, from grain 1, acts as a second incident beam in grain 2. However, although this second incident beam excites diffracted beams in grain 2 along the systematic row normal to the Kikuchi lines corresponding to $\boldsymbol{g}_2$, in general, none of these beams will be strongly excited. This is so because, provided there is an appreciable angular separation between $\boldsymbol{g}_1$ and $\boldsymbol{g}_2$, the deviation from the Bragg condition for such beams will be very large. This is illustrated by the relative positions of the diffracted beam 1 and the Kikuchi lines corresponding to $\boldsymbol{g}_2$ in figure 2.4(*d*). More importantly, because these diffracted beams in grain 2 lie in the systematic row normal to the Kikuchi lines of $\boldsymbol{g}_2$, they cannot diffract back into the central beam 0, or its strongly diffracted beam 2. Thus, for column j in grain 2 the Howie–Whelan two-beam theory can also be applied to the central beam 0, and the diffracted beam 2, with diffracting vector $\boldsymbol{g}_2$, from the interface down to the exit surface of the bicrystal.

Clearly, in the calculation of image contrast from a bicrystal, two pairs of Howie–Whelan equations are involved in the integration down a column such as j, one pair in grain 1 and the other in grain 2, and these pairs of equations have to be coupled by appropriate boundary conditions at the interface. The integration of the Howie–Whelan equations in grain 1 starts at the entrance surface with input boundary conditions $T = 1$, $S = 0$, giving wave amplitudes T_I^F and S_I^F at the interface. The integration of the Howie–Whelan equations in grain 2 starts at the interface with input boundary conditions $T = T_\mathrm{I}^\mathrm{F}$, $S = 0$, giving wave amplitudes T^F and S^F at the exit surface of the bicrystal. The simultaneous double two-beam image of the interface is then obtained as a bright-field image, with electron intensity $|T^\mathrm{F}|^2$, formed by the central beam common to both grains.

It should be noted that since the two beams in the lower crystal are excited only by the central beam from the upper crystal, no relative phase information between the central beam and the diffracted beam 1 from grain 1 can be transferred to the central beam and the diffracted beam 2 in grain 2. Thus, a rigid-body displacement between the two grains across the interface AA′ (which is analogous to the displacement associated with a stacking fault in a single crystal) cannot be detected in a simultaneous double two-beam image with different diffracting vectors in each grain. However, it will be seen later that such rigid-body displacements can be detected and determined using simultaneous double two-beam images with the same diffracting vector in both grains (diffraction condition (ii)).

As for the case of a simple two-beam image of a dislocation in a single crystal, the displacement field associated with a dislocation in a bicrystal enters the Howie–Whelan equations via the term β' in equations (2.6). Each pair of Howie–Whelan equations for the bicrystal has its own expression for β', namely

$$\beta'^{(1)} = \mathrm{d}(g_k^{(1)} u_k^{(1)})/\mathrm{d}Z$$

for grain 1 and

$$\beta'^{(2)} = \mathrm{d}(g_k^{(2)} u_k^{(2)})/\mathrm{d}Z$$

for grain 2, where $g_k^{(1)}$ and $g_k^{(2)}$ are the components of the diffracting vector in grains 1 and 2 respectively, and $u_k^{(1)}$ and $u_k^{(2)}$ are the components of the elastic displacement associated with the interfacial dislocation in grains 1 and 2 respectively. In order to obtain a suitable expression for $\beta'^{(1)}$ and $\beta'^{(2)}$ for substitution in equations (2.6), the case of an interfacial dislocation lying along a straight line in a planar interface is considered. A set of cartesian axes is defined in which the normal to the plane of the interface is along the axis Ox_2, pointing from grain 2 into grain 1 (where grain 1 is the upper grain in the sense already defined), and the line direction, $\boldsymbol{r}$, of the interfacial dislocation is along the axis Ox_3. Then, under the assumption that there is no elastic relaxation at the intersections of the interfacial dislocation with the surfaces of the bicrystal, the displacements $u_k^{(1)}$ and $u_k^{(2)}$ are given by equation (1.48) as

$$u_k^{(g)} = \mathscr{R}\left[(2\pi)^{-1}\sum_\alpha A_{k\alpha}^{(g)}M_{\alpha j}^{(g)}G_{ji}^{(g,h)}b_i\ln(z_\alpha^{(g)})\right].$$

Here the superscripts g and h refer to grains 1 and 2 in the same way as for the elastic half-spaces I and II discussed in section 1.5, and take the values $g = 1, h = 2$ when $x_2 > 0$ and $g = 2, h = 1$ when $x_2 < 0$. Thus, for each pair of Howie–Whelan equations β' can be written as

$$\beta'^{(g)} = \mathrm{d}(g_k^{(g)}u_k^{(g)})/\mathrm{d}Z = -\cos\psi\sum_\alpha \frac{P_\alpha^{(g)}x_1 + Q_\alpha^{(g)}x_2}{(x_1 + R_\alpha^{(g)}x_2)^2 + (S_\alpha^{(g)}x_2)^2} \tag{2.7}$$

where

$$P_\alpha^{(g)} = \mathscr{R}[g_k^{(g)}A_{k\alpha}^{(g)}M_{\alpha j}^{(g)}G_{ji}^{(g,h)}b_i(p_\alpha^{(g)} - \tan\theta)(1 + \bar{p}_\alpha^{(g)}\tan\theta)]/T_\alpha^{(g)}$$
$$Q_\alpha^{(g)} = \mathscr{R}[g_k^{(g)}A_{k\alpha}^{(g)}M_{\alpha j}^{(g)}G_{ji}^{(g,h)}b_i(p_\alpha^{(g)} - \tan\theta)(\bar{p}_\alpha^{(g)} - \tan\theta)]/T_\alpha^{(g)}$$
$$R_\alpha^{(g)} = [-2\tan\theta + 2p_\alpha^{(g)}\bar{p}_\alpha^{(g)}\tan\theta + (p_\alpha^{(g)} + \bar{p}_\alpha^{(g)})(1 - \tan^2\theta)]/2T_\alpha^{(g)}$$
$$S_\alpha^{(g)} = \{[\tan^2\theta + p_\alpha^{(g)}\bar{p}_\alpha^{(g)} - (p_\alpha^{(g)} + \bar{p}_\alpha^{(g)})\tan\theta]/T_\alpha^{(g)} - (R_\alpha^{(g)})^2\}^{1/2}$$
$$T_\alpha^{(g)} = 1 + p_\alpha^{(g)}\bar{p}_\alpha^{(g)}\tan^2\theta + (p_\alpha^{(g)} + \bar{p}_\alpha^{(g)})\tan\theta.$$

Here, θ is the angle of rotation around the Ox_3 axis necessary to bring the Ox_2 axis into the plane containing the Ox_3 axis and the direction Z down the column and ψ is the angle between the Ox_2 axis, after the rotation θ, and the $-Z$ direction. The expression $\beta'^{(g)}$ for an interfacial dislocation in equation (2.7) has the same form as β' for a dislocation in a single crystal, as given in section 2.3.1. Therefore, the methods developed to integrate the Howie–Whelan equations and display the theoretical contrast of a dislocation in a single crystal (Hirsch *et al* 1965, Head *et al* 1973) can also be applied to an interfacial dislocation in a bicrystal, and the following discussion will be concerned with the computation of theoretical images.

2.4.2 Application to Image Computation

The method for computing theoretical images which applies the concept of a generalised cross-section to the case of a bicrystal containing an interfacial dislocation will now be discussed for the simultaneous double two-beam diffraction condition (i).

Figure 2.5(a) is a schematic illustration of a bicrystal EFGHKLMN which is tilted with respect to the electron beam and contains an interfacial

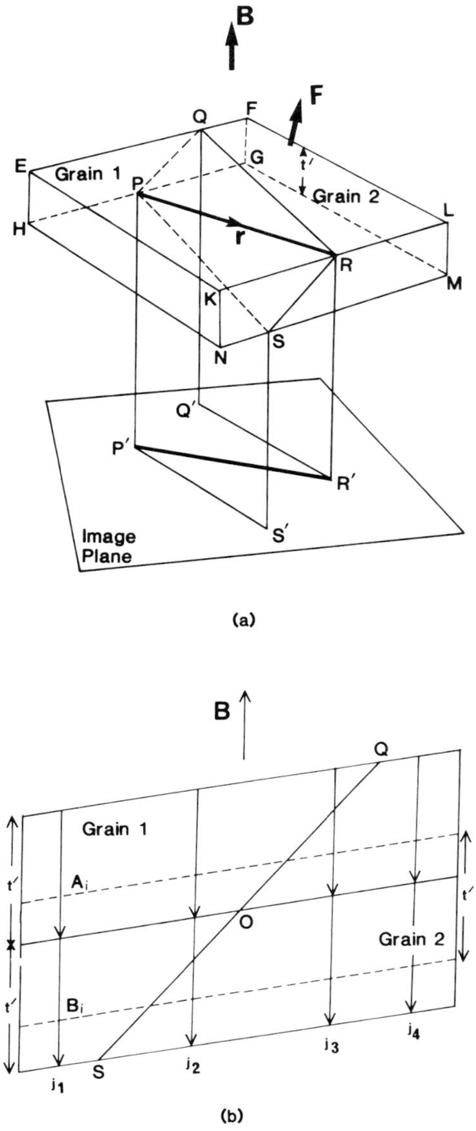

Figure 2.5 (*a*) Schematic illustration of the projection of an interfacial dislocation PR in a planar interface PQRS in a tilted bicrystal on to the fluorescent screen of an electron microscope. In this diagram the surfaces EFLK and HGMN are normal to ***F*** and the edges EH, FG, LM and KN are drawn parallel to ***B***, so that angles such as $\widehat{HEK}$ are not right angles. (*b*) Generalised cross-section of the bicrystal in (*a*).

dislocation PR in the plane of the interface PQRS. The beam direction $\boldsymbol{B}$ is defined to be anti-parallel to the direction of the incident electrons, and the direction $\boldsymbol{r}$ of the interfacial dislocation line is defined in the sense from P to R in figure 2.5(a), i.e. from the bottom to the top of the bicrystal. The normal to the surface of the bicrystal, $\boldsymbol{F}$, is defined as being upward-drawn, i.e. acute with $\boldsymbol{B}$. Included in figure 2.5(a) is a projection, in the direction of the incident electrons, of the interface and the interfacial dislocation on to the image plane which is normal to $\boldsymbol{B}$. This is the projection that applies to an experimental image and to the corresponding computed image which is calculated via the generalised cross-section of figure 2.5(b). The generalised cross-section of figure 2.5(b) is the projection along $\boldsymbol{r}$ of a region of the bicrystal in the vicinity of the interface on to the plane $(\boldsymbol{r} \wedge \boldsymbol{B}) \wedge \boldsymbol{B}$, i.e. on to a plane parallel to the plane KLMN in figure 2.5(a). In the generalised cross-section of figure 2.5(b) the line QS is the projection of the plane PQRS of figure 2.5(a), where the points Q and S in figure 2.5(b) correspond to the projections along $\boldsymbol{r}$ of the points Q and S from figure 2.5(a). The point O in figure 2.5(b) is the projection of the points P and R from figure 2.5(a) and marks the point of emergence of the interfacial dislocation in the generalised cross-section.

A theoretical image of the interfacial dislocation is computed by the numerical integration of the Howie–Whelan equations down successive columns such as j_1, j_2, j_3 and j_4 in the generalised cross-section of figure 2.5(b). Each column of integration then gives the intensity at the image points in one complete row, parallel to the line of the dislocation, in the theoretical image. For a column such as j_1 which is contained solely in grain 1 and does not cross the boundary, only one pair of Howie–Whelan equations is required. Whereas for columns such as j_2 and j_3, which cross the interface QS, two pairs of Howie–Whelan equations are required. The numerical integration of each pair of Howie–Whelan equations uses principle (ii) (section 2.3.2) which allows all solutions for the wave amplitudes T and S to be obtained as linear combinations of two independent solutions. For columns such as j_1 the computational procedure is identical with that described in detail in Chapter 4 of Head *et al* (1973) for a dislocation in a single crystal. In brief, two independent integrations of the Howie–Whelan equations for grain 1 are made down the column j_1 through the thickness $2t'$ of the generalised cross-section. The input boundary conditions for these two integrations are ${}_{(1)}T^{(1)} = 1$, ${}_{(1)}S^{(1)} = 0$ and ${}_{(2)}T^{(1)} = 0$, ${}_{(2)}S^{(1)} = 1$, where the subscripted prefixes (1) and (2) refer to the two independent integrations and the superscripts indicate the grain in which the integration is being carried out. During both integrations, $2n$ values of ${}_{(1)}T^{(1)}$, ${}_{(1)}S^{(1)}$ and ${}_{(2)}T^{(1)}$, ${}_{(2)}S^{(1)}$, at depths t'/n down the column, are stored. These values therefore correspond to n pairs of points which are t' apart, such as those labelled A_i and B_i in figure 2.5(b). The intensities of image points in the row corresponding to the j_1th column of integration are then obtained from linear combinations of these stored values in the following way. The two independent integrations at points such as A_i in the top half of the generalised cross-section are combined linearly by proportionality coefficients a_i and b_i to give the boundary condition $T^{(1)} = 1$, $S^{(1)} = 0$, appropriate to the electron beam entering a free surface at A_i, that is

$$\left.\begin{aligned} a_{i\,(1)}T^{(1)}_{A_i} + b_{i\,(2)}T^{(1)}_{A_i} &= 1 \\ a_{i\,(1)}S^{(1)}_{A_i} + b_{i\,(2)}S^{(1)}_{A_i} &= 0 \end{aligned}\right\} \quad \text{at } A_i. \tag{2.8}$$

The proportionality coefficients, a_i and b_i, obtained from equations (2.8) are used at the points B_i (t' below A_i) to obtain the wave amplitudes $T_i^{(1)F}$ at the exit surface of the specimen as

$$T_i^{(1)F} = a_{i\,(1)}T^{(1)}_{B_i} + b_{i\,(2)}T^{(1)}_{B_i} \qquad \text{at } B_i. \tag{2.9}$$

Thus the intensity in the ith image point in the row of the theoretical image corresponding to the j_1th column is given by $|T_i^{(1)F}|^2$. The other image points in the row are obtained from similar linear combinations of the remaining values for the other $(n-1)$ pairs of points A_i and B_i in the j_1th column stored during the integrations. The same procedure is adopted to obtain the row of the theoretical image corresponding to a column such as j_4 which is completely contained in grain 2, except in this case the Howie–Whelan equations appropriate to diffraction in grain 2 are used.

The above procedure is modified for columns such as j_2 and j_3 which cross the interface QS. For such a column, when the two integrations of the Howie–Whelan equations in grain 1 reach the interface, the wave amplitudes of the transmitted beams at the interface ${}_{(1)}T_{\rm I}^{(1)}$ and ${}_{(2)}T_{\rm I}^{(1)}$ are stored and two new independent integrations are started for the Howie–Whelan equations in grain 2, with ${}_{(1)}T^{(2)} = 1$, ${}_{(1)}S^{(2)} = 0$ and ${}_{(2)}T^{(2)} = 0$, ${}_{(2)}S^{(2)} = 1$, and continued through the remainder of the column. When A_i and B_i both lie in the same crystal, then the final wave amplitudes at the exit surface of the bicrystal are obtained from equations of the type (2.8) and (2.9) in the manner already described. However, when A_i and B_i are on opposite sides of the interface a different procedure is necessary in order to couple the two pairs of Howie–Whelan equations at the interface as discussed in section 2.4.1. In this case the proportionality coefficients a_i and b_i at A_i, obtained from equation (2.8), are used to determine the wave amplitude $T_{\rm I}^{(1)F}$ transmitted in the direct beam across the interface as

$$T_{\rm I}^{(1)F} = a_{i\,(1)}T_{\rm I}^{(1)} + b_{i\,(2)}T_{\rm I}^{(1)}. \tag{2.10}$$

Since only the direct beam $T_{\rm I}^{(1)F}$ from grain 1 excites the two-beam diffraction condition in grain 2, the input conditions at the interface for the integration of the Howie–Whelan equations in grain 2 are $T^{(2)} = T_{\rm I}^{(1)F}$, $S^{(2)} = 0$. Since the input conditions chosen for the independent integration (1) in grain 2 are ${}_{(1)}T^{(2)} = 1$, ${}_{(1)}S^{(2)} = 0$, the determination of the wave amplitude $T_i^{(2)F}$ of the direct beam at the exit surface of the bicrystal does not require linear combinations of the type given in equations (2.9), but is given simply as

$$T_i^{(2)\,\mathrm{F}} = T_{\mathrm{I}}^{(1)\,\mathrm{F}}\ {}_{(1)}T_{\mathrm{B}_i}^{(2)}. \tag{2.11}$$

Thus $|T_i^{(2)\,\mathrm{F}}|^2$ gives the intensities of those image points in a row of the theoretical images which are derived from stored values of the integrations for which the pairs of points A_i and B_i lie on different sides of the interface QS in the generalised cross-section. Such image points are the most important ones in a theoretical image, since it is these points which constitute the characteristic features of contrast associated with the interfacial dislocation.

2.4.3 Computer Program for Interfacial Dislocations

The structure of the computer program which has been developed to compute theoretical images of interfacial dislocations under simultaneous double two-beam diffraction conditions is given in the block diagram of figure 2.6. The program is called PCGBD and it computes the contrast of one interfacial dislocation, or two interfacial dislocations separated in the plane of the interface. It is available on a floppy disc from the CSIRO Division of Materials Science and Technology and is suitable for use on personal computers of the IBM type with the theoretical images printed on an EPSON FX/80 printer. It is this program which is used in this book for the image matching process in which theoretical images are compared with experimental electron micrographs for the determination of the Burgers vectors of interfacial dislocations. The program is a development of the one described in Chapter 10 of Head *et al* (1973) which is used for computing the contrast of a dislocation in a single crystal, and this chapter (of Head *et al*) should be consulted if details of the computational methods are required.

The essential data required to operate the program, 'Data in' of figure 2.6, are listed in table 2.1. Of the parameters in this table the elastic constants are usually known, $\mathscr{A}$ and ξ_g can be calculated for the particular crystal

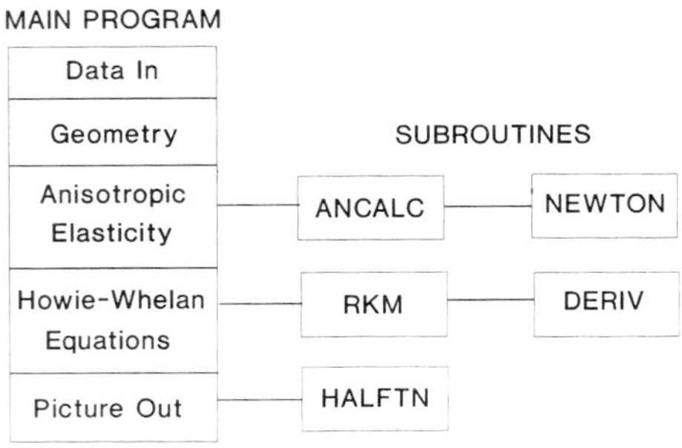

Figure 2.6 Structure of the computer program PCGBD.

Table 2.1 Essential data for operating interfacial dislocation image program.

Input data	Grain 1	Grain 2
Orientation relationship	$(\boldsymbol{u}, \theta)$	
Elastic constants	$c_{ij}^{(1)}$	$c_{ij}^{(2)}$
Diffracting vector	$\boldsymbol{g}_1$	$\boldsymbol{g}_2$
Deviation from the Bragg condition	w_1	w_2
Anomalous absorption coefficient	$\mathscr{A}_1$	$\mathscr{A}_2$
Extinction distance	ξ_{g_1}	ξ_{g_2}
Beam direction	$\boldsymbol{B}$	
Specimen surface normal	$\boldsymbol{F}$	
Specimen thickness	t	
Interface normal	$\boldsymbol{v}$	
Line direction of interfacial dislocation	$\boldsymbol{r}$	
Trial Burgers vector of interfacial dislocation	$\boldsymbol{b}$	

structures involved (see section 3.6) and the remaining parameters, including the orientation relationship between the two grains, are determined from experimental electron micrographs and diffraction patterns (see sections 3.4, 3.5 and 3.7). Some of the input parameters are defined in terms of the crystallographic axes of both grains 1 and 2 and others only in terms of the crystallographic axes of grain 1, as indicated in the table.

The section of the program labelled 'Geometry' in figure 2.6 is concerned with setting up the appropriate axis system required for determining the anisotropic displacement field of the interfacial dislocation (section 1.6), and the appropriate axis system for integrating the Howie–Whelan equations in the generalised cross-section (section 2.4.2).

In the section 'Anisotropic elasticity' the subroutine ANCALC is called, once for each grain, to calculate the elastic matrices $A_{k\alpha}^{(g)}$, $M_{\alpha j}^{(g)}$ and $G_{ji}^{(g,h)}$ required in equation (2.7) for β', and it does this by calling the subroutine NEWTON to obtain the roots p_α of the sextic equation (1.31). The remainder of 'Anisotropic elasticity' calculates the constants $P_\alpha^{(g)}$, $Q_\alpha^{(g)}$, $R_\alpha^{(g)}$ and $S_\alpha^{(g)}$ in equation (2.7).

The part of the program in the block labelled 'Howie–Whelan equations' first sets up the geometry necessary to define the grid of image points in the computed image which determines, for example, the spacing of the integration columns in the generalised cross-section. It then proceeds to the two independent numerical integrations of the Howie–Whelan equations to obtain the two independent solutions for each of the grains, as discussed in section 2.4.2. Numerical integration is done using the subroutine RKM which in turn calls the subroutine DERIV for evaluation of the terms in the Howie–Whelan equations (2.6).

The 'Picture out' section of the program, together with the subroutine HALFTN, is concerned with the production of the computed half-tone image which is printed a row at a time, with each row corresponding to a column of integration in the generalised cross-section.

The parameter in equations (2.6) which is not included in the input data in table 2.1 is the normal absorption coefficient $\mathscr{N}$. For the case of a single crystal of constant thickness, the normal absorption coefficient acts as a constant scaling factor for the intensities of all image points. Since (as pointed out in section 2.3.2) the intensities in a theoretical micrograph of a dislocation in a single crystal are expressed relative to a fixed background intensity, $\mathscr{N}$ plays no part in assigning shades of grey to the image points. However, in the case of a bicrystal imaged under the simultaneous double two-beam condition (i), the different diffracting conditions in each grain result in different background intensities and different values of normal absorption coefficient, $\mathscr{N}_1$ and $\mathscr{N}_2$. These coefficients, for a specimen of constant thickness t' in the beam direction, occur in expressions for the intensities transmitted by each grain as factors $\exp(-2\mathscr{N}_1 t')$ and $\exp(-2\mathscr{N}_2 t')$ for grains 1 and 2 respectively. In the computation of a theoretical micrograph of a bicrystal, the intensities, I, at the image points are always expressed relative to the lighter of the two background intensities I_0, so that the shade of grey at each image point is determined by the value of $\ln(I/I_0)$. Thus the effect of the two values of $\mathscr{N}$ is simply a linear transition in grey shades across the interface from the lighter background of one grain to the darker background of the other. Such a linear graduation in grey will not have a significant effect on the characteristic features of contrast associated with an interfacial dislocation and, therefore, the values used for $\mathscr{N}_1$ and $\mathscr{N}_2$ are not very critical for dislocation contrast. Experimental values for the parameter $\mathscr{N}$ are not well known, but the way in which $\mathscr{N}$ changes with the diffracting vector for a particular crystal is similar to the way in which the anomalous absorption coefficient $\mathscr{A}$ changes (Radi 1970). Thus, in practice, values of the normal absorption coefficients $\mathscr{N}_1$ and $\mathscr{N}_2$ are put equal to the values of the anomalous absorption coefficients $\mathscr{A}_1$ and $\mathscr{A}_2$ respectively.

2.4.4 Verification of Image Matching Procedure for Interfacial Dislocations

In this section the method of computing theoretical images described in sections 2.4.2 and 2.4.3 will be tested by comparing computed images with experimental images of interfacial dislocations which have known Burgers vectors. In practice an interfacial dislocation with a known Burgers vector can be obtained in an interphase interface or in a grain boundary by making use of a glide dislocation which has intersected the interface in such a way that it lies partly in one of the grains and partly as an extrinsic dislocation in the interface. In this way, two segments of a dislocation (which of course have the same Burgers vector) are obtained, one segment in one grain and the other in the interface. Using the segment of the dislocation in the grain, the Burgers vector can be determined by the routine method of image matching for a dislocation in a single crystal (Head *et al* 1973). Then, experimental micrographs of the segment of the dislocation in the interface can be compared with the corresponding theoretical images computed for the determined Burgers vector using the interfacial dislocation program. Good agreement between the experimental and theoretical micrographs would then verify that the program is a good approximation in the way that it treats the diffraction processes and image formation for interfaces and interfacial dislocations in the electron microscope. Such a test is only complete if the set of simultaneous double two-beam experimental images selected contains at least three non-coplanar diffracting vectors, so that the full three-dimensional displacement field of the interfacial dislocation is sampled. This set of three non-coplanar diffracting vectors can be made up of diffracting vectors from either grain.

Examples of this test are shown for an interphase interface in a two-phase copper–iron alloy in figures 2.7, 2.8 and 2.9, and for a high-angle grain boundary in austenitic stainless steel in figure 2.10. Figure 2.7 is an example of the type of interphase interface observed in a Cu–25 wt% Fe alloy (Forwood and Clarebrough 1989) where, in this case, the FCC copper matrix is the upper grain (grain 1) and the BCC iron phase is the lower grain (grain 2). The dislocation lying in the interface along RQ has been generated by the slip dislocation QP in the copper matrix. The segment QP, which runs from the interface to the top surface of the thin-foil specimen, was identified by routine image matching as having the Burgers vector $(1/2)[01\bar{1}]_{\mathrm{Cu}}$. An example of one of the comparisons made in this identification which establishes the sign of the Burgers vector is given in figure 2.8. In this example the experimental micrograph in figure 2.8(*a*) is compared with a matching computed image for the Burgers vector $(1/2)[01\bar{1}]_{\mathrm{Cu}}$ in figure 2.8(*b*) and the mismatching computed image for the Burgers vector of opposite sign $(1/2)[0\bar{1}1]_{\mathrm{Cu}}$ in figure 2.8(*c*). The Burgers vector of the dislocation lying in the interface along RQ must also be $(1/2)[01\bar{1}]_{\mathrm{Cu}}$. Figure 2.9 shows three experimental images, (*a*), and their corresponding theoretical images, (*b*), for the dislocation along RQ computed for the Burgers vector $(1/2)[01\bar{1}]_{\mathrm{Cu}}$. The set of experimental images (i), (ii) and (iii) involves three non-coplanar diffracting vectors in both grains 1 and 2. It can be seen from a comparison of the images in figure 2.9 that the $(1/2)[01\bar{1}]_{\mathrm{Cu}}$ Burgers vector gives excellent agreement between the experimental images and their corresponding theoretical images. This test demonstrates that the computer program described in section 2.4.3 can be used for the identification of the Burgers vectors of interfacial dislocations in interphase interfaces.

Figure 2.10 shows the test procedure applied to a high-angle grain boundary in austenitic stainless steel separating two grains U and L. This boundary is used in an analysis of slip transfer in section 6.5.2 and is shown

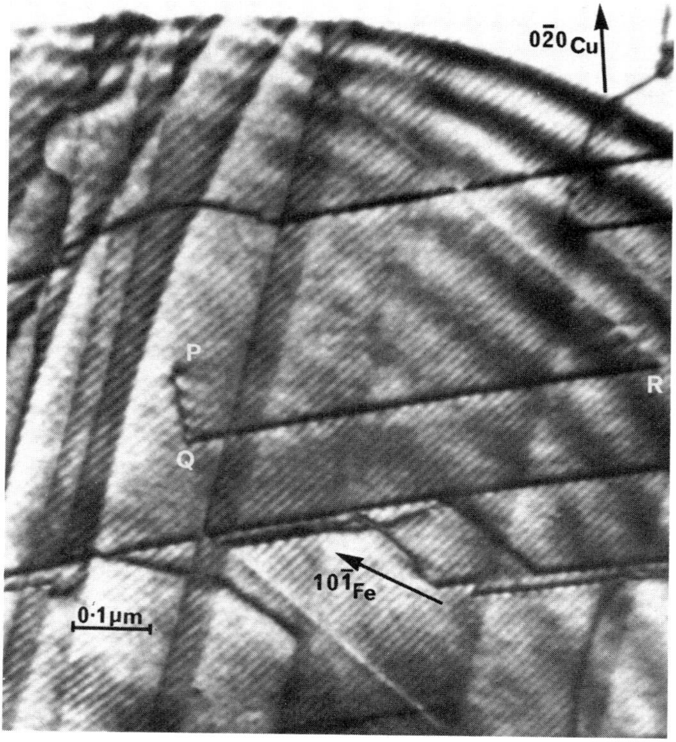

Figure 2.7 Double two-beam image of part of an interface between copper and iron phases in a two-phase Cu–Fe alloy. The diffracting vectors are indicated.

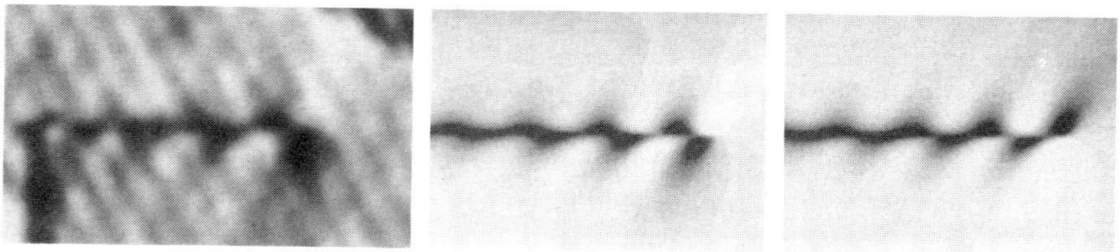

a b c

Figure 2.8 Comparison of an experimental two-beam image (*a*) of the segment QP (figure 2.7) of a slip dislocation in the copper matrix with computed images (*b*) and (*c*). The Burgers vectors are $(1/2)[01\bar{1}]_{Cu}$ in (*b*) and $(1/2)[0\bar{1}1]_{Cu}$ in (*c*), the line direction $\boldsymbol{r}$ of the dislocation is $[0\bar{1}1]_{Cu}$, the beam direction $\boldsymbol{B}$ is $[\bar{1}03]_{Cu}$, the diffracting vector $\boldsymbol{g}$ is $0\bar{2}0$ and w is 0.2.

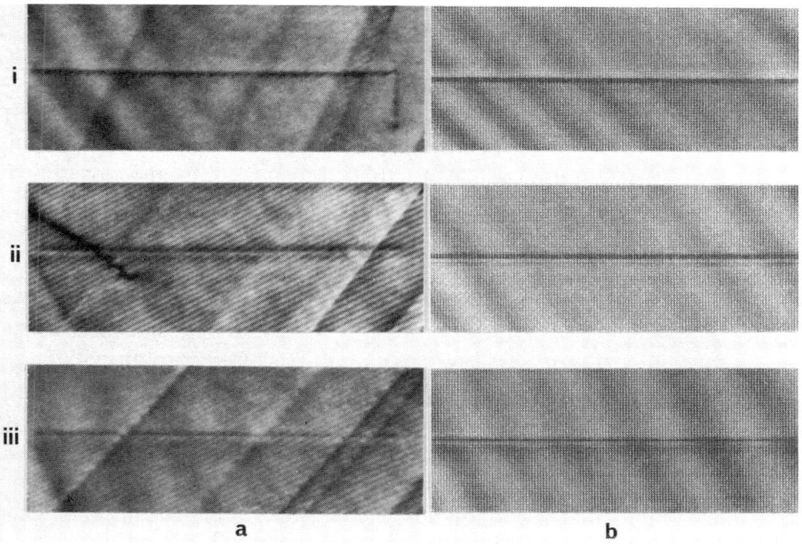

Figure 2.9 Comparison of double two-beam experimental images (*a*) with matching computed images (*b*) for the interfacial dislocation along RQ (figure 2.7). The thin-foil specimen is $10\xi_{\mathrm{Cu111}}$ thick with foil normal $[\bar{2}\bar{3}4]_{\mathrm{Cu}}$. The Burgers vector in the computed images is $(1/2)[01\bar{1}]_{\mathrm{Cu}}$ and $\boldsymbol{r}$ is $[25\ 3\ 22]_{\mathrm{Cu}}$. The values of $\boldsymbol{g}$; $\boldsymbol{B}$; and w are as follows:

(i) $0\bar{2}0_{\mathrm{Cu}}$, $10\bar{1}_{\mathrm{Fe}}$; $[\bar{1}03]_{\mathrm{Cu}}$; 0.2_{Cu}, 0.2_{Fe}
(ii) 111_{Cu}, $\bar{1}10_{\mathrm{Fe}}$; $[\bar{6}\bar{1}7]_{\mathrm{Cu}}$; 0.3_{Cu}, 0.3_{Fe}
(iii) 200_{Cu}, $0\bar{1}\bar{1}_{\mathrm{Fe}}$; $[0\bar{7}4]_{\mathrm{Cu}}$; 0.2_{Cu}, 0.2_{Fe}.

in more detail in figure 6.43. In this case the leading slip dislocation *D*2, from a pile-up of dislocations labelled 2 in figure 6.43, has entered the boundary from grain L, and the Burgers vector of the slip dislocations in the pile-up was determined by routine image matching as $(1/2)[110]_{\mathrm{L}}$. Figure 2.10(*a*) shows three experimental images of the leading slip dislocation in the boundary. These images involve three non-coplanar diffracting vectors. Their corresponding theoretical images, computed for the Burgers vector $(1/2)[110]_{\mathrm{L}}$, are shown in figure 2.10(*b*). In this case the excellent agreement between the experimental and theoretical images verifies the applicability of the computer program for the identification of grain boundary dislocations.

For simultaneous double two-beam images with different diffracting vectors operating in each grain, computation is necessary to interpret the image contrast and determine the Burgers vector of an interfacial dislocation because $\boldsymbol{g}\cdot\boldsymbol{b}$ criteria, of the type discussed in section 2.3.1, cannot be applied.

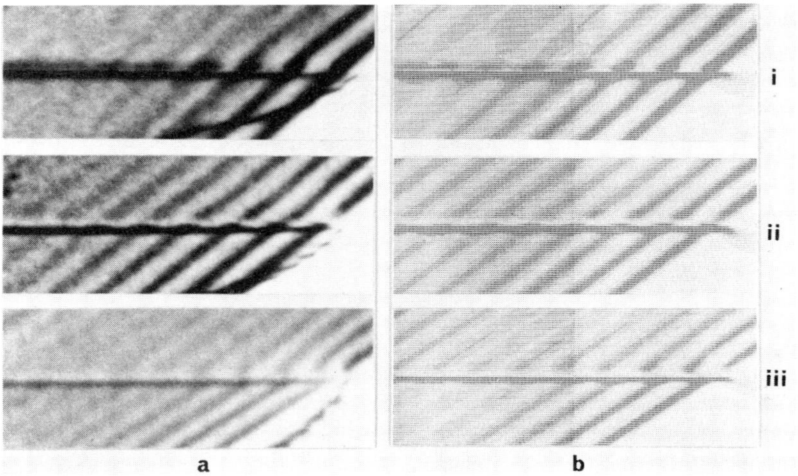

Figure 2.10 Three experimental double two-beam images (a) of a grain boundary dislocation in stainless steel with the matching set of computed images (b) for a Burgers vector of $(1/2)[110]_L$. In (i), $\boldsymbol{B}$ is close to $[13\bar{2}]_U$ with $\boldsymbol{g}$ $\bar{1}11_{U'}$ and $[703]_L$ with $\boldsymbol{g}$ $0\bar{2}0_L$. In (ii), $\boldsymbol{B}$ is close to $[01\bar{1}]_U$ with $\boldsymbol{g}$ $1\bar{1}\bar{1}_{U'}$ and $[5\bar{2}3]_L$ with $\boldsymbol{g}$ $\bar{1}\bar{1}1_L$. In (iii), $\boldsymbol{B}$ is close to $[\bar{2}3\bar{1}]_U$ with $\boldsymbol{g}$ $\bar{1}\bar{1}1_{U'}$ and $[6\bar{5}\bar{1}]_L$ with $\boldsymbol{g}$ $\bar{1}\bar{1}\bar{1}_L$.

2.4.5 Simultaneous Double Two-beam Imaging Condition (ii)

A special case, condition (ii), of the simultaneous double two-beam condition is when the same diffracting vector, $\boldsymbol{g}_c$, is present in each grain. This corresponds to two-beam electron diffraction from a set of crystallographic planes in each grain which are parallel and equally spaced. This situation commonly occurs across grain boundaries for particular orientation relationships between neighbouring grains and is associated with coincident site lattice (CSL) orientations. With reference to figure 2.4, the same-$\boldsymbol{g}_c$ condition corresponds to the situation where $\boldsymbol{g}_1$ and $\boldsymbol{g}_2$ are identical, and the two beams 0, 1 from grain 1 are completely coupled with and constitute the two beams 0, 2 in grain 2. Thus, in this case diffraction from the bicrystal reduces to that of two-beam diffraction from a single crystal but, of course, the elastic displacements of any interfacial dislocation are still those associated with the bicrystal. Clearly, for these diffraction conditions phase information between the central and diffracted beams is transmitted across the interface, so that any rigid-body displacement between the grains, which gives rise to an offset of the diffracting planes at the interface, can be detected. In the practical use of diffraction condition (ii), electron diffraction patterns from neighbouring grains often indicate a small difference in the deviation from the Bragg condition w associated with the same $\boldsymbol{g}_c$ in each grain. This situation is accommodated in the treatment of the diffraction process by including appropriate values of w in the Howie–Whelan equations for each grain but still considering that the two beams 0, 1 from grain 1 are still completely coupled with the beams 0, 2 in grain 2 as before.

When a same-$\boldsymbol{g}_c$ image is to be computed using the interfacial dislocation program this is indicated in the input data. In this case, new integrations of the Howie–Whelan equations are not started at the interface QS in the generalised cross-section of figure 2.5(b), as described previously for diffraction condition (i), but the integrations are continued with T being continuous across the interface, and S also being continuous apart from the inclusion of a phase factor, $\exp(2\pi i \boldsymbol{g}_c \cdot \boldsymbol{R})$, which takes account of any rigid-body displacement $\boldsymbol{R}$ of grain 2 relative to grain 1 across the interface. In the continuation of the integrations in grain 2 the appropriate value of w is used. Thus, the interfacial dislocation program allows the strain contrast associated with interfacial dislocation to be computed, not only for the simultaneous double two-beam diffraction condition with different diffracting vectors operating in both grains, but also for the simultaneous double two-beam diffraction condition with the same diffracting vector operating in both grains. As a result, same-$\boldsymbol{g}_c$ diffracting vectors can be part of the set of three non-coplanar diffracting vectors required for the identification of the Burgers vector of an interfacial dislocation by image matching. In addition, the fringe contrast which is associated with rigid-body displacements between grains in same-$\boldsymbol{g}_c$ images can be computed and, from the comparison of experimental and corresponding theoretical images from three non-coplanar same-$\boldsymbol{g}_c$ vectors, the rigid-body displacement $\boldsymbol{R}$ can be determined (see section 5.4).

Since, from a diffraction viewpoint, same-$\boldsymbol{g}_c$ images of interfacial dislocations are essentially equivalent to two-beam images of dislocations in single crystals (Barry and Mahajan 1971), $\boldsymbol{g} \cdot \boldsymbol{b} = 0$ invisibility criteria can be used subject to the same limitations already discussed in section 2.3.1. An example of the use of $\boldsymbol{g} \cdot \boldsymbol{b} = 0$ criteria is given in figure 2.11 where interfacial dislocations in a grain boundary show virtually no contrast in same-$\boldsymbol{g}_c$ images when $\boldsymbol{g}_c \cdot \boldsymbol{b} = 0$. The example is for a coherent twin boundary on $(111)_U/(111)_L$, separating an upper grain U from a lower grain L, in which the three $(1/6)\langle 11\bar{2}\rangle_U$ dislocations a, b, c form an hexagonal array as shown by the simultaneous double two-beam image of figure 2.11(a). Three different same-$\boldsymbol{g}_c$ images are shown in figures 2.11(b), (c) and (d), and in each of these images one component of the hexagonal network is out of contrast. By assigning $\boldsymbol{g}_c \cdot \boldsymbol{b} = 0$ to these invisibilities, the Burgers vectors of the interfacial dislocations can be designated, with an uncertainty as to sign, as follows: $\pm(1/6)[2\bar{1}\bar{1}]_U$ for a, $\pm(1/6)[1\bar{2}1]_U$ for b and $\pm(1/6)[\bar{1}\bar{1}2]_U$ for c.

An essential feature of the same-$\boldsymbol{g}_c$ images in figure 2.11, which enables invisibility criteria to be applied, is that no fringe contrast due to a rigid-body displacement between grains U and L is present. In cases where rigid-body displacement gives rise to strong fringe contrast, the $\boldsymbol{g}_c \cdot \boldsymbol{b} = 0$ invisibility

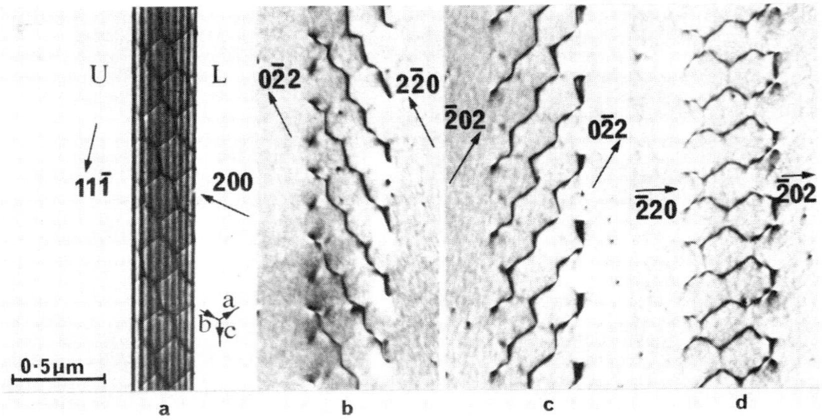

Figure 2.11 Double two-beam images of a near-Σ3 coherent twin-boundary on $111_{U/L}$ in a Cu − 6 at% Si alloy, the diffracting vectors are indicated. In (*a*), (*b*), (*c*) and (*d*) $\boldsymbol{B}$ is close to $[\bar{1}43]_U/[015]_L$, $[011]_U/[114]_L$, $[111]_U/[111]_L$ and $[111]_U/[111]_L$ respectively.

criteria fail, and this is illustrated in figure 2.12. Figure 2.12(*a*) is a same-$\boldsymbol{g}_c = 02\bar{2}_U/\bar{2}20_L$ experimental electron micrograph of an incoherent twin-boundary on the $(\bar{4}21)_U/(\bar{8}\ \bar{5}\ 10)_L$ plane. The fringe contrast in this image is very strong due to the presence of a rigid-body displacement. There are two arrays of dislocations in the boundary; in one array the individual dislocations are in strong contrast and run diagonally across the micrograph, whilst only one interfacial dislocation of the other array is present (which is horizontal in the figure and shows white contrast). Detailed image matching showed that the Burgers vector of this horizontal dislocation is $(1/3)[111]_{U/L}$, so that $\boldsymbol{g}_c \cdot \boldsymbol{b} = 0$, but, despite this, the dislocation is in contrast in the micrograph of figure 2.12(*a*). This failure to obtain invisibility for $\boldsymbol{g}_c \cdot \boldsymbol{b} = 0$ in the presence of displacement fringes is confirmed by the computed images in figures 2.12(*b*) and (*c*), which are computed for interfacial dislocations with Burgers vectors $(1/3)[111]_{U/L}$ and $(1/3)[\bar{1}\bar{1}\bar{1}]_{U/L}$ respectively. Both of these computed images are for $\boldsymbol{g}_c \cdot \boldsymbol{b} = 0$, but both show strong contrast.

The similarity between the diffraction process in single crystals and that in same-$\boldsymbol{g}_c$ images of bicrystals is used in the technique developed by Ishida *et al* (1980, 1986) for the determination of the Burgers vectors of interfacial dislocations. Ishida *et al* showed that at the intersection of an interfacial dislocation with either surface of the specimen there is an apparent change in the number of pendellösung fringes. They demonstrated that the Burgers vector of the interfacial dislocation is related to the integral number n of terminating pendellösung fringes by $\boldsymbol{g}_c \cdot \boldsymbol{b} = n$. This enables the Burgers vector $\boldsymbol{b}$, including its sign, to be determined. This method has not been widely used

scattering angle, θ_s, given by

$$\theta_s = 1.5(\lambda/C_s)^{1/4}.$$

Under these conditions, the point-to-point resolution obtained is given by

$$x = 0.66C_s^{1/4}\lambda^{3/4}.$$

The discussion so far has been for the case in which the crystal specimen is thin enough to be regarded essentially as a phase object. For thicker specimens, computation using the full *n*-beam theory of electron diffraction is necessary in order to interpret image contrast in terms of atomic structure. The procedure is one where an assumed structure for the crystal is tested by calculating the corresponding theoretical images and comparing them with experimental images. The most common method for calculating the image contrast arising from dynamical *n*-beam diffraction is based on the analytical multi-slice description put forward by Cowley and Moodie (1957). In the multi-slice description of the diffraction process the crystal is divided into planar slices which are normal to the direction of the incident electron beam and sufficiently thin for the phase object approximation to apply to each slice. The *n*-beam electron scattering from the crystal as a whole is then obtained by the sequential linking of the scattering from each of the slices by a suitable propagation function which accounts for the propagation of the electron waves from one slice to the next. In the computer programs based on the Cowley–Moodie multi-slice description of *n*-beam diffraction full account is taken of the degree of defocus, spherical aberration, astigmatism, chromatic aberration and beam divergence (see, for example, Lynch and O'Keefe 1972, O'Keefe 1973).

For a crystal containing a defect such as a dislocation or a stacking fault, the calculations are more complicated because not only must the crystal structure be periodic in the plane of the slice, but also the defect has to be made a periodic feature of the structure of the slice. This involves artificially replacing, in the computations, the single defect by a periodic set of defects which are sufficiently well-spaced so that the computed lattice image associated with each defect does not interfere with that of its neighbour. This is known as the 'periodic continuation approximation'. For the case of a bicrystal containing an interface, such a periodic continuation would involve constructing a periodic set of interfaces in each slice. Then, in using computed images to test trial models for the structure of the interface, the coordinates of all the atoms must be specified within a unit cell with dimensions defined by the thickness of the slice, the repeat distance of the trial periodic structure of the interface and the repeat distance of the artificial periodic structure introduced for the periodic continuation. An alternative procedure which also uses the *n*-beam multi-slice method, but does not involve the periodic continuation approximation, has been developed by Krakow (1981) and used by Krakow *et al* (1986) to investigate a tilt boundary in gold.

In practice, a major difficulty with using high-resolution *n*-beam images for the study of interface structure is that the technique can only be applied to a restricted class of interfaces. The types of interface which give high-resolution *n*-beam images that are amenable to interpretation are ones where the specimen can be oriented in such a way that the direction of the incident electron beam is in the plane of the interface and along a low-index crystallographic direction in each grain. Only very special interfaces with particular plane normals and misorientations satisfy these conditions. For example, in the case of grain boundaries these conditions virtually restrict the type of interface that can be readily studied by *n*-beam microscopy to pure tilt boundaries with the electron beam incident along the tilt axis which in turn must be a low-index crystallographic direction. Under these conditions the grain boundary dislocations are edge dislocations parallel to the tilt axis.

Following the original observations of Krivanek *et al* (1977) on the structure of a high-angle tilt boundary in germanium, most *n*-beam high-resolution studies of interfaces have been carried out, until recently, on grain boundaries in silicon and germanium. The initial concentration of effort on these materials was because they have larger unit cells than metals and alloys, and therefore require less stringent conditions for point-to-point resolution. However, the recent advent of a new generation of electron microscopes has enabled the resolution of $\{111\}$ and $\{200\}$ planes in FCC and $\{110\}$ planes in BCC metals and alloys. These microscopes operate at 300–500 kV with much improved lens characteristics which give a point-to-point resolution of approximately 1.7 Å at the Scherzer focus. In metals and alloys this enhanced resolution is now enabling detailed studies of the atomic structure of interfaces (see sections 4.5 and 5.5) to be undertaken. In addition to the advent of the new generation of electron microscopes, computer programs are also available for computing *n*-beam lattice images of interfaces (see, for example, Skarnulis 1979, Lynch and Qin 1987, Stadelmann 1987). An example of the type of agreement that can now be obtained between experimental and computed images of the atomic structure of grain boundaries is given in figure 4.16 which is reproduced from the work of Penisson *et al* (1988) and shows a grain boundary dislocation in a symmetric-tilt boundary in molybdenum.

A serious problem with the use of *n*-beam lattice imaging in studies of interface structure is associated with the necessity of using very thin specimens, usually with thicknesses of about 200 Å or less. For an interface in a specimen of such a thickness, the defect structure observed could be unrepresentative of the structure of the interface in the bulk specimen in that periodic defect structure with large repeat distances of 1000 Å or more is likely to be missing in the thin-foil specimen. In addition, in such thin specimens surface image forces are likely to play a dominant role and could bring about rearrangement and/or loss of defect structure.

2.6 DIFFRACTION PHENOMENA ASSOCIATED WITH INTERFACES

The discussion in this section will be concerned with special diffraction effects which are observed in electron diffraction patterns from interfaces in bicrystals. The three diffraction effects which will be considered are: double diffraction, diffraction due to the presence of an inclined interface and diffraction due to periodic structure in an interface.

The simplest diffraction effect which always occurs in association with electron diffraction from interfaces is double-diffraction. Double-diffraction arises when exit beams from the upper grain act as incident beams for the lower grain (see figure 2.3). In principle, each diffracted beam from the upper grain could reproduce the diffraction pattern which arises from diffraction of the central beam by the lower grain. Thus, extra diffraction spots occur in diffraction patterns from the interface which are not present in the patterns from either of the individual grains. One of the major disadvantages associated with double-diffraction is that under certain conditions it can give contrast in images which can be confused with contrast arising from the structure of an interface. For example, when double-diffraction spots are present in the neighbourhood of the central beam and are included in the objective aperture during bright-field imaging, fringe contrast can arise which is not related in any way to interface structure. Therefore, it is important to be able to recognise double-diffraction and its effects, and methods for doing this are discussed in section 4.3.

Diffraction effects due to the presence of an inclined interface and those due to periodic structure in an interface will be discussed in terms of the Ewald construction described in section 2.2. For a given direction of the incident electron beam the intensity distribution observed in the resulting electron diffraction pattern of a specimen is represented by that section in reciprocal space which results from the intersection of the Ewald sphere with the reciprocal lattice of the specimen. Thus, in discussing diffraction effects associated with an interface it is necessary to consider how the presence of the interface modifies the reciprocal lattice of each of the crystals making up the bicrystal.

As pointed out in section 2.2, the reciprocal lattice points of a crystal structure are modified by the geometry of the specimen, and for the case of a parallel-sided thin-foil specimen the reciprocal lattice points are spiked normal to the foil surface. For a bicrystal specimen each grain will be bounded on one side by a foil surface and on the other by the plane of the interface. As a result, the reciprocal lattice points of each grain will be spiked both normal to the foil plane and normal to the plane of the interface. In the simple case illustrated in figure 2.13(a), where it is assumed that the interface (which is parallel to the foil surface) does not have any structure, the spiking of the reciprocal lattice points for each grain is along a single direction which, for this case, is parallel to both the foil normal $\boldsymbol{F}$ and the normal to the

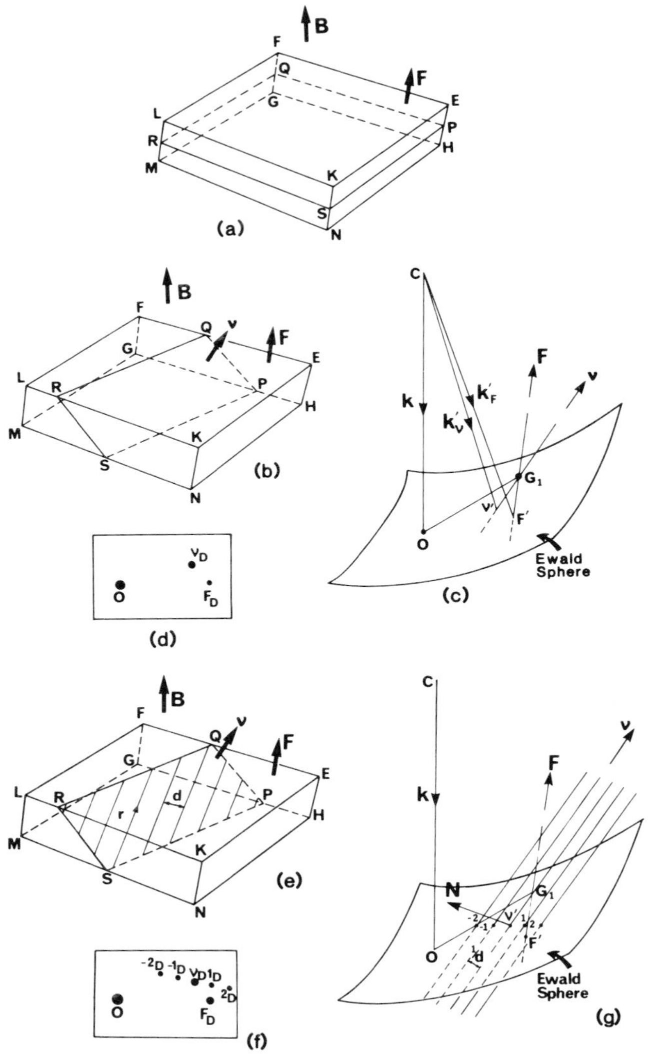

Figure 2.13 Schematic illustration of diffraction phenomena associated with interfaces.

plane of the interface. Thus, in diffraction from such a specimen the only complicating diffraction effects will be those associated with double-diffraction. The more general case of an inclined interface is illustrated in figure 2.13(*b*), where again it is assumed that the interface is without structure. Here the reciprocal lattice points of each grain will be spiked along two directions, i.e. parallel to the foil normal $\boldsymbol{F}$ and parallel to the normal $\boldsymbol{v}$ of the plane of the interface. The Ewald construction for one of the grains for this case is illustrated schematically in figure 2.13(*c*) for the origin O and the reciprocal lattice point G_1. The construction is for the specimen oriented relative to the incident electron beam (with wave vector $\boldsymbol{k}$) so that the departure from the exact Bragg condition is positive and the reciprocal lattice point G_1 lies inside the Ewald sphere. The spikes along $\boldsymbol{F}$ and $\boldsymbol{v}$ intersect the Ewald sphere at the points F' and v' respectively, so that the two diffracted beams with wave vectors $\boldsymbol{k}'_F$ and $\boldsymbol{k}'_v$ are generated.† Thus, for the reciprocal lattice point G_1 these two beams will give rise to two spots F_D and v_D in the diffraction pattern from the interface (as illustrated schematically in figure 2.13(*d*)), rather than a single spot F_D which would be the case for diffraction from the parallel-sided grain remote from the interface. The direction $F_D v_D$ in the schematic diffraction pattern of figure 2.13(*d*) is normal to the projection in the beam direction of the line of intersection QR of the plane of the interface with the foil surface, and the two spots F_D and v_D give rise to pendellösung fringes parallel to this line of intersection in an image of the interface. The occurrence of extra diffraction spots in diffraction patterns due to the presence of inclined interfaces was first pointed out and analysed by Whelan and Hirsch (1957) for a stacking fault in a single crystal. This work was extended by Amelinckx and his colleagues to include, in addition to stacking faults, diffraction effects from domain boundaries, wedge crystals and grain boundaries (see Gevers *et al* 1966, Van Landuyt *et al* 1966, Gevers *et al* 1968, De Ridder *et al* 1968, Gevers *et al* 1969).

In addition to the extra diffraction spots which arise from the geometry of the interface and the specimen, further diffraction effects can occur as a result of the presence of periodic structure in an interface. This was first pointed out by Spyridelis *et al* (1967) for a periodic array of dislocations in a domain boundary in tungsten trioxide. Following this work Balluffi *et al* (1972b), Forwood and Clarebrough (1977), Carter *et al* (1979) and Hall *et al* (1982) demonstrated the importance of periodic arrays of spots in diffraction patterns from grain boundaries in the interpretation of boundary structure in terms of independent periodic arrays of grain boundary dislocations. Similarly, Howell *et al* (1979) and Ecob and Ralph (1984) have used periodic spots in diffraction patterns to study the dislocation structure of FCC/BCC interphase interfaces. Figures 2.13(*e*), (*f*) and (*g*) illustrate schematically the way in which a periodic array of additional diffraction spots in a diffraction pattern is generated by the presence of a periodic array of straight dislocations in a planar interface. Figure 2.13(*e*) shows an inclined interface containing a regular periodic array of interfacial dislocations with line direction $\boldsymbol{r}$ and spacing d. In the diffraction process the displacement

† Although the spikes along $\boldsymbol{F}$ and $\boldsymbol{v}$ are shown passing through the reciprocal lattice point G_1, dynamical theory indicates that very close to G_1 the spikes curve away from G_1 (Whelan and Hirsch 1957).

fields associated with this set of dislocation lines act as a phase grating. As a result, the reciprocal lattice point G_1 in figure 2.13(g) is not only spiked parallel to $\boldsymbol{F}$ and $\boldsymbol{v}$ as before, but also has in association with it a periodic set of spikes parallel to $\boldsymbol{v}$ with a periodic spacing given in magnitude and direction by

$$\boldsymbol{N} = (\boldsymbol{v} \wedge \boldsymbol{r})/d.$$

The intersections of this set of periodic spikes with the Ewald sphere at $\ldots -2, -1, v', 1, 2\ldots$ correspond to a set of diffracted beams (not shown in figure 2.13(g)) connecting the centre of the Ewald sphere to these points. These diffracted beams give a corresponding periodic row of diffraction spots, $\ldots -2_D, -1_D, v_D, 1_D, 2_D \ldots$, in the diffraction pattern, as shown schematically in figure 2.13(f). The zeroth-order diffraction spot in this row is not the F_D diffraction spot associated with diffraction in the grain, but the v_D spot associated with the presence of the interface. For the simpler case of figure 2.13(a), where the interface is parallel to the foil surfaces, extra diffraction spots associated with a periodic array of dislocations in the interface will of course be centred on the F_D spot. The strength of the periodic set of spikes parallel to $\boldsymbol{v}$ will depend on the structure factor of the phase grating. Guan and Sass (1973) have shown that the structure factor is zero for the central transmitted beam but, in general, is non-zero for all other diffracted beams. However, the periodicities in diffracted intensity will be present in the central beam from multiple scattering and can thus give rise to the formation of fringes in bright-field images. It will be seen in Chapters 4 and 5 how these diffraction effects have been used in the analysis of the structure of both low-angle and high-angle grain boundaries.

3

Determination of Crystallographic and Diffraction Parameters

3.1 INTRODUCTION

This chapter will be concerned with the determination of the parameters that specify the diffraction conditions corresponding to electron micrographs of interfaces and the use of this information to determine the orientation relationship between the grains, the crystallographic geometry of the interface and the structural features in it. The emphasis will be on the determination of the necessary parameters for the general case where an interface is inclined to the surfaces of a thin-foil specimen (which is typical of specimens prepared by thinning bulk polycrystalline single-phase or two-phase material) rather than for the special case where an interface is parallel to the surfaces of a thin-foil specimen (which is typical of specimens prepared by epitaxial techniques). The discussion will be concerned with determining the data (such as that listed in table 2.1) which are essential for the quantitative analysis of interfacial structure. In addition, methods will be given for calibrating the characteristics of the electron microscope, such as rotations between images and diffraction patterns and the wavelength of the electrons at the nominal accelerating voltage.

3.2 RELATIVE ROTATION BETWEEN AN IMAGE AND A DIFFRACTION PATTERN

Image formation in the electron microscope involves the use of electromagnetic lenses so that there are rotations of ray paths, due to the magnetic fields of the lenses, in addition to the inversions due to crossing ray paths common to optical microscopes. Such rotations and inversions must be determined in order to relate crystallographic directions in a diffraction pattern to directions in the corresponding image. To correlate directions in diffraction patterns and images correctly, a consistent method must be adopted for viewing negatives and prints of both diffraction patterns and images. The method used by the authors is always to view negatives of images and diffraction patterns with the emulsion side uppermost and, to produce positive prints, the negatives are always printed with the emulsion side up. Thus, in viewing both negatives and prints, the orientation seen is the same as that which appears on the fluorescent screen of the electron microscope.

A schematic diagram illustrating the formation of an image and a diffraction pattern and intermediate image are magnified and projected on to the 3.1. The diffraction pattern is formed in the back focal plane of the objective lens and the image in the plane of the intermediate image. When the diffraction pattern and intermediate image are magnified and projected on to the fluorescent screen of the electron microscope there will be a relative rotation, ϕ, between the final diffraction pattern and the final image, due to the different settings of the intermediate lens required to focus each of these planes. The

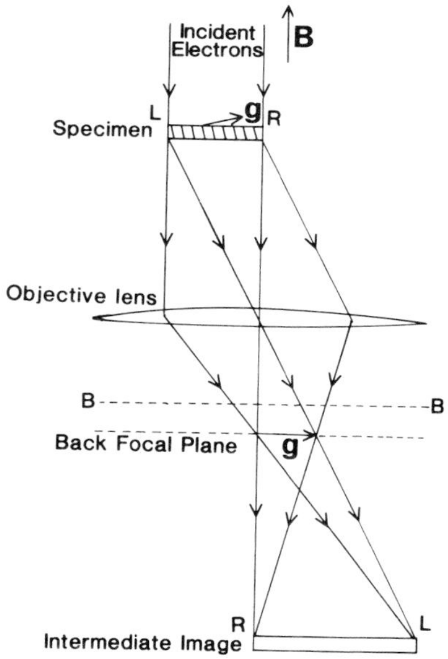

Figure 3.1 Schematic ray diagram illustrating the formation of a diffraction pattern and an intermediate image by the objective lens of an electron microscope.

angle ϕ can be determined, for particular magnification settings of the image and diffraction pattern, by using a specimen of molybdenum trioxide in the form of a small vapour-deposited crystal. Such crystals are long laths with their long dimension parallel to the [100] direction, and in an electron diffraction pattern from such a specimen this [100] direction corresponds to the larger spacing in the rectangular cross-grating pattern. Thus, when an image and a diffraction pattern of a crystal of molybdenum trioxide are recorded on the same negative, the required angle ϕ is the angle between the long edge of the image of the crystal and the [100] direction in the diffraction pattern. However, since $\pm$[100] directions cannot be distinguished, this determination of ϕ contains an uncertainty of 180°.

The way to eliminate this uncertainty of 180° can be seen by using figure 3.1 to consider in more detail how an image and a diffraction pattern of a specimen are formed by the objective lens. In figure 3.1 the diffracting vector $\boldsymbol{g}$ is defined as being normal to the diffracting planes in the specimen and acute with the beam direction $\boldsymbol{B}$. This vector appears in the diffraction pattern as the displacement of the diffracted beam from the transmitted beam in the back focal plane of the objective lens. It can be seen from the relative positions of L and R in the specimen and in the intermediate image in figure 3.1 that the objective lens produces an inversion between the specimen and the intermediate image, i.e. directions in the intermediate image are rotated by 180° relative to directions in the specimen. However, it can be seen that the diffracting vector $\boldsymbol{g}$ in the specimen is not inverted in the diffraction pattern. Thus, directions in the intermediate image are inverted with respect to corresponding directions in the diffraction pattern and this inversion occurs at the back focal plane. If the intermediate lens is focused so as to produce an image on the fluorescent screen of the plane BB†, then an out-of-focus diffraction pattern is obtained where each spot appears as a disc containing an image of the specimen. Since the inversion has not occurred at BB, the directions in each of these images of the specimen are in the same orientation as directions in the diffraction pattern, apart from the possibility of a small rotation arising from focusing on the plane BB rather than on the back focal plane. Thus, a comparison between directions in the normal image, formed by focusing the intermediate image, with corresponding directions in the images from the plane BB indicates whether a rotation of 180° needs to be added to the angle ϕ. Worked examples of this calibration procedure are not given here, but can be found in Head *et al* (1973) and Loretto and Smallman (1975).

† The plane BB is imaged by decreasing the current in the intermediate lens relative to the setting required for a focused diffraction pattern from the back focal plane of the objective lens.

3.3 SPECIMEN TILTING TO COLLECT CRYSTALLOGRAPHIC INFORMATION

Experimental electron microscopy of a bicrystal involves tilting the specimen and recording images of the interface and diffraction patterns from each grain and (when required) from the interface for a number of different orientations of the specimen over a wide angular range of tilt. In other words, it is necessary to bring a considerable number of crystallographic directions into alignment with the electron beam in order to obtain a sufficient number of two-dimensional projections so that the three-dimensional geometry of the bicrystal and of the interface can be determined. In addition, the information recorded in the images and diffraction patterns enables the beam directions and diffracting vectors used to be indexed specifically, and hence the corresponding crystallography of the bicrystal and of the interface to be established. For an example of all the procedures involved a high-angle grain boundary in an FCC Cu -6 at% Si alloy will be used throughout, but exactly the same methods would be involved for an interphase interface.

3.3.1 Tilting Procedure and Orientation Maps

During the tilting procedure particular diffracting conditions can be set up, such as simultaneous double two-beam diffraction with different diffracting vectors in each grain, or with the same diffracting vector in each grain, and the following description will be given with reference to the setting up of these simultaneous double two-beam conditions.

Figure 3.2 shows an image of the boundary separating two grains marked U and L which contains an hexagonal network of interfacial dislocations. The nature of this interfacial dislocation structure will be discussed later in section 5.4.2. The bicrystal specimen was examined in an electron microscope operating at 200 kV with a double-tilting goniometer stage capable of giving $\pm 30^\circ$ of tilt about two orthogonal axes. In order to carry out the tilting experiment systematically, use is made of the Kikuchi lines in the diffraction patterns from the two grains as discussed in sections 2.2 and 2.4.1. The different crystallographic directions in the two grains along which the electron beam was aligned, in order to set up simultaneous double two-beam diffraction, are shown by the encircled numbers ①–⑩ in the stereographic projection for grain U in figure 3.3(*a*) and for grain L in figure 3.3(*b*). These directions are shown in relation to the poles corresponding to low-index crystallographic directions which are indicated with non-specific indices in diamond brackets. The pairs of Kikuchi lines in the diffraction patterns, which enabled the controlled tilting of the specimen, are indicated by the single lines in figure 3.3 joining the low-index poles and are labelled with non-specific indices to indicate the type of diffracting vector associated with them. The two 'orientation maps' in figure 3.3 look rather complex at first

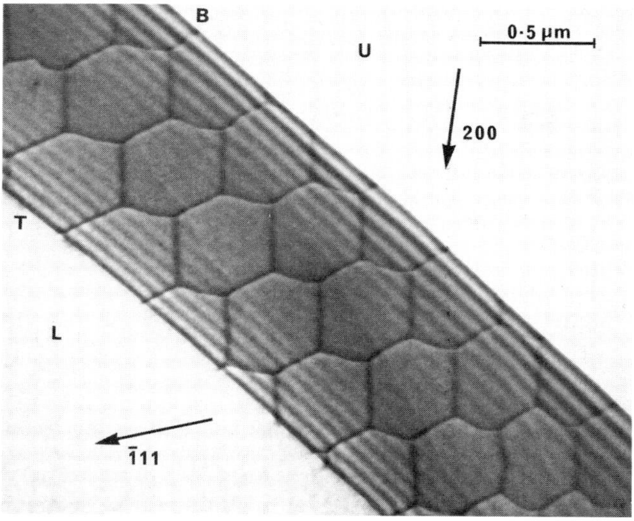

Figure 3.2 Double two-beam image of a near-Σ9 grain boundary in a Cu − 6 at% Si alloy separating grains U and L. The diffracting vectors are indicated.

sight, but they are obtained quite simply by tilting the specimen in a sequential way and recording the results appropriately. For example, the first double two-beam condition at map position ① in figures 3.3(*a*) and (*b*) involved a diffracting vector of the 311_U type in grain U and of the 220_L type in grain L. After recording images of the interface and diffraction patterns from each grain at ①, the specimen was rotated around the 220_L diffracting vector in

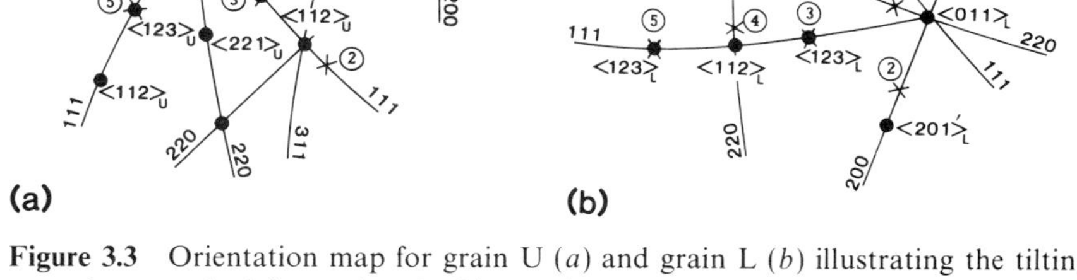

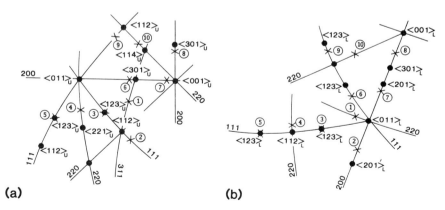

Figure 3.3 Orientation map for grain U (*a*) and grain L (*b*) illustrating the tilting procedure required for setting double two-beam conditions for the grain boundary in figure 3.2.

grain L, corresponding to movement along the 220_L Kikuchi lines, until the electron beam was sufficiently close to the $\langle 011 \rangle_L$ direction to enable rotation to be transferred to an axis about the diffracting vector of the 200_L type. Rotation about this axis towards $\langle 201 \rangle_L'$ gives position ② in which a second double two-beam condition arises where the diffracting vectors are of the type 200_L and 111_U. As before, images of the interface and diffraction patterns from each grain are recorded. In addition, the small portion of the orientation map defined so far is recorded. By similar rotations along Kikuchi lines in grains U or L, the remaining double two-beam conditions ③–⑩ are set up, appropriate images and diffraction patterns recorded and the orientation maps of figure 3.3 completed.

3.3.2 Specific Indexing of Beam Directions and Diffracting Vectors

The beam directions in diffraction patterns from each grain are very important parameters to be determined because not only do they specify diffraction conditions and projection directions for images, but also the orientation relationship between the grains in a bicrystal is obtained from them by finding the transformation matrix which, for each position of tilt, relates the beam direction indexed in one grain with that indexed in the other. Similarly, the diffracting vectors in each grain must be determined not only to define the operative diffracting conditions, but also to enable the beam directions to be specified.

In order to index specifically the beam directions and the diffracting vectors for the diffraction patterns from both grains in positions ①–⑩ of figure 3.3, it is first necessary to give specific indices to the low-index poles in each of the orientation maps. This is done by locating the maps on a standard stereographic projection of crystal directions. For a bicrystal composed of crystals of the lowest symmetry there is only one way in which each orientation map can be located on such a stereographic projection. However, for a bicrystal composed of crystals of higher symmetry there is a choice of crystallographically equivalent regions of the stereographic projection on which each orientation map can be located. For the current example, an FCC bicrystal, there are twelve crystallographically equivalent locations on a standard stereographic projection of crystal directions for each of the orientation maps of figure 3.3. In figures 3.4(*a*) and (*b*) each of these maps has been located on an appropriate region of a standard [001] stereographic projection of crystal directions in a cubic crystal, and the low-index poles (in diamond brackets in figure 3.3) have been assigned the specific indices in square brackets accordingly. In standard stereographic projections, such as the [001] projection used in figure 3.4, the indices are those of upward-drawn directions. The beam directions in diffraction patterns are also defined as upward-drawn. Thus, the beam directions for positions ①–⑩ can be

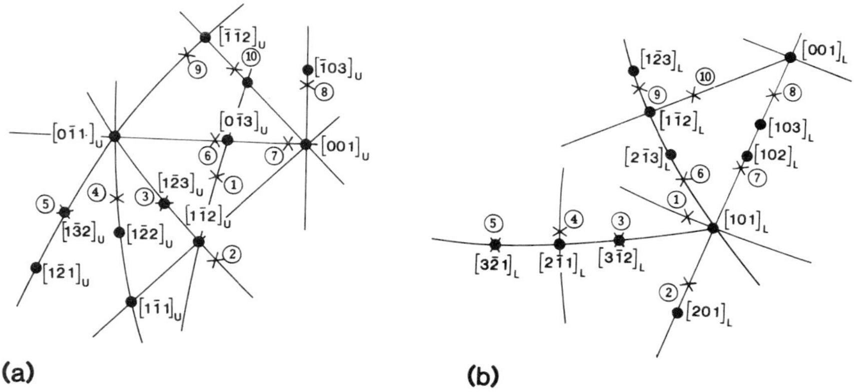

Figure 3.4 Orientation maps of figure 3.3 after specific indexing.

given specific indices in accordance with their relative positions in the specifically indexed orientation maps of figure 3.4.

In order to give specific indices to the operative diffracting vectors in the diffraction patterns, it is first necessary to consider the relation between the diffraction patterns and the specifically indexed orientation maps. Each orientation map, as recorded from observations of diffraction patterns, represents a projection in the direction of electron flow of the relative positions of crystallographic directions in the specimen, i.e. each orientation map is a downward projection of crystallographic directions. Thus, in representing such a downward projection of directions on a standard stereographic projection, for which the sense of projection is upwards, a rotation of 180° has been introduced. In other words, a rotation of 180° has been introduced between the recorded diffraction patterns and the specifically indexed orientation maps. Therefore, in assigning specific indices to the operative diffracting vector in a diffraction pattern, this 180° rotation has to be removed. In practice, specific indices are assigned to the operative diffracting vector in a diffraction pattern by using a standard projection of plane normals which is first placed in the same orientation as the diffraction pattern and the specifically indexed orientation map. To remove the 180° rotation, the projection of plane normals is then rotated by 180° relative to the diffraction pattern. The diffracting vector, which is in the direction from the central spot in the diffraction pattern to the spot corresponding to the diffracted beam, is then identified with the appropriate plane normal and the diffracting vector specifically indexed with the indices of this plane normal. The direction of this diffracting vector can then be marked on the corresponding image using the angle ϕ (see section 3.2) to orient the image relative to the diffraction pattern.

The two-beam diffraction pattern in figure 3.5(a) will be used to illustrate specific indexing, first for a diffracting vector and then for a beam direction. This pattern is from grain U of the FCC Cu–Si alloy of figure 3.2. The pattern was recorded as one of the pair of diffraction patterns for the bicrystal at position ① on the specifically indexed orientation maps of figure 3.4 and has an operative diffraction vector of the 311 type. For crystals with cubic

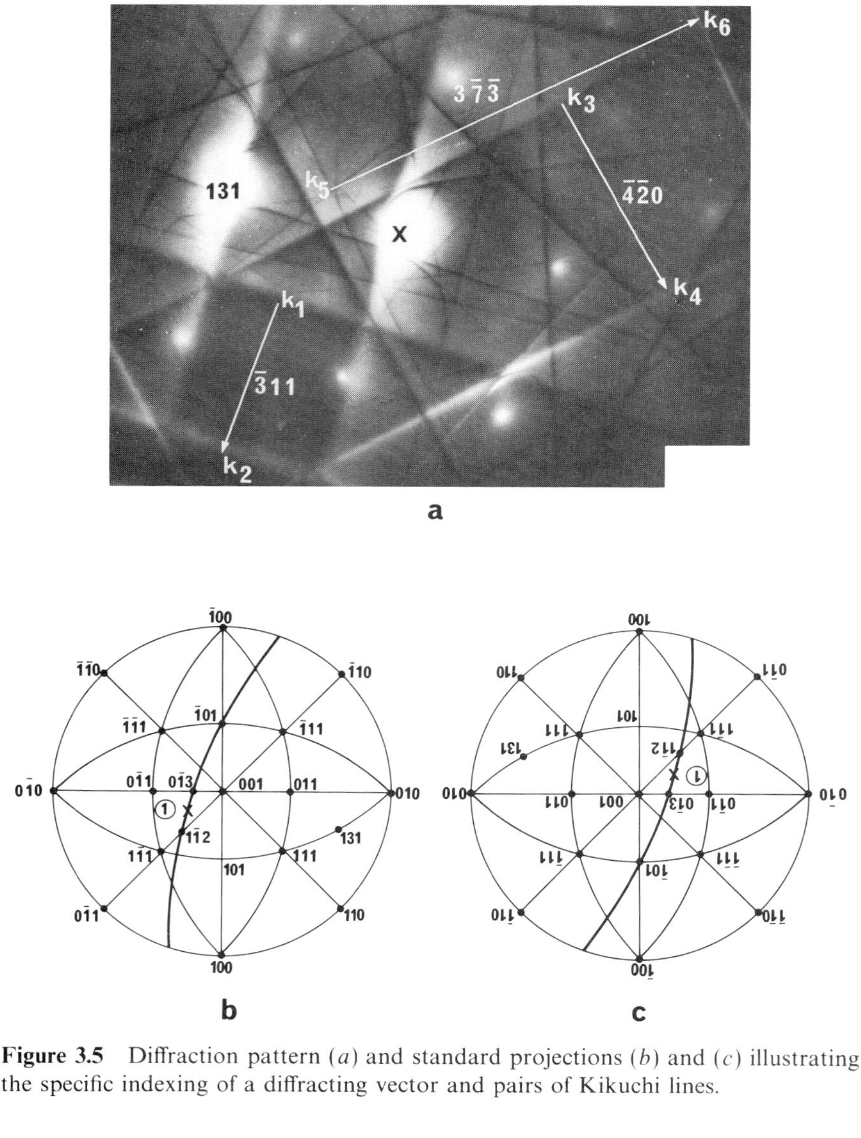

Figure 3.5 Diffraction pattern (a) and standard projections (b) and (c) illustrating the specific indexing of a diffracting vector and pairs of Kikuchi lines.

symmetry the same standard stereographic projection specifies both crystallographic directions and plane normals, so that the one projection can be used to determine both the specific beam direction and the specific diffracting vector from the diffraction pattern of figure 3.5(*a*). The standard stereographic projection used for indexing the beam directions is shown in figure 3.5(*b*) in its correct orientation with respect to the diffraction pattern of figure 3.5(*a*), and the beam direction corresponding to position ① is indicated between $[0\bar{1}3]_U$ and $[1\bar{1}2]_U$ on the $(131)_U$ zone. Clearly the operative diffracting vector is either 131_U or $\bar{1}\bar{3}\bar{1}_U$. In figure 3.5(*c*) the standard projection is rotated through 180° for indexing the diffracting vector. In the diffraction pattern the operative diffracting vector points to the left so that, with reference to figure 3.5(*c*), it corresponds to the plane normal $(131)_U$. Thus the operative diffracting vector has the specific indices 131_U.

In a similar way, other diffracting vectors associated with weaker diffraction spots or with the normals to pairs of Kikuchi lines in the diffraction pattern can be given specific indices. For example, in figure 3.5(*a*), from measurements of spacings and directions relative to the 131_U diffracting vector, and with reference to the standard stereographic projection of cubic crystals in the orientation of figure 3.5(*c*), the normal $\overrightarrow{k_1k_2}$ to the pair of Kikuchi lines k_1, k_2 has the indices $\bar{3}11_U$, the normal $\overrightarrow{k_3k_4}$ to the pair of Kikuchi lines k_3, k_4 has the indices $\bar{4}\bar{2}0_U$ and the normal $\overrightarrow{k_5k_6}$ to the pair of Kikuchi lines k_5, k_6 has the indices $3\bar{7}\bar{3}_U$.

In the diffraction pattern of figure 3.5(*a*) the beam direction is the upward-drawn direction normal to the pattern that passes through the central spot marked X. This direction is obtained by determining its angular separation from a low-index crystallographic direction which is defined by crossing pairs of Kikuchi lines in the pattern. For example, in the pattern of figure 3.5(*a*) the 131_U, $\bar{3}11_U(k_1, k_2)$ and $\bar{4}\bar{2}0_U(k_3, k_4)$ pairs of Kikuchi lines cross over to define such a low-index crystallographic direction. The centre, P, of the Kikuchi line pattern formed by these crossovers has been constructed in figure 3.6 from the intersection of the dashed lines drawn midway between each pair of Kikuchi lines. The upward-drawn crystallographic direction through the point P is obtained as $[1\bar{2}5]_U$ by taking the cross product of the indices of any two pairs of the crossing Kikuchi lines. The angular separation of the beam direction corresponding to the central spot X and the direction $[1\bar{2}5]_U$ is obtained by making use of the approximation that there is a linear relationship between distances in the diffraction pattern and small angles of tilt as discussed in section 2.2. There, it is shown that the separation x of a pair of Kikuchi lines in a diffraction pattern corresponds to an angular separation of $2\theta_B$, i.e. twice the Bragg angle for the diffracting vector $\boldsymbol{g}$ associated with the pair of Kikuchi lines. For a particular diffraction pattern this can be written as

$$2\theta_B = Cx$$

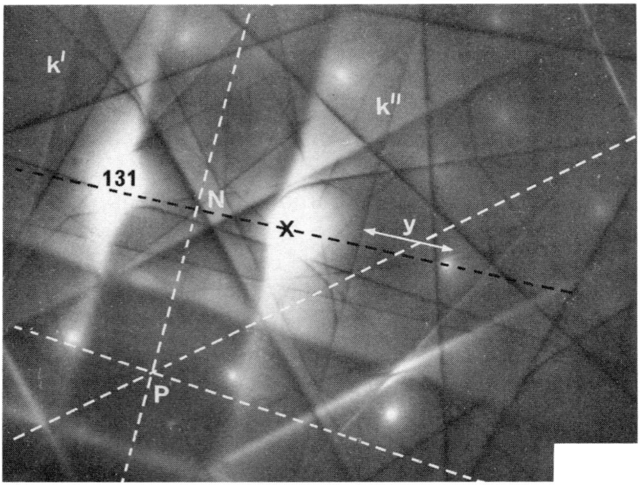

Figure 3.6 Diffraction pattern illustrating the determination of the beam direction $\boldsymbol{B}$ corresponding to the central spot X.

where C is the calibration constant of the diffraction pattern. For small values of θ_B, Bragg's law can be written as

$$2\theta_B = \lambda|\boldsymbol{g}|$$

so that the calibration constant of the diffraction pattern is given by

$$C = \lambda|\boldsymbol{g}|/x. \tag{3.1}$$

In figure 3.6 the angular separation between X and P can be expressed as a rotation through an angle corresponding to the distance $\overrightarrow{PN}$ around the $[131]_U$ axis from $[1\bar{2}5]_U$ towards $[0\bar{1}3]_U$, as defined by the specifically indexed orientation map, followed by a rotation through an angle corresponding to the distance $\overrightarrow{NX}$ towards the direction $[131]_U$†, again as defined by the specifically indexed orientation map. In practice, since the point X is an ill-defined, diffuse diffraction spot, the distance $\overrightarrow{NX}$ is obtained by measuring the equivalent distance y between the diffraction spot $\bar{1}\bar{3}\bar{1}_U$ and the Kikuchi line k″. In this way the beam direction X is specified as the direction which departs from $[1\bar{2}5]_U$ by 1.38° towards $[0\bar{1}3]_U$ and 0.67° towards $[131]_U$.

Angles such as $\overrightarrow{PN}$ and $\overrightarrow{NX}$ can be determined from a typical diffraction pattern with an accuracy of $\pm 0.02°$ and this leads to a similar accuracy in the determination of the orientation relationship between the grains in a

† The sense of this direction is opposite to the sense defined by the indices of the diffracting vector, because indexing the diffracting vector involves a rotation of 180°.

Table 3.4 (*Continued*)

$\mathbf{U}_1$	$\mathbf{U}_2$		u		θ (deg.)
010 001 100	001 010 $\bar{1}00$	−0.707 549	−0.706 664	−0.000 433	141.00
$00\bar{1}$ 100 $0\bar{1}0$	$00\bar{1}$ $\bar{1}00$ 010	0.970 115	0.242 643	0.000 608	152.66
010 100 $00\bar{1}$	010 $\bar{1}00$ 001	−0.969 975	−0.243 202	−0.000 607	152.73
$0\bar{1}0$ $00\bar{1}$ 100	$00\bar{1}$ 100 $0\bar{1}0$	−0.728 157	−0.484 955	−0.484 361	152.75
001 100 010	010 001 100	0.727 443	0.485 491	0.484 897	152.82
$00\bar{1}$ $\bar{1}00$ 010	$\bar{1}00$ $00\bar{1}$ $0\bar{1}0$	0.845 201	0.506 993	0.169 094	160.71
100 010 001	010 100 $00\bar{1}$	−0.845 556	−0.506 544	−0.168 669	160.76
$\bar{1}00$ 001 010	001 $\bar{1}00$ $0\bar{1}0$	−0.845 080	−0.506 920	−0.169 917	160.81
$0\bar{1}0$ 100 001	100 $0\bar{1}0$ $00\bar{1}$	0.845 435	0.506 471	0.169 491	160.85
$\bar{1}00$ 010 $00\bar{1}$	$\bar{1}00$ $0\bar{1}0$ 001	0.942 631	0.236 346	0.235 769	179.93
$00\bar{1}$ 010 100	010 $00\bar{1}$ $\bar{1}00$	0.666 958	0.666 123	0.333 836	179.95

Although the different values of $\boldsymbol{u}$ and θ obtained by re-indexing describe the same physical situation, a listing of them is useful as it can reveal a more suitable form for $\boldsymbol{u}$ and θ (see section 1.4) than that obtained from the initial positioning of the orientation maps on the standard projection. If a different $\boldsymbol{u}$ and θ are chosen, the beam directions and diffracting vectors for grains 1 and 2 must, of course, be re-indexed accordingly.

3.4.2 Coincident Site Lattice Orientations

The misorientations across grain boundaries in metals and alloys with cubic crystal structure are often observed to be close to special orientation relationships which belong to a class known as coincident site lattice (CSL) orientations. These CSL orientations play an important role in the structure of high-angle grain boundaries and this will be discussed in detail in Chapter 5. In this section the concept of a coincident site lattice for crystals with cubic symmetry will be introduced, orientation relationships associated with CSLs will be listed, and the use of these orientation relationships in transmission electron microscopy for setting up particular diffraction conditions, such as double two-beam diffraction with the same diffracting vector in each grain, will be demonstrated.

The concept of a CSL can be considered in the following way. If two identical but misoriented cubic lattices are allowed to interpenetrate, then, for particular misorientations of these lattices, there will be a number of sites in the interpenetrating lattices which are common to both of the misoriented lattices. A particular CSL is specified by a parameter Σ where $1/\Sigma$ of the lattice sites are common to both lattices and Σ is an odd integer. An example of a CSL is given in figure 3.9 which is a [001] projection showing the atomic arrangement for a $\Sigma 5$ CSL formed by two interpenetrating FCC lattices, 1 and 2, misoriented by a rotation $\theta = 36.87^\circ$ around a common axis $\boldsymbol{u} = [001]$. Lattice 1 is represented by the symbols + and □, where + is in the plane of the page and □ is $(1/2)[00\bar{1}]$ below the plane of the page. Similarly,

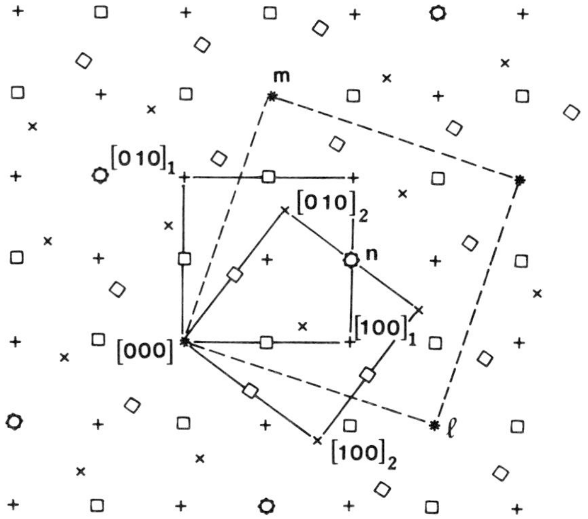

Figure 3.9 [001] projection of two interpenetrating FCC lattices at the $\Sigma 5$ CSL orientation.

lattice 2 is represented by the symbols $\times$ and $\diamond$ where $\times$ is in the plane of the page and $\diamond$ is $(1/2)[00\bar{1}]$ below the plane of the page. The coincident sites are indicated by coincident symbols and four of these coincident sites in the plane of the page are linked by dashed lines. The basis CSL vectors are from the origin [000] to the coincident points *l*, *m* and *n*, and are $(1/2)[3\bar{1}0]_1/(1/2)[310]_2$, $(1/2)[130]_1/(1/2)[\bar{1}30]_2$ and $(1/2)[21\bar{1}]_1/(1/2)[12\bar{1}]_2$ respectively.

For values of $\Sigma > 11$, more than one CSL can exist for a given value of Σ and these involve non-equivalent values of $\boldsymbol{u}$ and θ. When more than one CSL exists for a particular value of Σ, they are distinguished by subscripts *a*, *b*, *c*,... where the subscript *a* is given to the smallest angle of misorientation, *b* to the next smallest etc. For different CSLs with the same values of Σ and angle of misorientation alphabetic precedence is given to the CSL with the smallest sum of the squares of the indices of the rotation axis. A CSL is defined by a rotation matrix **R** of the form

$$\mathbf{R} = 1/\Sigma \begin{pmatrix} R_{11} & R_{12} & R_{13} \\ R_{21} & R_{22} & R_{23} \\ R_{31} & R_{32} & R_{33} \end{pmatrix}$$

where the elements R_{ij} are integers. The theory of CSL lattices and their properties is contained in a basic paper by Grimmer *et al* (1974), in which they list values of $\boldsymbol{u}$ and θ for CSLs up to $\Sigma 49$. More recently, Mykura (1980) has calculated the values of $\boldsymbol{u}$ and θ for CSLs up to $\Sigma 101$ together with the 24 equivalent values of $\boldsymbol{u}$ and θ for each CSL up to $\Sigma 43$. Mykura's results are reproduced in tables 3.5 and 3.6 and the usefulness of these tables will be demonstrated in the following example.

In the transmission electron microscopy of grain boundaries in cubic metals and alloys which are close to CSL orientations a quick identification of the type of CSL orientation involved for a particular boundary can be made prior to engaging in the full determination of the orientation relationship in the way described in section 3.4.1. This is done by first finding, for a particular beam direction, a pair of Kikuchi lines which is common to the diffraction patterns from neighbouring grains and is therefore associated with a same-$\boldsymbol{g}_c$ diffracting vector. Then, by specifically indexing these patterns, a prediction can be made for the CSL and therefore for other same-$\boldsymbol{g}_c$ vectors. The prediction for the CSL orientation is confirmed if these other same-$\boldsymbol{g}_c$ vectors are found.

This procedure will now be described for the grain boundary of figure 3.2. In the initial stages of tilting the specimen in the electron microscope a common pair of Kikuchi lines of the 220 type was found to pass through position ④ in grains U and L of the orientation maps of figure 3.3. Further tilting of the specimen established the positions of a sufficient number of other beam directions in grains U and L, relative to position ④, to enable position ④ in each grain and the other beam directions in the grains, obtained

Table 3.5 Minimum angle of rotation θ and axis of rotation $\boldsymbol{u}$ for cubic CSLs up to $\Sigma 101$ (after Mykura 1980).

Σ	θ	$\boldsymbol{u}$	Σ	θ	$\boldsymbol{u}$	Σ	θ	$\boldsymbol{u}$	Σ	θ	$\boldsymbol{u}$
1	0°	Any	43_b	27.91°	⟨210⟩	67_b	36.31°	⟨320⟩	87_a	19.51°	⟨210⟩
3	60°	⟨111⟩	43_c	60.77°	⟨332⟩	67_c	47.23°	⟨533⟩	87_b	30.44°	⟨211⟩
5	36.87°	⟨100⟩	45_a	28.62°	⟨311⟩	67_d	60.49°	⟨433⟩	87_c	48.64°	⟨731⟩
7	38.21°	⟨111⟩	45_b	36.87°	⟨221⟩	69_a	31.23°	⟨210⟩	87_d	48.64°	⟨553⟩
9	38.94°	⟨110⟩	45_c	53.13°	⟨221⟩	69_b	41.09°	⟨410⟩	89_a	25.99°	⟨100⟩
11	50.48°	⟨110⟩	47_a	37.07°	⟨331⟩	69_c	41.09°	⟨322⟩	89_b	25.99°	⟨221⟩
13_a	22.62°	⟨100⟩	47_b	43.66°	⟨320⟩	69_d	50.92°	⟨551⟩	89_c	34.88°	⟨110⟩
13_b	27.80°	⟨111⟩	49_a	43.58°	⟨111⟩	71_a	45.23°	⟨421⟩	89_d	51.43°	⟨733⟩
15	48.19°	⟨210⟩	49_b	43.58°	⟨511⟩	71_b	54.23°	⟨553⟩	89_e	51.83°	⟨433⟩
17_a	28.07°	⟨100⟩	49_c	49.22°	⟨332⟩	73_a	11.64°	⟨111⟩	91_a	10.42°	⟨111⟩
17_b	61.93°	⟨221⟩	51_a	16.10°	⟨110⟩	73_b	13.44°	⟨100⟩	91_b	31.00°	⟨320⟩
19_a	26.53°	⟨110⟩	51_b	22.84°	⟨110⟩	73_c	41.11°	⟨100⟩	91_c	38.70°	⟨310⟩
19_b	46.83°	⟨111⟩	51_c	48.94°	⟨531⟩	73_d	41.11°	⟨221⟩	91_d	53.99°	⟨111⟩
21_a	21.79°	⟨111⟩	53_a	27.52°	⟨211⟩	73_e	48.88°	⟨430⟩	91_e	53.99°	⟨751⟩
21_b	44.40°	⟨211⟩	53_b	31.89°	⟨100⟩	75_a	22.08°	⟨311⟩	93_a	19.79°	⟨311⟩
23	40.45°	⟨311⟩	53_c	53.53°	⟨533⟩	75_b	23.07°	⟨211⟩	93_b	35.19°	⟨410⟩
25_a	16.25°	⟨100⟩	55_a	35.10°	⟨310⟩	75_c	45.03°	⟨311⟩	93_c	35.19°	⟨322⟩
25_b	51.68°	⟨331⟩	55_b	38.57°	⟨211⟩	75_d	52.17°	⟨432⟩	93_d	42.10°	⟨111⟩

27_a	31.58°	⟨110⟩	55_c	57.55°	⟨551⟩	77_a	28.76°	⟨331⟩	93_e	56.37°	⟨753⟩
27_b	35.42°	⟨210⟩	57_a	13.17°	⟨111⟩	77_b	29.52°	⟨310⟩	95_a	25.84°	⟨331⟩
29_a	43.61°	⟨100⟩	57_b	41.03°	⟨321⟩	77_c	48.52°	⟨320⟩	95_b	38.84°	⟨421⟩
29_b	46.39°	⟨221⟩	57_c	43.99°	⟨110⟩	77_d	55.15°	⟨441⟩	95_c	45.15°	⟨321⟩
31_a	17.90°	⟨111⟩	57_d	61.16°	⟨553⟩	79_a	33.99°	⟨111⟩	97_a	30.59°	⟨111⟩
31_b	52.19°	⟨211⟩	59_a	24.93°	⟨311⟩	79_b	33.99°	⟨511⟩	97_b	30.59°	⟨511⟩
33_a	20.05°	⟨110⟩	59_b	45.98°	⟨110⟩	79_c	34.63°	⟨321⟩	97_c	42.07°	⟨100⟩
33_b	33.55°	⟨311⟩	59_c	45.98°	⟨411⟩	81_a	38.37°	⟨531⟩	97_d	42.07°	⟨430⟩
33_c	58.98°	⟨110⟩	61_a	10.38°	⟨100⟩	81_b	38.94°	⟨411⟩	97_e	60.68°	⟨771⟩
35_a	34.04°	⟨211⟩	61_b	32.42°	⟨331⟩	81_c	54.52°	⟨322⟩	97_f	60.68°	⟨755⟩
35_b	43.23°	⟨331⟩	61_c	50.25°	⟨332⟩	81_d	60.41°	⟨443⟩	99_a	11.53°	⟨110⟩
37_a	18.92°	⟨100⟩	61_d	52.64°	⟨111⟩	83_a	17.86°	⟨110⟩	99_b	34.59°	⟨531⟩
37_b	43.13°	⟨310⟩	63_a	22.98°	⟨210⟩	83_b	42.19°	⟨533⟩	99_c	45.01°	⟨520⟩
37_c	50.57°	⟨111⟩	63_b	38.21°	⟨511⟩	83_c	42.70°	⟨332⟩	99_d	45.01°	⟨432⟩
39_a	32.21°	⟨111⟩	63_c	54.03°	⟨431⟩	83_d	57.15°	⟨331⟩	99_e	62.62°	⟨773⟩
39_b	50.13°	⟨321⟩	65_a	14.25°	⟨100⟩	85_a	8.79°	⟨100⟩	101_a	11.40°	⟨100⟩
41_a	12.68°	⟨100⟩	65_b	30.51°	⟨100⟩	85_b	25.05°	⟨100⟩	101_b	19.85°	⟨211⟩
41_b	40.88°	⟨210⟩	65_c	30.51°	⟨221⟩	85_c	45.57°	⟨711⟩	101_c	38.08°	⟨533⟩
41_c	55.88°	⟨110⟩	65_d	43.05°	⟨531⟩	85_d	45.57°	⟨551⟩	101_d	47.68°	⟨522⟩
43_a	15.18°	⟨111⟩	67_a	24.43°	⟨111⟩	85_e	46.04°	⟨431⟩	101_e	47.68°	⟨441⟩

Table 3.6 The set of 24 angle–axis pairs for cubic CSLs up to Σ43 (after Mykura 1980).

Σ1	0° on any axis	6 of 90° on ⟨100⟩	3 of 180° on ⟨100⟩
		8 of 120° on ⟨111⟩	6 of 180° on ⟨110⟩
Σ3	2 of 60°, 180° } on same ⟨111⟩	3 of 70°, 3 of 109.47° } on same ⟨110⟩	6 of 131.81° on ⟨210⟩
			6 of 146.44° on ⟨311⟩
			3 of 180° on ⟨211⟩
Σ5	36.87°, 53.13°, 126.87°, 143.13° } on same ⟨100⟩	4 of 95.74° on ⟨311⟩	4 of 154.16° on ⟨331⟩
		4 of 101.54° on ⟨211⟩	2 of 180° on ⟨210⟩
		4 of 143.13° on ⟨221⟩	2 of 180° on ⟨310⟩
Σ7	38.21°, 81.79°, 158.21° } on same ⟨111⟩	3 of 73.40° on ⟨210⟩	3 of 135.58° on ⟨211⟩
		3 of 110.92° on ⟨331⟩	3 of 149.00° on ⟨320⟩
		3 of 115.38° on ⟨310⟩	3 of 158.21° on ⟨511⟩
			3 of 180° on ⟨321⟩
Σ9	38.94°, 141.06° } on same ⟨110⟩	2 of 67.11° on ⟨311⟩	2 of 152.73° on ⟨410⟩
		2 of 90° on ⟨221⟩	2 of 152.73° on ⟨322⟩
		2 of 96.38° on ⟨210⟩	4 of 160.81° on ⟨531⟩
		2 of 120° on ⟨511⟩	180° on ⟨221⟩
		4 of 123.75° on ⟨321⟩	180° on ⟨411⟩
Σ11	50.48°, 129.52° } on same ⟨110⟩	2 of 62.96° on ⟨211⟩	2 of 144.90° on ⟨310⟩
		2 of 82.16° on ⟨331⟩	4 of 155.38° on ⟨421⟩
		2 of 100.48° on ⟨320⟩	2 of 162.66° on ⟨533⟩
		4 of 126.22° on ⟨531⟩	180° on ⟨311⟩
		2 of 129.52° on ⟨411⟩	180° on ⟨332⟩
$\Sigma 13_a$	22.62°, 67.38°, 12.62°, 157.38° } on same ⟨100⟩	4 of 92.20° on ⟨511⟩	4 of 164.06° on ⟨551⟩
		4 of 107.92° on ⟨322⟩	2 of 180° on ⟨320⟩
		4 of 133.81° on ⟨332⟩	2 of 180° on ⟨510⟩
$\Sigma 13_b$	27.80°, 92.20°, 147.80° } on same ⟨111⟩	3 of 76.66° on ⟨310⟩	3 of 130.83° on ⟨533⟩
		3 of 107.92° on ⟨410⟩	3 of 157.38° on ⟨430⟩
		3 of 112.62° on ⟨221⟩	3 of 164.06° on ⟨711⟩
			3 of 180° on ⟨431⟩
Σ15	48.19° on ⟨210⟩	117.82° on ⟨311⟩	2 of 158.96° on ⟨432⟩
	50.70° on ⟨311⟩	134.43° on ⟨551⟩	158.96° on ⟨520⟩
	78.46° on ⟨211⟩	134.43° on ⟨711⟩	165.16° on ⟨553⟩
	2 of 86.18° on ⟨321⟩	2 of 137.17° on ⟨431⟩	2 of 165.16° on ⟨731⟩
	2 of 99.59° on ⟨531⟩	137.17° on ⟨510⟩	180° on ⟨521⟩
	2 of 113.58° on ⟨421⟩	2 of 150.07° on ⟨321⟩	
$\Sigma 17_a$	28.07°, 61.93°, 118.07°, 151.93° } on same ⟨100⟩	4 of 93.37° on ⟨411⟩	4 of 160.25° on ⟨441⟩
		4 of 105.35° on ⟨533⟩	2 of 180° on ⟨410⟩
		4 of 137.33° on ⟨553⟩	2 of 180° on ⟨530⟩

$\Sigma 17_b$	2 of 61.93° on ⟨221⟩	2 of 118.07° on ⟨430⟩	2 of 160.25° on ⟨522⟩
	2 of 63.82° on ⟨331⟩	2 of 121.97° on ⟨320⟩	2 of 166.07° on ⟨733⟩
	86.63° } on same	4 of 137.33° on ⟨731⟩	180° on ⟨322⟩
	93.37° } ⟨110⟩	4 of 139.88° on ⟨521⟩	180° on ⟨433⟩
$\Sigma 19_a$	26.53° } on same	2 of 73.17° on ⟨511⟩	2 of 142.14° on ⟨433⟩
	153.47° } ⟨110⟩	2 of 93.02° on ⟨310⟩	2 of 161.33° on ⟨610⟩
		2 of 99.08° on ⟨332⟩	4 of 166.83° on ⟨751⟩
		2 of 110.01° on ⟨711⟩	180° on ⟨331⟩
		4 of 121.76° on ⟨432⟩	180° on ⟨611⟩
$\Sigma 19_b$	46.83° }	3 of 71.59° on ⟨320⟩	3 of 139.74° on ⟨733⟩
	73.17° } on same	3 of 110.01° on ⟨551⟩	3 of 142.14° on ⟨530⟩
	166.83° } ⟨111⟩	3 of 121.76° on ⟨520⟩	3 of 153.47° on ⟨411⟩
			3 of 180° on ⟨532⟩
$\Sigma 21_a$	21.79° }	3 of 79.02° on ⟨410⟩	3 of 128.25° on ⟨322⟩
	98.21° } on same	3 of 103.77° on ⟨510⟩	3 of 162.25° on ⟨540⟩
	141.79° } ⟨111⟩	3 of 113.87° on ⟨553⟩	3 of 167.48° on ⟨911⟩
			3 of 180° on ⟨541⟩
$\Sigma 21_b$	44.40° on ⟨211⟩	124.84° on ⟨441⟩	154.80° on ⟨210⟩
	58.40° on ⟨210⟩	124.84° on ⟨522⟩	162.25° on ⟨443⟩
	79.02° on ⟨322⟩	128.25° on ⟨410⟩	2 of 162.25° on ⟨621⟩
	2 of 80.41° on ⟨531⟩	2 of 141.79° on ⟨751⟩	2 of 167.48° on ⟨753⟩
	2 of 103.77° on ⟨431⟩	2 of 144.05° on ⟨532⟩	180° on ⟨421⟩
	2 of 113.87° on ⟨731⟩	144.05° on ⟨611⟩	
$\Sigma 23$	40.45° on ⟨311⟩	117.16° on ⟨733⟩	155.94° on ⟨332⟩
	55.56° on ⟨310⟩	127.49° on ⟨610⟩	163.04° on ⟨210⟩
	2 of 85.01° on ⟨421⟩	130.71° on ⟨331⟩	2 of 163.04° on ⟨542⟩
	86.25° on ⟨533⟩	143.56° on ⟨911⟩	2 of 168.04° on ⟨931⟩
	2 of 102.55° on ⟨321⟩	2 of 143.56° on ⟨753⟩	180° on ⟨631⟩
	2 of 107.72° on ⟨521⟩	2 of 145.70° on ⟨541⟩	
$\Sigma 25_a$	16.25° }	4 of 91.13° on ⟨711⟩	4 of 168.53° on ⟨771⟩
	73.75° } on same	4 of 111.10° on ⟨433⟩	2 of 180° on ⟨430⟩
	106.25° } ⟨100⟩	4 of 129.80° on ⟨443⟩	2 of 180° on ⟨710⟩
	163.75° }		
$\Sigma 25_b$	51.68° on ⟨331⟩	2 of 120° on ⟨751⟩	156.93° on ⟨211⟩
	2 of 63.88° on ⟨321⟩	129.80° on ⟨540⟩	2 of 163.75° on ⟨632⟩
	73.75° on ⟨221⟩	2 of 129.80° on ⟨621⟩	168.53° on ⟨311⟩
	90° on ⟨430⟩	2 of 132.86° on ⟨421⟩	168.53° on ⟨755⟩
	91.13° on ⟨551⟩	2 of 145.09° on ⟨931⟩	180° on ⟨543⟩
	111.10° on ⟨530⟩	2 of 147.15° on ⟨631⟩	
$\Sigma 27_a$	31.58° } on same	2 of 70.51° on ⟨411⟩	2 of 146.44° on ⟨755⟩
	148.42° } ⟨110⟩	2 of 94.25° on ⟨520⟩	2 of 157.82° on ⟨510⟩
		2 of 95.30° on ⟨553⟩	4 of 164.36° on ⟨641⟩
		2 of 114.05° on ⟨611⟩	180° on ⟨511⟩
		4 of 122.50° on ⟨753⟩	180° on ⟨552⟩

Table 3.6 (*Continued*)

$\Sigma 27_b$	35.42° on ⟨210⟩	2 of 114.05° on ⟨532⟩	2 of 157.82° on ⟨431⟩
	60° on ⟨511⟩	122.50° on ⟨911⟩	164.36° on ⟨720⟩
	79.32° on ⟨311⟩	2 of 131.81° on ⟨542⟩	2 of 168.97° on ⟨951⟩
	2 of 94.25° on ⟨432⟩	146.44° on ⟨771⟩	168.97° on ⟨773⟩
	2 of 95.30° on ⟨731⟩	148.42° on ⟨710⟩	180° on ⟨721⟩
	109.47° on ⟨411⟩	2 of 148.42° on ⟨543⟩	
$\Sigma 29_a$	43.61°, 46.39°, 133.61°, 136.39° on same ⟨100⟩	4 of 97.93° on ⟨522⟩	4 of 149.55° on ⟨552⟩
		4 of 98.92° on ⟨733⟩	2 of 180° on ⟨520⟩
		4 of 147.65° on ⟨773⟩	2 of 180° on ⟨730⟩
$\Sigma 29_b$	46.39° on ⟨221⟩	2 of 116.62° on ⟨541⟩	2 of 149.55° on ⟨721⟩
	2 of 66.63° on ⟨531⟩	2 of 124.68° on ⟨931⟩	164.92° on ⟨722⟩
	76.02° on ⟨332⟩	2 of 133.61° on ⟨632⟩	164.92° on ⟨544⟩
	84.07° on ⟨320⟩	136.40° on ⟨430⟩	2 of 169.36° on ⟨953⟩
	97.93° on ⟨441⟩	2 of 147.65° on ⟨951⟩	180° on ⟨432⟩
	112.29° on ⟨210⟩	149.55° on ⟨211⟩	
$\Sigma 31_a$	17.90°, 102.10°, 137.90° on same ⟨111⟩	3 of 80.70° on ⟨510⟩	3 of 126.62° on ⟨755⟩
		3 of 101.16° on ⟨610⟩	3 of 165.41° on ⟨650⟩
		3 of 114.79° on ⟨332⟩	3 of 169.70° on ⟨11 1 1⟩
			3 of 180° on ⟨651⟩
$\Sigma 31_b$	52.19° on ⟨211⟩	126.62° on ⟨311⟩	150.57° on ⟨730⟩
	54.49° on ⟨320⟩	126.62° on ⟨771⟩	2 of 159.33° on ⟨521⟩
	72.15° on ⟨533⟩	135.20° on ⟨720⟩	2 of 165.41° on ⟨643⟩
	2 of 80.70° on ⟨431⟩	2 of 135.20° on ⟨641⟩	169.70° on ⟨775⟩
	2 of 102.10° on ⟨751⟩	137.91° on ⟨511⟩	180° on ⟨732⟩
	2 of 118.93° on ⟨631⟩	2 of 148.74° on ⟨953⟩	
$\Sigma 33_a$	20.05°, 159.95° on same ⟨110⟩	2 of 76.86° on ⟨711⟩	2 of 136.67° on ⟨544⟩
		2 of 91.73° on ⟨410⟩	2 of 165.87° on ⟨810⟩
		2 of 104.02° on ⟨443⟩	4 of 170.02° on ⟨971⟩
		2 of 104.95° on ⟨911⟩	180° on ⟨441⟩
		4 of 121.01° on ⟨543⟩	180° on ⟨811⟩
$\Sigma 33_b$	33.55° on ⟨311⟩	117.04° on ⟨211⟩	2 of 151.51° on ⟨651⟩
	60.98° on ⟨410⟩	121.01° on ⟨710⟩	2 of 165.87° on ⟨652⟩
	2 of 84.78° on ⟨521⟩	128.41° on ⟨773⟩	165.87° on ⟨740⟩
	91.73° on ⟨322⟩	2 of 139.25° on ⟨432⟩	2 of 170.02° on ⟨11 3 1⟩
	2 of 104.02° on ⟨621⟩	149.72° on ⟨775⟩	180° on ⟨741⟩
	2 of 104.95° on ⟨753⟩	149.72° on ⟨11 1 1⟩	
$\Sigma 33_c$	58.98°, 121.02° on same ⟨110⟩	2 of 60.98° on ⟨322⟩	2 of 139.25° on ⟨520⟩
		2 of 76.86° on ⟨551⟩	4 of 151.51° on ⟨732⟩
		2 of 104.02° on ⟨540⟩	2 of 170.02° on ⟨955⟩
		4 of 128.41° on ⟨951⟩	180° on ⟨522⟩
		2 of 136.67° on ⟨722⟩	180° on ⟨554⟩

$\Sigma 35_a$	34.04° on ⟨211⟩	119.05° on ⟨510⟩	152.35° on ⟨811⟩
	64.63° on ⟨310⟩	122.89° on ⟨211⟩	160.54° on ⟨530⟩
	2 of 80.95° on ⟨731⟩	122.89° on ⟨552⟩	2 of 166.28° on ⟨821⟩
	88.37° on ⟨433⟩	2 of 137.98° on ⟨643⟩	2 of 170.31° on ⟨973⟩
	2 of 106.60° on ⟨542⟩	2 of 150.63° on ⟨971⟩	180° on ⟨531⟩
	2 of 107.46° on ⟨931⟩	152.35° on ⟨554⟩	
$\Sigma 35_b$	43.23° on ⟨331⟩	2 of 119.05° on ⟨431⟩	2 of 152.35° on ⟨741⟩
	2 of 66.40° on ⟨421⟩	2 of 122.89° on ⟨721⟩	160.54° on ⟨433⟩
	80.95° on ⟨553⟩	2 of 130.00° on ⟨953⟩	2 of 166.28° on ⟨742⟩
	88.37° on ⟨530⟩	137.98° on ⟨650⟩	170.31° on ⟨11 3 3⟩
	94.90° on ⟨331⟩	150.63° on ⟨955⟩	180° on ⟨653⟩
	106.60° on ⟨210⟩	2 of 150.63° on ⟨11 3 1⟩	
$\Sigma 37_a$	18.92°	4 of 91.55° on ⟨611⟩	4 of 166.66° on ⟨661⟩
	71.08° on same	4 of 109.75° on ⟨755⟩	2 of 180° on ⟨610⟩
	108.92° ⟨100⟩	4 of 131.46° on ⟨775⟩	2 of 180° on ⟨750⟩
	161.08°		
$\Sigma 37_b$	43.13° on ⟨310⟩	109.75° on ⟨311⟩	2 of 153.12° on ⟨653⟩
	50.57° on ⟨511⟩	131.46° on ⟨11 1 1⟩	161.09° on ⟨221⟩
	84.56° on ⟨733⟩	2 of 139.18° on ⟨652⟩	166.66° on ⟨830⟩
	2 of 91.55° on ⟨532⟩	139.18° on ⟨810⟩	2 of 170.58° on ⟨11 5 1⟩
	2 of 97.75° on ⟨421⟩	141.62° on ⟨441⟩	180° on ⟨831⟩
	2 of 108.92° on ⟨632⟩	2 of 151.45° on ⟨973⟩	
$\Sigma 37_c$	50.57° on same	3 of 71.06° on ⟨430⟩	3 of 139.18° on ⟨740⟩
	69.43° ⟨111⟩	3 of 109.75° on ⟨771⟩	3 of 141.62° on ⟨522⟩
	170.57°	3 of 124.59° on ⟨730⟩	3 of 151.45° on ⟨11 3 3⟩
			3 of 180° on ⟨743⟩
$\Sigma 39_a$	32.21° on same	3 of 75.14° on ⟨520⟩	3 of 132.80° on ⟨955⟩
	87.79° ⟨111⟩	3 of 111.02° on ⟨720⟩	3 of 153.83° on ⟨750⟩
	152.21°	3 of 111.83° on ⟨773⟩	3 of 161.57° on ⟨611⟩
			3 of 180° on ⟨752⟩
$\Sigma 39_b$	50.13° on ⟨321⟩	122.58° on ⟨521⟩	152.22° on ⟨11 5 1⟩
	56.53° on ⟨531⟩	126.15° on ⟨651⟩	153.83° on ⟨743⟩
	73.62° on ⟨321⟩	126.15° on ⟨732⟩	153.83° on ⟨831⟩
	75.14° on ⟨432⟩	132.80° on ⟨971⟩	161.57° on ⟨532⟩
	87.80° on ⟨751⟩	132.80° on ⟨11 3 1⟩	167.01° on ⟨654⟩
	94.40° on ⟨541⟩	140.29° on ⟨742⟩	167.01° on ⟨832⟩
	111.02° on ⟨641⟩	140.29° on ⟨821⟩	170.82° on ⟨975⟩
	111.83° on ⟨951⟩	142.65° on ⟨531⟩	170.82° on ⟨11 5 3⟩
$\Sigma 41_a$	12.68°	4 of 90.68° on ⟨911⟩	4 of 171.05° on ⟨991⟩
	77.32° on same	4 of 112.95° on ⟨544⟩	2 of 180° on ⟨540⟩
	102.68° ⟨100⟩	4 of 127.56° on ⟨554⟩	2 of 180° on ⟨910⟩
	167.32°		

Table 3.6 (*Continued*)

$\Sigma 41_b$	40.88° on ⟨210⟩	2 of 113.73° on ⟨953⟩	2 of 154.48° on ⟨752⟩
	55.88° on ⟨411⟩	127.56° on ⟨811⟩	162.03° on ⟨310⟩
	78.75° on ⟨522⟩	2 of 134.05° on ⟨973⟩	167.32° on ⟨221⟩
	2 of 90.68° on ⟨753⟩	141.32° on ⟨661⟩	2 of 167.32° on ⟨841⟩
	2 of 97.00° on ⟨631⟩	143.60° on ⟨610⟩	180° on ⟨621⟩
	112.95° on ⟨722⟩	2 of 152.91° on ⟨975⟩	
$\Sigma 41_c$	55.88° } on same	2 of 61.61° on ⟨533⟩	2 of 141.32° on ⟨830⟩
	124.12° } ⟨110⟩	2 of 78.75° on ⟨441⟩	4 of 152.91° on ⟨11 5 3⟩
		2 of 102.68° on ⟨430⟩	2 of 167.32° on ⟨744⟩
		4 of 127.56° on ⟨741⟩	180° on ⟨443⟩
		2 of 134.05° on ⟨11 3 3⟩	180° on ⟨833⟩
$\Sigma 43_a$	15.18° }	3 of 81.97° on ⟨610⟩	3 of 125.57° on ⟨433⟩
	104.82° } on same ⟨111⟩	3 of 99.68° on ⟨710⟩	3 of 167.62° on ⟨760⟩
	135.82° }	3 of 115.47° on ⟨775⟩	3 of 171.27° on ⟨13 1 1⟩
			3 of 180° on ⟨761⟩
$\Sigma 43_b$	27.91° on ⟨210⟩	2 of 114.74° on ⟨643⟩	2 of 162.47° on ⟨541⟩
	65.99° on ⟨711⟩	115.47° on ⟨11 1 1⟩	167.62° on ⟨920⟩
	80.62° on ⟨411⟩	2 of 128.89° on ⟨653⟩	171.27° on ⟨331⟩
	2 of 93.34° on ⟨931⟩	2 of 142.25° on ⟨654⟩	2 of 171.27° on ⟨11 7 1⟩
	2 of 99.68° on ⟨543⟩	153.56° on ⟨991⟩	180° on ⟨921⟩
	104.81° on ⟨511⟩	155.09° on ⟨910⟩	

at this stage, to be located on a standard stereographic projection of crystal directions (as described in section 3.3.2). Thus, the beam directions in both grains, U and L, obtained at this stage were specifically indexed with the indices in figure 3.4. From the relative locations of position ④ in grains U and L on the standard projection, as shown in figure 3.10, it was found that the angle between them was about 5° greater than the angle between $[2\bar{1}1]$ and $[1\bar{2}2]$, i.e. an angle of approximately 40°. Thus the misorientation between the grains corresponds to a rotation θ of approximately 40° about an axis $\boldsymbol{u}$ parallel to the same diffracting vector $\boldsymbol{g}_c$ with the same indices in both grains of $022_U/022_L$, i.e. $\theta \approx 40°$, $\boldsymbol{u} = [011]$. From table 3.6 it was found that there is only one CSL orientation for which the angle–axis pair is close to the experimental estimate, namely a $\Sigma 9$ CSL which corresponds to a rotation of 38.94° around a [011] axis common to both grains. The rotation matrix for this exact $\Sigma 9$ CSL, which re-indexes directions in grain U to directions in grain L, was obtained from equation (3.8) as

$$1/9\begin{pmatrix} 7 & -4 & 4 \\ 4 & 8 & 1 \\ -4 & 1 & 8 \end{pmatrix}. \qquad (3.15)$$

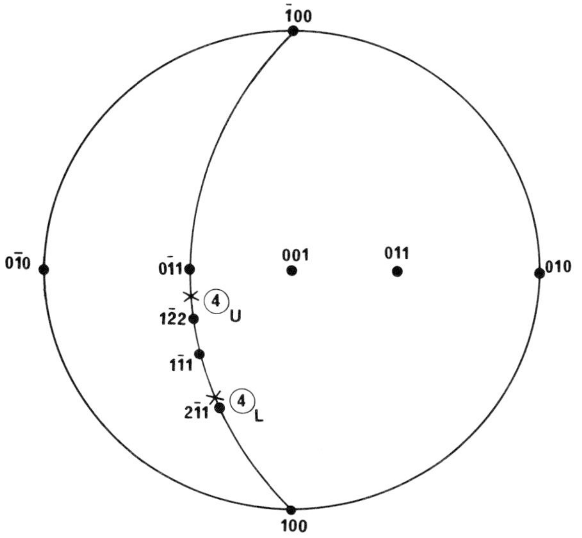

Figure 3.10 Stereographic projection showing the relative location of orientation map position (4) in grains U and L.

Additional low-order same-$\boldsymbol{g}_c$ diffracting vectors were then obtained from this matrix by inspection as $\bar{1}\bar{3}1_U/1\bar{3}1_L$, $\bar{1}\bar{1}3_U/1\bar{1}3_L$, $\bar{4}20_U/\bar{4}02_L$ and $402_U/420_L$.

It should be noted that in cases where the additional same-$\boldsymbol{g}_c$ diffracting vectors cannot be readily found by inspection of a rotation matrix, they can be found from table 3.6. For example, for the present case of a $\Sigma9$ CSL it can be seen from table 3.6 that in addition to the $\langle 110 \rangle$ rotation axis there are 22 other equivalent rotation axes, of which two are of the type $\langle 311 \rangle$ and two of the type $\langle 210 \rangle$. These equivalent rotation axes will correspond to the same-$\boldsymbol{g}_c$ diffracting vectors given above.

Beam directions containing the additional same-$\boldsymbol{g}_c$ diffracting vectors for the boundary of figure 3.2 were located on the standard projection and a check was made as to whether the tilt available in the microscope would enable them to be excited. It was found that only the $\bar{4}20_U/\bar{4}02_L$ same-$\boldsymbol{g}_c$ was available within the range of tilt, and appropriate tilt of the specimen showed the presence of this predicted same-$\boldsymbol{g}_c$ diffracting vector. In this way the boundary of figure 3.2 was shown in the early stages of examination to be close to the exact $\Sigma9$ CSL orientation given by the matrix (3.15).

A quick identification of the nearest exact CSL by this method is best carried out in the initial stages of the study of a boundary in the electron microscope so that further tilting to set up other same-$\boldsymbol{g}_c$ vectors can be done in a systematic way.

3.5 DETERMINATION OF CRYSTALLOGRAPHIC DATA FOR THE ANALYSIS OF INTERFACE STRUCTURE

For the quantitative analysis of the structure of an interface by transmission electron microscopy it is necessary to determine the line direction, $\boldsymbol{r}$, of individual structural features in the interface, the normal to the plane of the interface, $\boldsymbol{v}$, and the normal, $\boldsymbol{F}$, to the surface of the specimen. In this section an outline will be given of how these parameters are determined and how the bottom and top surfaces of the specimen are established so that the upper grain, i.e. the grain which is nearer the electron source, can be identified. It is necessary to know which is the upper grain because, when analysing diffraction effects and image contrast from an interface and the defects in it, it is necessary to know which grain the electron beam entered first (see section 2.4). However, a knowledge of which grain is the upper grain in a bicrystal is not needed at the outset when determining $\boldsymbol{r}$, $\boldsymbol{v}$ and $\boldsymbol{F}$, as the indexing of these directions can be done with respect to either one of the grains, and then, if necessary, they can be readily re-indexed with respect to the other grain using the determined orientation relationship.

3.5.1 Determination of a Line Direction r

The line direction $\boldsymbol{r}$ of a defect, such as one of the segments in the network of interfacial dislocations in figure 3.2, is determined by using images taken in at least three different beam directions, so as to obtain at least three different projections of the line direction of the defect. The problem then is to determine the direction $\boldsymbol{r}$ of the defect from these projections of its direction. For each image i, where $i = 1, \ldots, n$, the projected line direction $\boldsymbol{r}_{\boldsymbol{B}_i}$ in the plane normal to the beam direction $\boldsymbol{B}_i$ is obtained from the measured angle between it and the direction of the diffracting vector marked on the image for one of the grains. The line direction $\boldsymbol{r}$ of the defect must lie on a plane with plane normal $\boldsymbol{P}_i$ given by

$$\boldsymbol{P}_i = \boldsymbol{B}_i \wedge \boldsymbol{r}_{\boldsymbol{B}_i}.$$

The direction $\boldsymbol{r}$ of the defect is then defined by the common line of intersection of the different planes $\boldsymbol{P}_i$ and is given by

$$\boldsymbol{r} = \boldsymbol{P}_i \wedge \boldsymbol{P}_j$$

where $i, j = 1, \ldots, n$ and $i \neq j$. In practice n needs to be greater than 2 to obtain a meaningful experimental result and such a determination of the line direction of a defect is most readily carried out using standard stereographic techniques (see, for example, Head *et al* 1973).

3.5.2 Determination of Interface Normal $\boldsymbol{v}$

The normal $\boldsymbol{v}$ to the plane of an interface is obtained from the determined directions of two lines lying in the interface, such as line directions $\boldsymbol{r}_i$ and $\boldsymbol{r}_j$ of interfacial dislocations, and is given by

$$\boldsymbol{v} = \boldsymbol{r}_i \wedge \boldsymbol{r}_j.$$

As with the determination of $\boldsymbol{r}$, the determination of $\boldsymbol{v}$ is best achieved by standard stereographic methods.

3.5.3 Determination of Specimen Surface Normal $\boldsymbol{F}$

The normal to the surface of the specimen, $\boldsymbol{F}$, is found from the determined directions of traces lying in the surface, and one such trace, for example, is the line of intersection of the plane of the interface with the specimen surface. For two traces with directions $\boldsymbol{t}_i$ and $\boldsymbol{t}_j$, $\boldsymbol{F}$ is given by

$$\boldsymbol{F} = \boldsymbol{t}_i \wedge \boldsymbol{t}_j$$

and again is determined most readily using standard stereographic techniques.

3.5.4 Identification of Upper Grain

When a direction $\boldsymbol{r}$ in an interface and the specimen normal $\boldsymbol{F}$ have been determined, the intersections of the plane of the interface with the top and bottom surfaces of the specimen can be found and, for a given image, the upper and lower grains identified. For this purpose, the sense of $\boldsymbol{F}$ is defined as being upward-drawn (i.e. acute with $\boldsymbol{B}$) and the sense of $\boldsymbol{r}$ is defined from the bottom to the top of the specimen (i.e. acute with $\boldsymbol{F}$). Then, the projection of $\boldsymbol{r}$ with this sense on the image plane points from the line of intersection of the interface with the bottom of the specimen to the line of intersection of the interface with the top of the specimen, and this identifies the upper grain. For the example in figure 3.2 the intersections of the boundary with the top and bottom of the specimen are marked T and B respectively so that the upper grain, labelled U, is on the right.

3.6 ADDITIONAL DIFFRACTION PARAMETERS

In order to calculate the diffraction contrast in electron microscope images of interfaces it is necessary to determine the diffraction parameters listed in table 2.1. Methods for determining the beam direction $\boldsymbol{B}$ and the diffracting vector $\boldsymbol{g}$ have already been discussed in section 3.3, and methods for determining the additional parameters—extinction distance $\xi_{\boldsymbol{g}}$, the deviation from the Bragg condition, w, and the anomalous absorption coefficient, $\mathscr{A}$—will be described in this section.

3.6.1 Extinction Distance ξ_g

The extinction distance ξ_g, associated with a particular diffracting vector $\boldsymbol{g}$, was introduced in section 2.3.1, for two-beam diffracting conditions, as the distance within the crystal in the beam direction between neighbouring intensity maxima of either of the beams. The extinction distance depends on the operative diffracting vector $\boldsymbol{g}$ and can be obtained from the relation:

$$\xi_g = [(h^2E/2me)^{1/2} \cos \theta_B]/V_g \tag{3.16}$$

where V_g is the Fourier coefficient for the reciprocal lattice vector $\boldsymbol{g}$ of the periodic scattering potential V of the crystal, h is Planck's constant, E is the energy of the incident electrons, m is the relativistic mass of the electron, e is the electronic charge and θ_B is the Bragg angle. Values of V_g have been calculated by Radi (1970) and his tables list these values for specific diffracting vectors in a variety of materials of different crystal structure which can be used to obtain values of ξ_g from equation (3.16).

For materials involving compositions and crystal structures not listed by Radi, the extinction distance ξ_g can be calculated using the relation

$$\xi_g = (\pi V_c \cos \theta_B)/\lambda F_g \tag{3.17}$$

where V_c is the volume of the unit cell for the crystal structure concerned and F_g is the structure factor for the unit cell. The structure factor F_g can be expressed as

$$F_g = \sum_{j=1}^{n} f_j \exp[2\pi i(\boldsymbol{r}_j \cdot \boldsymbol{g})]\exp(-M_g)_j. \tag{3.18}$$

The sum in equation (3.18) is over the n atoms in the unit cell, and for the jth atom f_j is the atomic scattering amplitude for electrons, $\boldsymbol{r}_j$ its position in the unit cell, and $\exp(-M_g)_j$ is its Debye–Waller factor. The atomic scattering amplitude f is a function of $\sin \theta_B/\lambda = \boldsymbol{g}/2$, and values of f for different atomic species are given by Doyle and Turner (1968). Values of M_g can be obtained from values of the thermal Debye parameter B given in the *International Tables of X-ray Crystallography* (1962) using the relation

$$M_g = B(\sin \theta_B/\lambda)^2.$$

Examples of this type of calculation of ξ_g for alloys using equations (3.17) and (3.18) are given in Head *et al* (1973).

The discussion so far in this section has been concerned with the theoretical two-beam extinction distance. In practical electron microscopy the ideal two-beam diffraction condition of section 2.3 can never be exactly satisfied, as other weak beams, particularly those in a systematic row, are excited to some small degree, and this has the effect that the actual experimental extinction distance ξ_{ge} departs slightly from the theoretical two-beam extinction distance ξ_g. This difference between experimental and theoretical extinction distances is only of importance when dimensions in computed images are being related to real dimensions, and this point will be discussed in section 3.7 in connection with the determination of the thickness of an electron microscope specimen.

3.6.2 Deviation from the Bragg Condition w

In section 2.2, with reference to figure 2.1, it was shown that when a crystal was rotated through a small angle $\Delta\theta$ from the exact Bragg condition, then the deviation from the Bragg condition is specified by a scalar parameter s given by

$$s = |\boldsymbol{g}|\,\Delta\theta \tag{3.19}$$

where s is positive when the reciprocal lattice point G_1 of figure 2.1(b) lies inside the Ewald sphere. This deviation from the exact Bragg condition is usually expressed as the dimensionless parameter w where

$$w = \xi_g s. \tag{3.20}$$

The parameter w is most conveniently determined from the displacements of Kikuchi lines in diffraction patterns caused by small angular tilts of the specimen, and such displacements are illustrated schematically in figures 3.11(a) and (b). Figure 3.11(a) shows the position of the pairs of Kikuchi lines E_1D_1 and E_2D_2, associated with the diffraction spots $\boldsymbol{g}$ and $2\boldsymbol{g}$ respectively, when the exact Bragg condition ($s = 0$) is satisfied for the diffracting vector $\boldsymbol{g}$. Figure 3.11(b) shows the positions of the same pairs of Kikuchi lines after a small tilt $\Delta\theta$ of the specimen, making $s > 0$. As pointed out in section 2.2, this small angular tilt $\Delta\theta$ of the specimen can be related to the displacement of the Kikuchi lines from their positions at $s = 0$ to their positions at $s > 0$ by

$$\Delta\theta/2\theta_B = \Delta x_1/x \tag{3.21}$$

where Δx_1 is the displacement of the Kikuchi line E_1 from the diffraction spot $\boldsymbol{g}$. From equations (3.19)–(3.21) it can be seen that the deviation from the Bragg condition w is given by

$$w = 2\xi_g|\boldsymbol{g}|\theta_B(\Delta x_1/x).$$

In practice it is difficult to measure Δx_1 because the contrast in the diffraction pattern in the region of the diffraction spot $\boldsymbol{g}$ is too strong and diffuse. However, the diffraction spot $2\boldsymbol{g}$ and the Kikuchi line E_2 in its vicinity are usually clearly defined, and in practice w is determined from a measurement

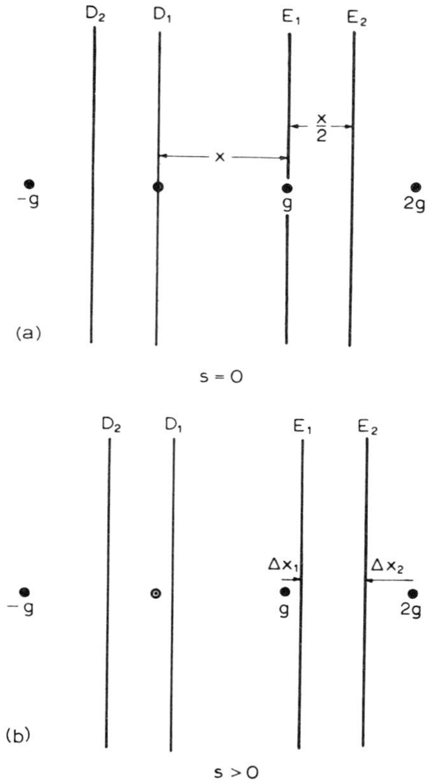

Figure 3.11 Schematic illustration of the parameters required for the determination of *w*.

of Δx_2 which is related to Δx_1 by

$$\Delta x_1 + \Delta x_2 = x/2$$

so that

$$w = \xi_{\boldsymbol{g}} |\boldsymbol{g}| \theta_{\mathrm{B}}(1 - 2\Delta x_2/x).$$

3.6.3 Anomalous Absorption Coefficient

The anomalous absorption coefficient $\mathscr{A}$ was introduced in the discussion of contrast calculations in section 2.3.1 to take account, in a phenomenological way, of an apparent loss of some of the electrons by inelastic scattering processes in two-beam diffraction. In section 2.3.1 $\mathscr{A}$ is defined in terms of a complex extinction distance with a real part $\xi_{\boldsymbol{g}}$ and an imaginary part $\xi'_{\boldsymbol{g}}$ as

$$\mathscr{A} = \xi_{\boldsymbol{g}}/\xi'_{\boldsymbol{g}}.$$

An alternative way of expressing $\mathscr{A}$ derives from a description of scattering of electrons in a crystal in terms of the interactions of the electrons with a periodic lattice potential which is complex. This lattice potential is expressed in terms of real Fourier coefficients V_g, associated with elastic scattering, and imaginary Fourier coefficients V'_g, associated with inelastic scattering, for the reciprocal lattice vectors $\boldsymbol{g}$ (Yoshioka 1957, Humphreys and Hirsch 1968 and Radi 1970). In these terms $\mathscr{A}$ can be re-expressed as

$$\mathscr{A} = V'_g / V_g.$$

Humphreys and Hirsch (1968) have calculated the variation of V'_g/V_g with $|\boldsymbol{g}|$ for a range of elements with atomic number Z varying from $Z = 6$ for carbon to $Z = 79$ for gold. Their results are reproduced in figure 3.12, where it is clear that there is a considerable variation in V'_g/V_g with $|\boldsymbol{g}|$ and Z. The values in figure 3.12 have been calculated for 'zero aperture' and thus only give relative values of anomalous absorption $\mathscr{A}$. However, theoretical values of V'_g as well as V_g have been calculated by Radi (1970) for a finite aperture size, and these can be used to obtain values of $\mathscr{A}$ for the materials that he considered. For other materials, values of $\mathscr{A}$ can be calculated in the way described by Radi, or alternatively an estimate of $\mathscr{A}$ can be obtained experimentally (Clarebrough 1969) by matching the contrast in experimental images of a known defect with theoretical images computed for a variation in $\mathscr{A}$.

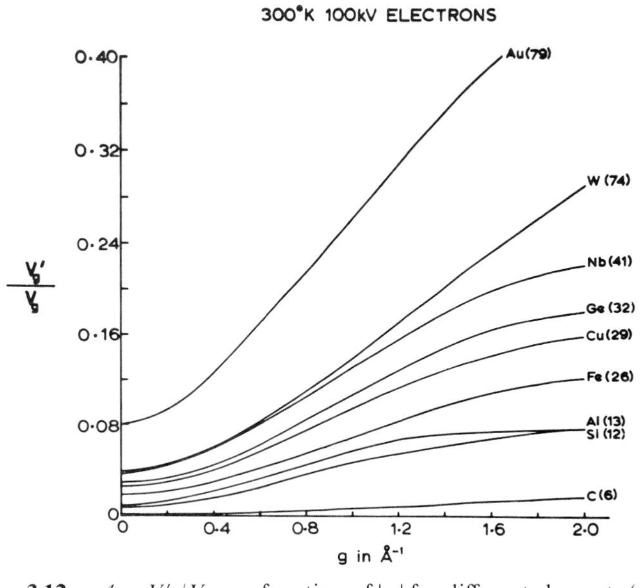

Figure 3.12 $\mathscr{A} = V'_g/V_g$ as a function of $|\boldsymbol{g}|$ for different elements (after Humphreys and Hirsch 1968).

3.7 DETERMINATION OF SPECIMEN THICKNESS

The thickness of a thin-foil specimen containing the interface under investigation in the electron microscope needs to be determined before any calculations of image contrast associated with the interface can be made. A common method for determining the thickness of a foil is from the measurement on an image of the projected length of a feature which extends from the bottom to the top surface of the foil. For a known magnification, this length l, after correction for projection, gives the thickness t of the specimen directly as

$$t = l \cos \alpha / \sin \beta \tag{3.22}$$

where α is the angle between the line direction of the measured feature and the foil normal and β is the angle between the line direction of the measured feature and the beam direction. These angles α and β are most readily determined using standard stereographic techniques.

The value of foil thickness obtained in this way will be expressed in units of length such as ångströms (Å), but for theoretical calculations of two-beam image contrast all dimensions are required in units of extinction distance (see equations (2.5)). Furthermore, the appropriate unit of length that is required in a computation, when comparing dimensions in computed and experimental images, is the experimental extinction distance $\xi_{\boldsymbol{g}e}$. Thus, the most convenient determination of foil thickness for use in calculations of image contrast is one which gives the foil thickness directly in units of $\xi_{\boldsymbol{g}e}$. A suitable way of making this determination is to use the oscillatory character of contrast in images of features that extend throughout the foil, such as the fringes in an interface, stacking fault fringes or oscillations in contrast along the length of a dislocation. Then, for an image taken in a given beam direction, with a known diffracting vector and value of w, the number of these oscillations n gives the projected thickness t' of the foil in terms of $\xi_{\boldsymbol{g}e}$ as

$$t' = [n/(1 + w^2)^{1/2}]\xi_{\boldsymbol{g}e}$$

so that the thickness t measured normal to the surface of the foil is

$$t = [n/(1 + w^2)^{1/2}]\xi_{\boldsymbol{g}e} \cos \delta$$

where δ is the angle between the foil normal $\boldsymbol{F}$ and the beam direction $\boldsymbol{B}$. For a given diffracting vector, the number of contrast oscillations can only be estimated from an image to about half an oscillation, but, when averaged over a series of images in different beam directions with different values of w, the foil thickness can be obtained to an accuracy of $\pm 0.1\xi_{\boldsymbol{g}e}$. In practice, when estimating foil thickness by this method, the images used should be ones for a diffracting vector with a small extinction distance, and the foil thickness for other diffracting vectors is then obtained with sufficient accuracy by scaling this value in the ratio of the theoretical two-beam extinction distances.

When a numerical value of ξ_{ge} is required in order to relate dimensions in a theoretical image to real dimensions it can be obtained by comparing the value of foil thickness determined by counting contrast oscillations with that determined by direct measurement using equation (3.22).

3.8 COMMENT

Determinations of all the diffraction and crystallographic parameters which have been discussed in this chapter are essential for the quantitative analysis of interface structure by transmission electron microscopy. The particular methods described for obtaining the basic diffraction data, orientation relationship and crystallography of the interface require the analysis of a considerable number of images and diffraction patterns covering a wide range of beam directions and diffracting conditions. However, if the electron microscopy is carried out in a systematic way, these methods are straightforward and provide a basis which enables a quantitative interpretation of the structure of interfaces to be made.

4

Low-angle Grain Boundaries

4.1 INTRODUCTION

When the dislocation model of low-angle grain boundaries was first introduced in Chapter 1, it was discussed as being applicable in the approximate range of misorientation $\theta \lesssim 15^\circ$, and no distinction was made between different boundaries in this range. However, when discussing the electron microscopy of low-angle grain boundaries within this broad range, a distinction will be made between boundaries with $\theta \lesssim \frac{1}{2}^\circ$ and those with $\frac{1}{2}^\circ \lesssim \theta \lesssim 15^\circ$. For those misorientations where $\theta \lesssim \frac{1}{2}^\circ$, the individual dislocations making up the structure of the boundary can be clearly resolved by two-beam electron microscopy and the boundary can be considered as a dislocation network in a single crystal. In these cases the Burgers vectors of the grain boundary dislocations, and therefore the nature of the boundary, can be determined from image contrast using the same procedures that apply to dislocations in single crystals. However, for misorientations with increasing values of θ in the range $\frac{1}{2}^\circ \lesssim \theta \lesssim 15^\circ$ the spacings of the dislocations become small enough for the contrast in images of the boundary to change from that associated with diffraction by the displacement fields of individual dislocations to that associated with the interference of diffracted beams which arise from the periodic displacement field of the dislocation array as a whole. When this situation develops, $\boldsymbol{g} \cdot \boldsymbol{b}$ invisibility criteria and image matching techniques for isolated dislocations cannot be used to identify the Burgers vectors of the individual grain boundary dislocations. Under these conditions alternative methods such as geometric analysis, as discussed in section 1.2, or n-beam lattice-imaging techniques are required to determine the Burgers vectors of the grain boundary dislocations.

Detailed investigations of the structure of low-angle grain boundaries by transmission electron microscopy have been made on boundaries which have formed in very different ways. Many investigations have been concerned with low-angle boundaries formed during annealing of bulk polycrystalline metals and alloys, and these boundaries have been studied in thin-foil specimens prepared by thinning the bulk polycrystalline material. A very different approach, which is confined almost exclusively to gold, has involved the study of special low-angle boundaries in fabricated thin-film bicrystal specimens suitable for examination in the electron microscope without thinning. Another approach has involved the preparation of special bicrystals for lattice imaging. In this chapter examples of the analysis of the structure of low-angle boundaries in specimens obtained by thinning bulk polycrystals will be discussed. Also, the work on low-angle boundaries in fabricated thin-film bicrystals and on low-angle boundaries in bicrystals prepared so as to be suitable for lattice imaging will be reviewed.

4.2 LOW-ANGLE BOUNDARIES IN POLYCRYSTALS WITH $\theta \lesssim \frac{1}{2}°$

The first observations, using transmission electron microscopy, of coarsely spaced dislocations in low-angle grain boundaries with $\theta \lesssim \frac{1}{2}°$ were those of Hirsch *et al* (1956) in polycrystalline specimens of aluminium. Following these observations, similar low-angle boundaries have been studied in a variety of metals and alloys. One example is the work of Carrington *et al* (1960) who examined dislocation networks consisting of two independent arrays of dislocations in low-angle boundaries in BCC iron. Although they could not make any detailed analysis from dislocation contrast, due to limited tilting facilities in the electron microscope, they concluded from the geometry of the networks that each network had formed from two arrays of dislocations with Burgers vectors of the type $(1/2)\langle 111 \rangle$, which had interacted to form segments with Burgers vectors of the type $\langle 100 \rangle$ to give an hexagonal array. With the development of double-tilting goniometers it became possible to determine the Burgers vectors of dislocations using $\boldsymbol{g} \cdot \boldsymbol{b}$ invisibility criteria, and these criteria were successfully applied to dislocations in low-angle boundary networks by treating them as if they were dislocations in single crystals. One example of this is the work of Lindroos and Miekk-oja on low-angle boundaries in FCC aluminium–magnesium alloys (see, for example, Lindroos and Miekk-oja 1967, 1969). They analysed networks consisting of three independent arrays of dislocations and showed that the Burgers vectors involved were of the $(1/2)\langle 110 \rangle$ type. However, in order to specify completely the dislocations in low-angle boundary networks it is necessary to determine the sense as well as the magnitude of the Burgers vectors. This cannot be done using $\boldsymbol{g} \cdot \boldsymbol{b}$ invisibility criteria alone, but requires image-contrast calculations.

An example in which low-angle grain boundary dislocations are analysed using $\boldsymbol{g} \cdot \boldsymbol{b}$ invisibility criteria and image matching is shown in figures 4.1–4.4.

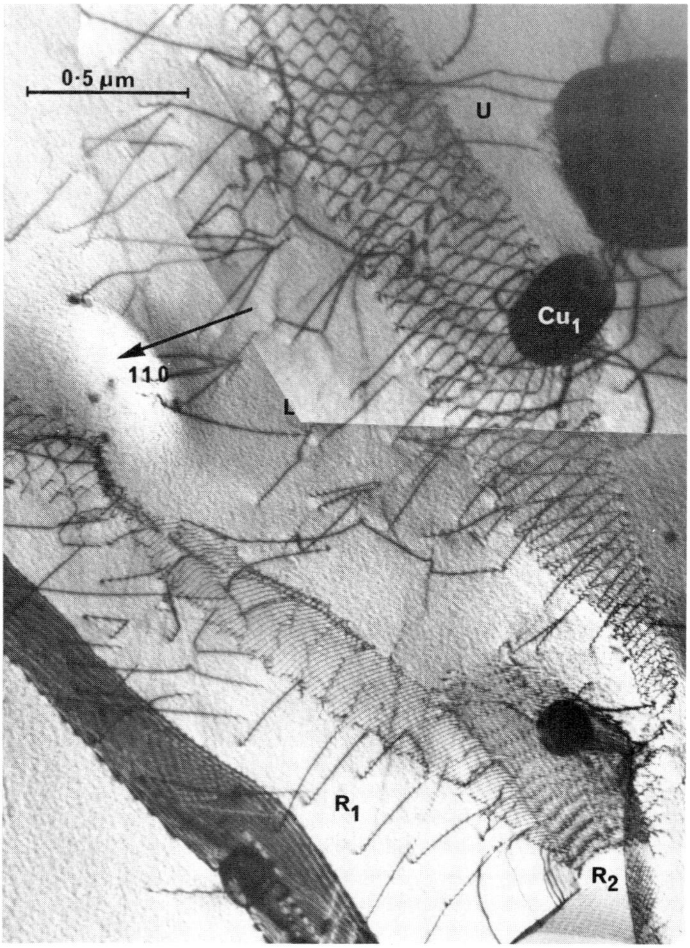

Figure 4.1 Electron micrograph of low-angle boundaries in the BCC phase of a two-phase Cu − 25 wt% Fe alloy.

Figure 4.1 is a low-magnification electron micrograph showing several low-angle grain boundaries in the BCC phase of a two-phase Cu − 25 wt % Fe alloy. The boundaries separating grains U, L, R_1 and R_2 are all general low-angle boundaries and the misorientations θ are approximately 0.3° for the boundary separating grains U and L, and approximately 1° for the boundaries separating grains L and R_1 and grains L and R_2. The dislocation arrays in the portion of the boundary between grains U and L immediately below the crystal of copper labelled Cu_1 are shown for different diffracting vectors in figure 4.2 and illustrated schematically in figure 4.3. In figure 4.3

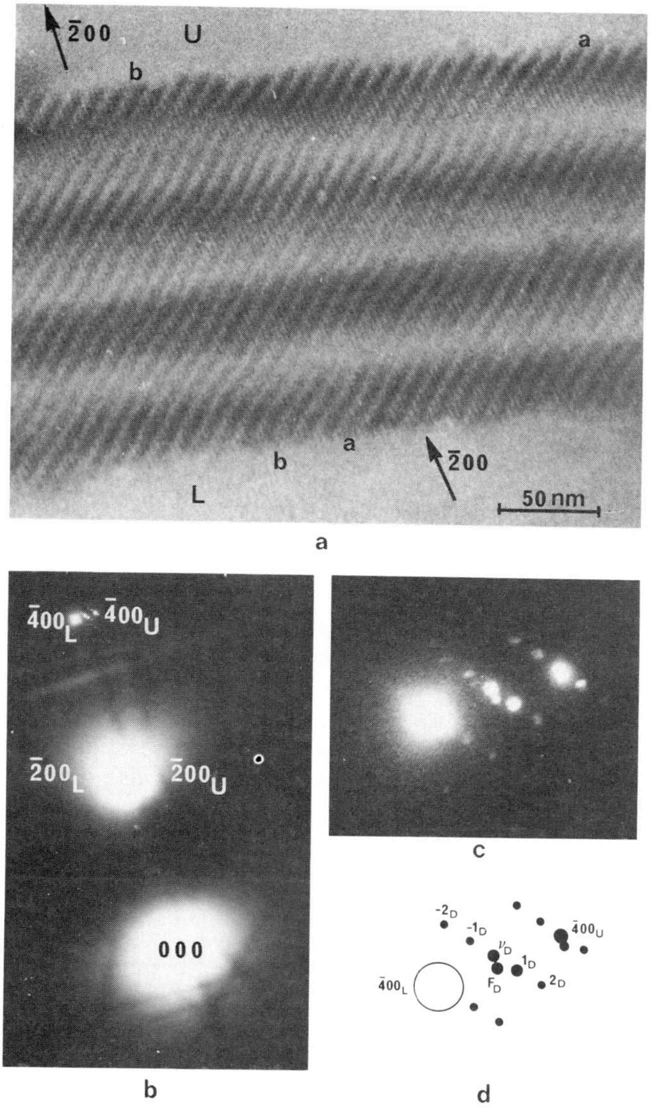

Figure 4.5 Double two-beam image (*a*) and diffraction patterns (*b*)–(*d*) used to distinguish between structural and non-structural periodicities in a low-angle boundary in copper.

these can only be seen in figure 4.5(*b*) in the region of the $\bar{4}00_U$ and $\bar{4}00_L$ spots. This region of the diffraction pattern is shown at higher magnification in figure 4.5(*c*) and represented schematically in figure 4.5(*d*). The additional rows of spots all exhibit the same periodicity and only occur in the diffraction pattern taken from the boundary itself, i.e. they are not present in the diffraction patterns taken from either of the grains.

The row of diffraction spots which is present in the vicinity of the central beam of figure 4.5(*b*) can be seen in figure 4.6(*a*) which shows this region at higher magnification. In the schematic diagram of figure 4.6(*b*) this row of spots is labelled −1, 0, 1, 2 and the central spot is labelled X. These diffraction spots are included in the objective aperture and contribute to the bright-field image. The coarsely spaced fringes *aa* in the bright-field image of figure 4.5(*a*) correlate with the row of diffraction spots −1, 0, 1, 2 in that the direction of this row is perpendicular to the fringes and the reciprocal spacing of the spots corresponds to the measured fringe spacing of 83 Å. Thus, the fringes *aa* arise as a result of the interference of the electron waves that generate the diffraction spots −1, 0, 1, 2. Similarly, the finely spaced fringes *bb* arise from the interference of the electron waves that generate the diffraction spots X and 0 in that the direction joining these spots is perpendicular to the fringes *bb* and the reciprocal spacing of the spots corresponds to the measured fringe spacing of 29 Å. The direction and spacing of the spots X and 0 are the same as those between the $\bar{2}00_L$ and $\bar{2}00_U$ spots, so that spot 0 and its associated spots −1, 1 and 2 are formed by the $\bar{2}00_U$ spot, with its associated row of spots, being doubly diffracted by the 200_L diffracting vector. Spot 0 thus arises simply as a result of double-diffraction of a diffraction spot in the upper grain, namely the $\bar{2}00_U$ spot, so that the

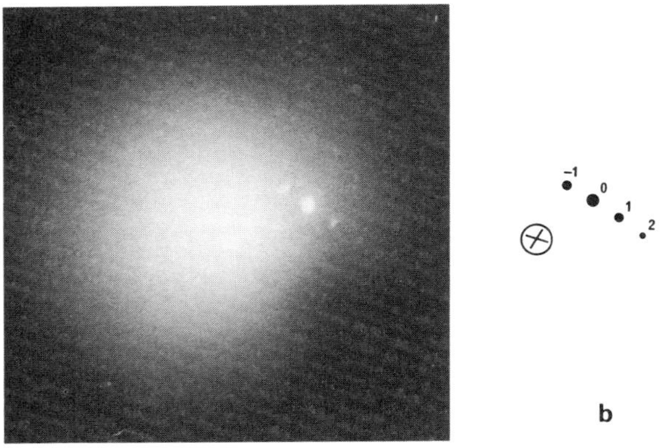

a b

Figure 4.6 (*a*) Row of diffraction spots in the vicinity of the 000 spot of figure 4.5(*b*). In (*b*) the 000 spot is marked X and the diffraction spots in the row are numbered.

fringes *bb* in the image of figure 4.5(*a*) are just a diffraction effect and are not associated with any structure in the boundary. In contrast, diffraction spots −1, 1 and 2 in the row −1, 0, 1, 2, and other equivalent rows of spots with the same periodicity that occur in association with diffraction spots from the grains in the diffraction pattern from the boundary, do not have any counterparts in either of the diffraction patterns from the grains, nor can they be generated by double-diffraction of any of the spots present in the diffraction patterns from the grains. Thus, the diffraction spots −1, 1 and 2 in the row −1, 0, 1, 2 and the corresponding fringes *aa*, in the bright-field image of figure 4.5(*a*), arise as a result of genuine periodic structure in the boundary.

The analysis of periodicities in diffraction patterns associated with boundary structure involves the measurement of their spacings and directions and this requires clearly visible arrays of diffraction spots. The arrays of spots associated with intense beams, such as the central beam and the diffracted beams for the double two-beam condition, seldom satisfy this requirement, but other arrays formed in association with more weakly excited beams are generally clearly visible. In the present case the row of diffraction spots in association with the $\bar{2}00_U$ diffraction spot, which can be seen on the original diffraction pattern, is not visible in the positive print of figure 4.5(*b*), but is clearly visible in the position where the $\bar{2}00_U$ diffraction spot has been doubly diffracted by the $\bar{2}00_L$ diffracting vector to give the row of spots between the $\bar{4}00_U$ and $\bar{4}00_L$ spots. This row of spots is shown at higher magnification in figure 4.5(*c*) and labelled -2_D to 2_D in figure 4.5(*d*). This clearly visible array of spots is identical with that associated with the $\bar{2}00_U$ spot and will be used as a simple illustration of the diffraction processes that occur at an interface as discussed in section 2.6. All the spots in the array -2_D to 2_D arise (in the manner illustrated in figure 2.13) from the intersections of the Ewald sphere with spikes in reciprocal space associated with the $\bar{2}00_U$ reciprocal lattice point. The intersection of the Ewald sphere with the spike parallel to the specimen normal gives the spot F_D, and the intersection with the spike parallel to the normal to the inclined boundary gives the spot ν_D, and these two spots are separated along the direction normal to the line of intersection of the boundary with the surface of the specimen†. The intersections of the Ewald sphere with the periodic set of spikes parallel to the boundary normal, arising from linear periodic structure in the boundary, give the row of spots -2_D, -1_D, ν_D, 1_D, 2_D.

The distinction between diffraction effects associated with genuine boundary structure and other diffraction effects is simplified when the grain boundary

† While the F_D and ν_D spots are clearly resolved in association with the $\bar{2}00_U$ spot when it is doubly diffracted into the region of the $\bar{4}00_L$ and $\bar{4}00_U$ spots (figure 4.5(*c*)), they are not clearly resolved when doubly diffracted into the vicinity of the central beam (figure 4.6(*a*)).

under investigation has a marked change in boundary plane along its length. For such boundaries, genuine structural periodicities can be readily distinguished from other diffraction effects from the way in which the different diffraction effects discussed in section 2.6 are separately influenced by changes in the boundary plane. For a given orientation of a bicrystal relative to the electron beam, the diffraction spots in diffraction patterns taken at different positions along a curved boundary will be affected by change in the boundary plane in the following ways.

(i) The positions of diffraction spots which arise from each of the grains, such as the spot F_D in figure 2.13, and the positions of spots which are due to double-diffraction are unaffected by changes in the boundary plane. This is because the positions of such spots are determined by the normal to the specimen surface and the direction of the electron beam in each of the grains.

(ii) The positions of diffraction spots such as the spot ν_D in figure 2.13, which arise because the boundary is inclined to the surface of the specimen, will change with boundary plane. This occurs because such spots are separated from their associated single-grain diffraction spots, such as the spot F_D, by an amount determined by the angle between the boundary normal and the specimen normal, in the direction normal to the line of intersection of the boundary with the specimen surface.

(iii) Rows of diffraction spots associated with diffraction from genuine periodic structure in the boundary will vary markedly with change of boundary plane, but in a manner unrelated to the intersection of the boundary with the foil surface.

The differing characteristics (i)–(iii) enable the separate diffraction effects to be readily recognised and correlated with images in the analysis of boundary structure by the geometric method.

4.3.2 Geometric Analysis of Low-angle Boundaries

For the case of a general grain boundary with boundary normal $\boldsymbol{v}$ it has been shown in section 1.2.1 (expressions (1.9) and (1.10)) that a misorientation θ about an axis $\boldsymbol{u}$ with common indices in both grains is accommodated by three independent arrays of grain boundary dislocations with Burgers vectors $\boldsymbol{b}_1$, $\boldsymbol{b}_2$, $\boldsymbol{b}_3$, line directions $\boldsymbol{r}_1$, $\boldsymbol{r}_2$, $\boldsymbol{r}_3$ and spacings d_1, d_2, d_3 when

$$\boldsymbol{r}_1 \parallel [\boldsymbol{u} \wedge (\boldsymbol{b}_2 \wedge \boldsymbol{b}_3)] \wedge \boldsymbol{v} \tag{4.1}$$

and

$$d_1 = \left(2 \sin(\theta/2) \left| \frac{[\boldsymbol{u} \wedge (\boldsymbol{b}_2 \wedge \boldsymbol{b}_3)] \wedge \boldsymbol{v}}{\boldsymbol{b}_1 \cdot (\boldsymbol{b}_2 \wedge \boldsymbol{b}_3)} \right| \right)^{-1} \tag{4.2}$$

with similar expressions applying to $\boldsymbol{r}_2$, d_2 and $\boldsymbol{r}_3$, d_3. These expressions provide the basis for the geometric analysis of boundary structure in that they show that changes in boundary plane can be accommodated simply by changes in the directions and spacings of the grain boundary dislocations without any change in their Burgers vectors.

An illustration of the marked changes in dislocation spacing and direction that can take place with change in boundary plane is given in figure 4.7 which shows a portion of a general low-angle grain boundary in a polycrystalline specimen of copper (99.999% Cu). The micrograph shows that the boundary separating grains 1 and 2 has four limbs and takes the form of a saddle in the central region, as can be seen from the lines of intersection of the different regions of the boundary with the top T and the bottom B of the specimen. Clearly, there are very marked changes in boundary plane between the four limbs. For this boundary three arrays of grain boundary dislocations accommodate the misorientation between the grains as can be seen in the region 1–2, where the dislocations in two of the arrays are coarsely spaced and those in the third array are finely spaced and parallel to the pendellösung fringes. As the boundary plane changes from the region 1–2 to any of the other three limbs the same dislocations remain, but their spacings and directions alter to accommodate the change in boundary plane and become closely spaced in three of the limbs.

Analysis of closely spaced arrays of grain boundary dislocations relies on the following two properties of expression (4.1) which hold under the

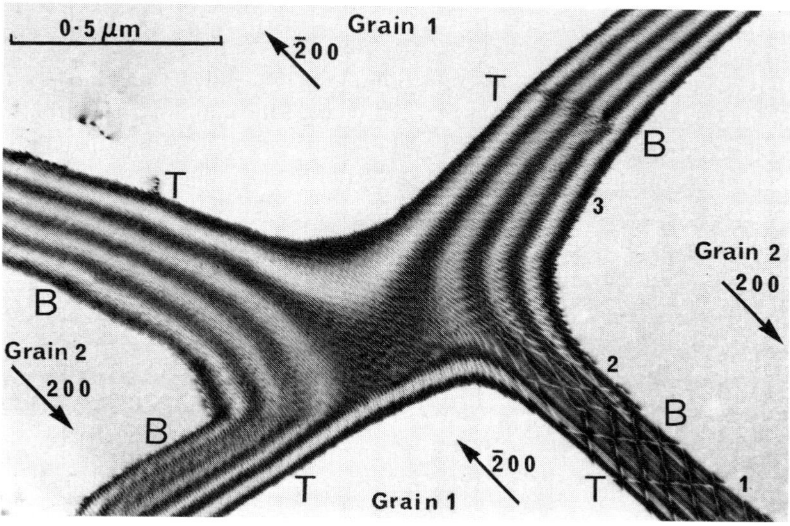

Figure 4.7 Double two-beam image of a low-angle boundary in copper.

condition that the Burgers vectors of the grain boundary dislocations remain constant with change in boundary plane.

(a) The different directions of $\boldsymbol{r}_1$, for different boundary plane normals $\boldsymbol{v}$, must be coplanar with the rotation axis $\boldsymbol{u}$, and so also for $\boldsymbol{r}_2$ and $\boldsymbol{r}_3$.
(b) The different directions of $\boldsymbol{r}_1$, for different boundary plane normals $\boldsymbol{v}$, must also be coplanar with the direction $(\boldsymbol{b}_2 \wedge \boldsymbol{b}_3)$, and so also for $\boldsymbol{r}_2$ with $(\boldsymbol{b}_3 \wedge \boldsymbol{b}_1)$ and $\boldsymbol{r}_3$ with $(\boldsymbol{b}_1 \wedge \boldsymbol{b}_2)$.

In practice, these two properties can be used to analyse the periodic structure of a boundary, provided there is a distinct change in direction and/or spacing of this periodic structure with change in boundary plane. If, for such a boundary, measurements of the directions of the periodicities present show that property (a) holds, in that they specify a rotation axis $\boldsymbol{u}$ in agreement with that determined independently in the way described in section 3.4, then the measured periodicities can be identified with dislocations accommodating the misorientation between the grains, which have Burgers vectors which do not change with boundary plane. At this stage, property (b) is used to give a set of possible Burgers vectors, from which the actual Burgers vectors $\boldsymbol{b}_1$, $\boldsymbol{b}_2$, $\boldsymbol{b}_3$ can be determined as those which give, from equation (4.2), spacings d_1, d_2, d_3 which are in agreement with the measured spacings.

Two low-angle boundaries will be used to illustrate the geometric analysis of boundary structure (Clarebrough and Forwood 1980a). These two examples are general boundaries of mixed tilt–twist character with misorientations θ of approximately 6° and 11°, and the structure of both boundaries involves three independent arrays of grain boundary dislocations.

6° Boundary in Copper

Figure 4.8 shows a low-angle grain boundary in a polycrystalline specimen of copper (99.999% Cu) with a marked change in boundary plane between the regions labelled 1, 2 and 3. The boundary separates an upper grain R from a lower grain L and the intersections of the boundary with the top and bottom of the specimen are marked T and B respectively. The misorientation between the two grains is such that grain L is rotated through an angle $\theta = 6.16°$ with respect to grain R in a right-handed sense around an axis $\boldsymbol{u}$ with common indices in both grains given by [cos 48.9° cos 65.5° cos 51.1°]. For this example the RMS error associated with the determined misorientation is $\pm 0.25°$, and this comparatively large uncertainty in the orientation relationship is probably due to the development of bend contours during the course of the experiment.

High-magnification images from the regions 1, 2 and 3 of figure 4.8(*a*) are shown in figures 4.8(*b*), (*e*) and (*h*) respectively. Apart from the pendellösung fringes, there is only one fringe system in the images and this has a spacing of approximately 100 Å and shows a marked change in direction between

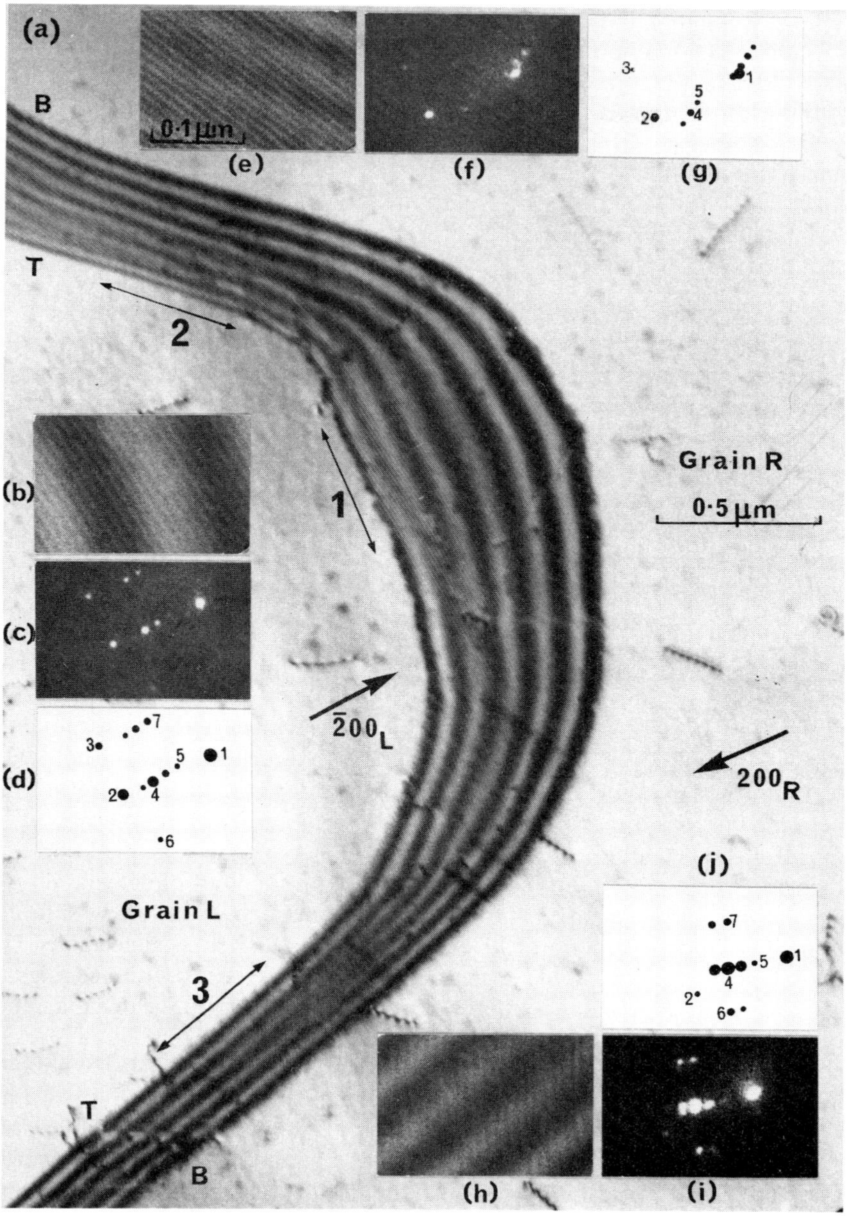

Figure 4.8 Low- (a) and higher-magnification (b), (e) and (h) double two-beam images and diffraction patterns (f), (c) and (i) of a low-angle boundary in copper with $\boldsymbol{B}$ close to [013] in both grains R and L and $\boldsymbol{g}_R = 200_R$ and $\boldsymbol{g}_L = \bar{2}00_L$.

regions 1, 2 and 3, indicating that it is probably associated with a structural periodicity. Diffraction patterns from regions 1, 2 and 3 show this periodicity and indicate the presence of two additional structural periodicities. An example of the diffraction effects obtained from each region of the boundary is shown in the patterns of figures 4.8(*c*), (*f*) and (*i*), and in the corresponding schematic diagrams in figures 4.8(*d*), (*g*) and (*j*). The arrays of diffraction spots are portions of diffraction patterns from the boundary corresponding to a beam direction close to [013] in both grains, and are centred on the spot labelled 1 which is the $\bar{1}3\bar{1}_L$ diffraction spot from grain L. The pattern of spots 1, 2 and 3 in figure 4.8(*c*) is unchanged in figures 4.8(*f*) and (*i*) (but in figure 4.8(*i*) spot 3 is too weak for reproduction), i.e. this pattern of spots remains constant with changes in boundary plane. Spot 2 is the $\bar{1}3\bar{1}_R$ diffraction spot from grain R, and spot 3 arises from the $13\bar{1}_R$ diffraction spot in grain R after double-diffraction by the strong $\bar{2}00_L$ operative diffracting vector in grain L. The relative positions of the remaining spots in each pattern change with boundary plane, and these spots can be interpreted in terms of periodic structures in the boundary.

In each of the three diffraction patterns of figure 4.8 there is a periodic repeat distance along the direction 4–5 and this is designated periodicity x. For each position on the boundary in figure 4.8 the direction and spacing of periodicity x correlates with the direction and spacing of the fringes in the corresponding image. Row of spots corresponding to periodicity x can be identified in diffraction patterns for other beam directions over the full range of tilt of $\pm 30°$ and can be correlated with fringes in the corresponding images. In addition, diffraction patterns in beam directions over the full range of tilt show two other periodicities, designated y and z. In figure 4.8 the periodicity y has a repeat distance along 6–4–7 and the periodicity z has a repeat distance along 1–4. For figure 4.8(*f*) at position 2 on the boundary the spots corresponding to periodicity y were on the limit of detectability in association with the $\bar{1}3\bar{1}_L$ diffraction spot, but in association with other diffraction spots the periodicity y at position 2 was readily detectable. The periodicities y and z should correspond to fringe systems with spacings of approximately 20 Å, but these were not observed in images at the magnification used. Periodicity z, like that of x, shows little change in spacing, but a marked change in direction with position on the boundary, while periodicity y shows little change in direction, but a marked change in spacing.

Periodicity z has been defined with reference to figure 4.8 as the inter-spot spacing 1–4 which makes use of spot 1, the $\bar{1}3\bar{1}_L$ diffraction spot. However, as pointed out in sections 2.6 and 4.3.1, the zeroth-order spot ν_D of the pattern associated with intrinsic structure is expected to be slightly displaced from its associated diffraction spot F_D in the direction normal to the intersection of the boundary with the specimen surface, and the magnitude of this displacement increases with the angle between the specimen and boundary normals. In cases where this displacement is resolved, periodicities are defined in relation to this zeroth-order spot ν_D. For this example, the ν_D and F_D spots should show the largest separation at position 2 on the boundary since the angle between the specimen and boundary normals at position 2 is more than twice that at positions 1 and 3 and, in fact, position 2 is the only position at which the ν_D and F_D spots are resolved. This resolution can be seen in figure 4.8(f) where there is a displacement between spot 1, the $\bar{1}3\bar{1}_L$ diffraction spot corresponding to F_D, and the row of spots associated with periodicity x, so the zeroth-order spot ν_D of the pattern associated with intrinsic structure is the nearest spot to spot 1 displaced in the direction normal to the intersection of the boundary (at position 2) with the specimen surface. Thus, in figure 4.8(f) it is this spot which is used to define periodicity z. In figures 4.8(c) and (i) the F_D and ν_D spots are not resolved and spot 1 is taken as the zeroth-order spot for determining periodicity z.

Standard stereographic techniques are used to analyse the geometry of the periodicities x, y and z in the diffraction patterns to determine the nature of the periodic structure in the boundary for the three regions 1, 2 and 3 with boundary normals $\boldsymbol{\nu}_1$, $\boldsymbol{\nu}_2$ and $\boldsymbol{\nu}_3$. The analysis will only be described for periodicity x, but exactly the same type of analysis is used for periodicities y and z. The stereographic projection in figure 4.9, which is indexed with respect to grain R, shows the local boundary planes corresponding to regions 1, 2 and 3 as the zones labelled 1, 2 and 3 normal to $\boldsymbol{\nu}_1$, $\boldsymbol{\nu}_2$ and $\boldsymbol{\nu}_3$. These zones have been determined by independent stereographic analysis of directions lying in each of the boundary planes, which are represented in the projection of figure 4.9 by the crosses (see section 3.5.2). For each beam direction the projected directions of the periodicity x for the three different positions 1, 2 and 3 on the boundary are labelled x_1, x_2 and x_3. For example, at position 1 on the boundary the directions x_1 in the diffraction patterns for several different beam directions are indicated by short arcs whose lengths are estimates of the uncertainty associated with each measurement. It can be seen that these directions x_1 lie on a great circle containing the boundary normal $\boldsymbol{\nu}_1$ and intersect the boundary plane at $\boldsymbol{X}_1$. This shows that the directions x_1 of the periodicity x in the different diffraction patterns are projections of $\boldsymbol{X}_1$ along the boundary plane normal $\boldsymbol{\nu}_1$. When the spacings of periodicity x in different beam directions are projected on to boundary plane 1 along its normal $\boldsymbol{\nu}_1$ a constant value for the spacing of a periodicity in the boundary along $\boldsymbol{X}_1$ is obtained. It follows, therefore, for position 1 on the boundary, that the periodicity x in each beam direction arises from the intersection of the Ewald sphere with a row of rods in reciprocal space parallel to the boundary normal, the spacing and direction of the row being that of a single periodicity along $\boldsymbol{X}_1$ in the boundary. A similar analysis of the directions x_2 and x_3 of the periodicity x for positions 2 and 3 on the boundary shows that they arise from a single periodicity along $\boldsymbol{X}_2$ and a single periodicity along $\boldsymbol{X}_3$ respectively. The structure of rods in reciprocal space, given by the analysis in figure 4.9, corresponds in real space to a structure

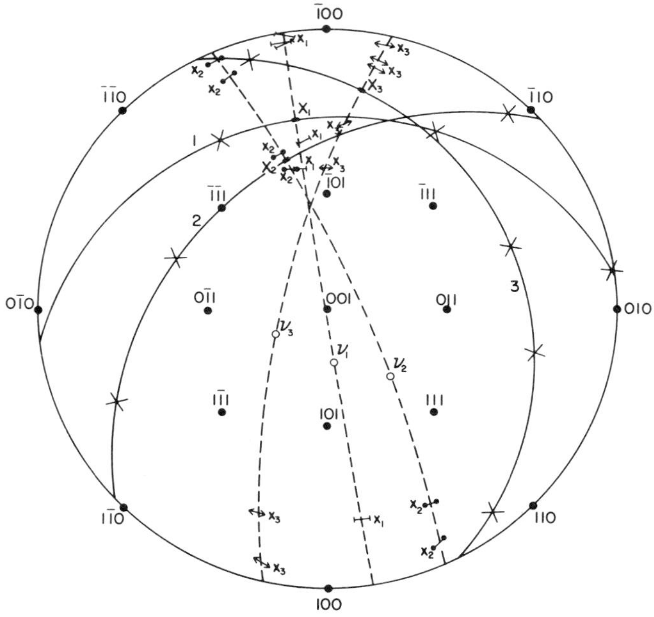

Figure 4.9 Stereographic projection, indexed with respect to grain R, showing the analysis of periodicity x for the boundary normals $\boldsymbol{v}_1$, $\boldsymbol{v}_2$ and $\boldsymbol{v}_3$ given by the direction cosines [cos 68.4°, cos 87.5°, cos 21.8°], [cos 64.4°, cos 66.6°, cos 35.9°] and [cos 80.4°, cos 110.0°, cos 22.4°] respectively, to an estimated accuracy of $\pm 2°$.

in the boundary consisting of a grating composed of an array of lines whose spacing and direction alter with change in boundary plane. The same type of analysis for periodicities y and z defines these periodicities in the three positions on the boundary as $\boldsymbol{Y}_1$, $\boldsymbol{Y}_2$, $\boldsymbol{Y}_3$ and $\boldsymbol{Z}_1$, $\boldsymbol{Z}_2$, $\boldsymbol{Z}_3$ respectively. Thus, the complete structure of the boundary in real space consists of a cross-grating composed of three independent arrays of lines whose spacings and directions change with boundary plane.

The next step is to determine whether the cross-grating structure satisfies property (a), i.e. whether the cross-grating structure represents arrays of grain boundary dislocations which accommodate the observed misorientation between the two grains. This step is illustrated in the stereographic projection of figure 4.10, which again is indexed with respect to grain R. Using the subscript i to denote the different regions of the boundary ($i = 1, 2, 3$), the spacing of the lines in the cross-grating structure in real space are reciprocally related to the spacings of the periodicities along $\boldsymbol{X}_i$, $\boldsymbol{Y}_i$ and $\boldsymbol{Z}_i$, and the directions of the lines are normal to the directions of $\boldsymbol{X}_i$, $\boldsymbol{Y}_i$ and $\boldsymbol{Z}_i$. These directions of the lines in the cross-grating structure for the three positions

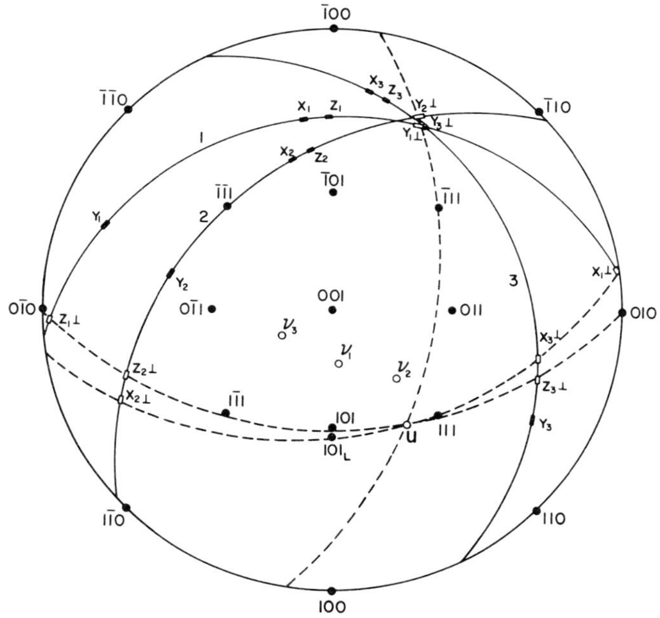

Figure 4.10 Stereographic projection indexed with respect to grain R, showing the determination of the common line of intersection (identifiable as $\boldsymbol{u}$) of the zones containing $\boldsymbol{b}_y \wedge \boldsymbol{b}_z$, $\boldsymbol{b}_z \wedge \boldsymbol{b}_x$ and $\boldsymbol{b}_x \wedge \boldsymbol{b}_y$ indicated by the dashed lines through $\boldsymbol{X}_{i\perp}$, $\boldsymbol{Y}_{i\perp}$ and $\boldsymbol{Z}_{i\perp}$ respectively. Also included is the direction $[101]_L$.

on the boundary are indicated in figure 4.10 by directions $\boldsymbol{X}_{i\perp}$, $\boldsymbol{Y}_{i\perp}$ and $\boldsymbol{Z}_{i\perp}$. It can be seen from figure 4.10 that the line directions of the cross-grating do satisfy property (a) in that the directions $\boldsymbol{X}_{i\perp}$ are coplanar, as are $\boldsymbol{Y}_{i\perp}$ and $\boldsymbol{Z}_{i\perp}$, and the three planes are consistent with having a common line of intersection which coincides (within the limits of experimental error) with the experimentally determined rotation axis marked $\boldsymbol{u}$ in the figure. Therefore, the cross-grating structure consists of three independent arrays of dislocations which accommodate the entire misorientation between the two grains.

The Burgers vectors $\boldsymbol{b}_x$, $\boldsymbol{b}_y$ and $\boldsymbol{b}_z$ of these grain boundary dislocations are likely to be simple $(1/2)\langle 110\rangle$ lattice vectors in either grain, or simple linear combinations of these vectors. Property (b) is now used to decide first whether simple $(1/2)\langle 110\rangle$ Burgers vectors are compatible with the observations. The possible values of $(\boldsymbol{b}_y \wedge \boldsymbol{b}_z)$ in the zone containing $\boldsymbol{X}_{i\perp}$ and $\boldsymbol{u}$, of $(\boldsymbol{b}_z \wedge \boldsymbol{b}_x)$ in the zone containing $\boldsymbol{Y}_{i\perp}$ and $\boldsymbol{u}$, and of $(\boldsymbol{b}_x \wedge \boldsymbol{b}_y)$ in the zone containing $\boldsymbol{Z}_{i\perp}$ and $\boldsymbol{u}$ are listed in table 4.2. From a consideration of all the possibilities in table 4.2, including the reversals of sign and the order of

Table 4.2 Possible values of $(\boldsymbol{b}_y \wedge \boldsymbol{b}_z)$, $(\boldsymbol{b}_z \wedge \boldsymbol{b}_x)$ and $(\boldsymbol{b}_x \wedge \boldsymbol{b}_y)$.

$\pm(\boldsymbol{b}_y \wedge \boldsymbol{b}_z)$ in zone containing $\boldsymbol{X}_{i\perp}$ and $\boldsymbol{u}$	$\pm(\boldsymbol{b}_z \wedge \boldsymbol{b}_x)$ in zone containing $\boldsymbol{Y}_{i\perp}$ and $\boldsymbol{u}$	$\pm(\boldsymbol{b}_x \wedge \boldsymbol{b}_y)$ in zone containing $\boldsymbol{Z}_{i\perp}$ and $\boldsymbol{u}$
$(1/2)[1\bar{1}0]_R \wedge (1/2)[0\bar{1}1]_R$	$(1/2)[110]_R \wedge (1/2)[101]_R$	$(1/2)[101]_R \wedge (1/2)[\bar{1}01]_R$
$(1/2)[1\bar{1}0]_R \wedge (1/2)[\bar{1}01]_R$	$(1/2)[110]_R \wedge (1/2)[0\bar{1}1]_R$	$(1/2)[101]_L \wedge (1/2)[\bar{1}01]_R$
$(1/2)[0\bar{1}1]_R \wedge (1/2)[\bar{1}01]_R$	$(1/2)[101]_R \wedge (1/2)[0\bar{1}1]_R$	$(1/2)[101]_L \wedge (1/2)[101]_R$
	$(1/2)[101]_L \wedge (1/2)[0\bar{1}1]_R$	

the cross products, the only self-consistent sets of Burgers vectors are $\boldsymbol{b}_x = \pm(1/2)[101]_R$ or $\pm(1/2)[101]_L$, $\boldsymbol{b}_y = \pm(1/2)[\bar{1}01]_R$ and $\boldsymbol{b}_z = \pm(1/2)[0\bar{1}1]_R$.

The determined sense of the misorientation between the grains enables the signs of the Burgers vectors $\boldsymbol{b}_x$, $\boldsymbol{b}_y$ and $\boldsymbol{b}_z$ to be specified for a given sense of the line directions of the dislocations. For the present example, where grain L is rotated in a right-handed sense with respect to grain R, expression (4.1) satisfies the FS/RH convention when the sense taken for the boundary normals $\boldsymbol{v}_1$, $\boldsymbol{v}_2$ and $\boldsymbol{v}_3$ is from grain R to grain L, i.e. from the upper grain to the lower grain (see section 1.2). In the stereographic projections of figures

Table 4.3 Comparison of observed and calculated periodicities for the low-angle grain boundary of figure 4.8†.

Position on boundary	$\boldsymbol{X}_\perp$ Spacing (Å)	$\boldsymbol{Y}_\perp$ Spacing (Å)	$\boldsymbol{Z}_\perp$ Spacing (Å)
		Observed periodicities	
1	109 ± 11	26 ± 3	23 ± 3
2	104 ± 10	22 ± 3	22 ± 3
3	118 ± 12	39 ± 4	25 ± 3
		Calculated periodicities for $\boldsymbol{b}_x = (1/2)[\bar{1}0\bar{1}]_L$, $\boldsymbol{b}_y = (1/2)[10\bar{1}]_R$, $\boldsymbol{b}_z = (1/2)[01\bar{1}]_R$	
1	116 ± 5	23.1 ± 0.9	20.4 ± 0.8
2	108 ± 4	20.1 ± 0.8	19.3 ± 0.8
3	131 ± 5	35.3 ± 1.4	22.2 ± 0.9

† The values in this table are for the values of $\boldsymbol{v}_i$ $(i = 1, 2, 3)$ given in the caption to figure 4.9. The error limits given for the spacings of the calculated periodicities arise from the uncertainty in the determined values of $\boldsymbol{u}$ and θ. Uncertainties, not included in the table, due to error limits of $\pm 2°$ in $\boldsymbol{v}_i$ affect spacings of observed and calculated periodicities in a coupled way, causing a relative change in agreement of $\pm 2\%$.

4.9 and 4.10 all directions are upward-drawn so that in these figures $\boldsymbol{v}_1$, $\boldsymbol{v}_2$ and $\boldsymbol{v}_3$ have the opposite sense, and this has to be taken into account when relating Burgers vectors and line directions. For the correct downward-drawn sense of boundary normals expression (4.1) and figure 4.10 show that $\boldsymbol{b}_x = (1/2)[\bar{1}0\bar{1}]_R$ or $(1/2)[\bar{1}0\bar{1}]_L$, $\boldsymbol{b}_y = (1/2)[10\bar{1}]_R$ and $\boldsymbol{b}_z = (1/2)[01\bar{1}]_R$ when the line directions of the dislocations $\boldsymbol{X}_{i\perp}$, $\boldsymbol{Y}_{i\perp}$ and $\boldsymbol{Z}_{i\perp}$ have the sense given in figure 4.10, with the exception of $\boldsymbol{X}_{2\perp}$ and $\boldsymbol{Z}_{2\perp}$ for which the opposite sense of line direction applies.

The final check on whether these Burgers vectors can be identified with the observed grain boundary dislocation structure is made by using them in equation (4.2) to determine whether they predict the observed spacings of the dislocation lines for the three positions on the boundary. For both sets of Burgers vectors the calculations agree equally well with the experimental results within the limits of experimental error, and table 4.3 shows the comparison of experimental and calculated results for the case where $\boldsymbol{b}_x = (1/2)[\bar{1}0\bar{1}]_L$. Thus the observed structure can be explained in terms of three arrays of grain boundary dislocations, two of which have Burgers vectors $(1/2)[10\bar{1}]_R$ and $(1/2)[01\bar{1}]_R$ in the upper grain, and the other either a Burgers vector of $(1/2)[\bar{1}0\bar{1}]_R$ in the upper grain or $(1/2)[\bar{1}0\bar{1}]_L$ in the lower grain.

11° Boundary in Copper

The analysis of the 6° boundary was straightforward because the three basis periodicities present were readily recognisable. However, for the 11° boundary it will be seen that this is not the case and it is necessary to distinguish the three basis periodicities associated with the actual dislocation structure of the boundary from a large number of other periodicities which are generated from combinations of these basis periodicities.

Figure 4.11(*a*) is a bright-field electron micrograph of the 11° boundary which, like the 6° boundary, is in copper (99.999% Cu) and shows a marked change in boundary plane along its length. The boundary separates an upper grain U from a lower grain L. The misorientation between the two grains is such that grain L is rotated through an angle of 10.65° with respect to grain U in a right-handed sense about an axis $\boldsymbol{u}$, with common indices in both grains given by [cos 122.4° cos 96.6° cos 33.2°]. For this boundary the RMS error associated with the determined misorientation is $\pm 0.20°$ and again reflects the development of bend contours during the course of the experiment. The analysis of the structure is carried out using the positions on the boundary marked 1, 2 and 3 in figure 4.11(*a*) and in each of these positions five different fringe systems are present in images with spacings in the range 20–200 Å. Examples of these fringe systems are shown for position 1 on the boundary in figures 4.11(*b*), (c) and (*d*) for different diffraction conditions. Figure 4.11(*b*) indicates the more widely spaced fringes along WW (~200 Å) and XX (~60 Å), and figures 4.11(*c*) and (*d*) show the finer fringe systems (~20 Å)

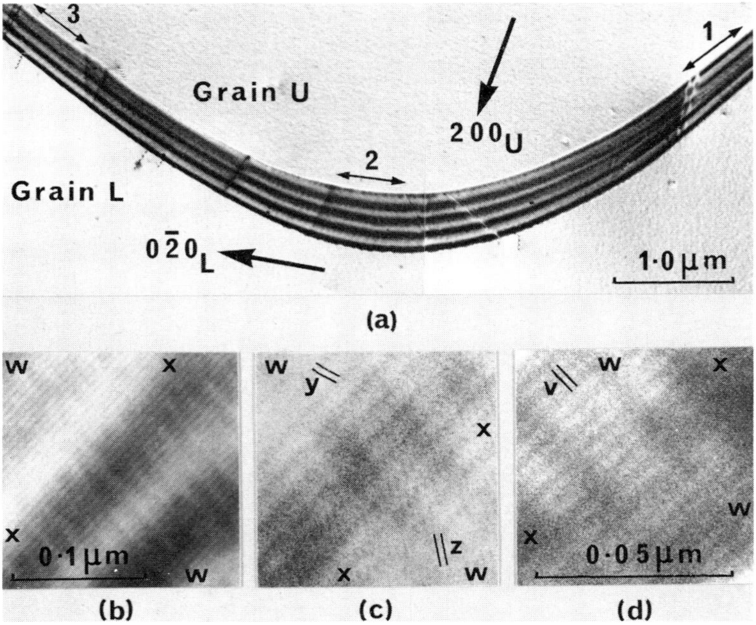

Figure 4.11 Double two-beam images of a low-angle boundary in copper. In (*a*) ***B*** is close to $[019]_U$ and $[001]_L$ are the ***g*** vectors are 200_U and $0\bar{2}0_L$; (*b*), (*c*) and (*d*) are higher magnification images from position 1 with equal magnifications for (*c*) and (*d*). For (*b*) ***B*** is close to $[\bar{1}12]$ in U and L and the ***g*** vectors are $\bar{1}1\bar{1}_U$ and $\bar{2}\bar{2}0_L$. For (*c*) ***B*** is close to $[\bar{1}01]$ in U and L and the ***g*** vectors are 111_U and $0\bar{2}0_L$. For (*d*) ***B*** is close to $[\bar{1}\ 3\ 10]_U$ and $[015]_L$ and the ***g*** vectors are $1\bar{3}1_U$ and $\bar{2}00_L$.

indicated along *y*, *z* and *v*. The detectability of the finer fringe systems is sensitive to the operative diffracting conditions, e.g. systems *y* and *z* are not detected in figure 4.11(*d*) and the system *v* is barely discernible in figure 4.11(*c*). The diffraction pattern in figure 4.12(*a*), which is taken from position 1 on the boundary and corresponds to the image in figure 4.11(*a*), is an illustration of the additional rows of weak diffraction spots, generated by the boundary structure, occurring in association with diffraction spots from the grains. Figure 4.12(*d*) is a magnified portion of figure 4.12(*a*) and shows the pattern of weak spots that occurs in the vicinity of the $\bar{2}00_L$ and $\bar{2}00_U$ diffraction spots for position 1 on the boundary. Figures 4.12(*c*) and (*b*) are similar magnified portions of patterns for positions 2 and 3 respectively on the boundary. All the diffraction patterns in figure 4.12 are in the correct orientation relative to the images in figure 4.11. The pattern of spots in figure 4.12(*c*) is shown schematically in figure 4.12(*e*), where the labelling of the spots 1–8 will also be used to describe the patterns in figures 4.12(*b*) and

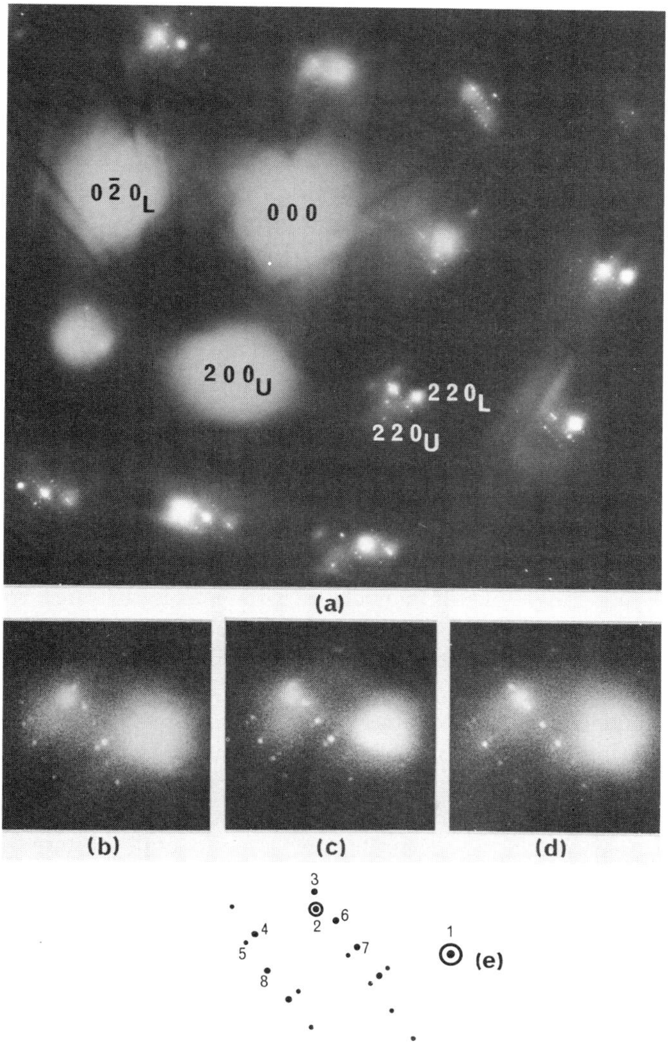

Figure 4.12 Diffraction patterns corresponding to the image of figure 4.11(*a*). The pattern in (*a*) is from position 1 on the boundary and those in (*b*), (*c*) and (*d*) are enlarged portions of patterns around the $\bar{2}00_U$ (spot 1) and $\bar{2}00_L$ (spot 2) crystal reflections for positions 3, 2 and 1 on the boundary respectively; (*e*) schematic representation of (*c*).

(*d*). In figures 4.12(*b*), (*c*) and (*d*), spots 1 and 2 are the $\bar{2}00_U$ and $\bar{2}00_L$ diffraction spots from the grains and their relative positions remain constant for the three positions on the boundary. However, the remaining spots in the patterns have spacings (e.g. 4–5, 3–6, 3–5) which are shorter than the smallest spacing between double-diffraction spots, and their relative positions change for the three boundary planes, so that this array of spots must be considered in terms of boundary structure.

In figure 4.12(*e*) spot 2 is the F_D spot $\bar{2}00_L$, and spot 3 is its associated ν_D spot. This can be seen from a comparison of figures 4.12(*d*), (*c*) and (*b*) which show the change in the position of spot 3 relative to the fixed F_D spot for the different lines of intersection of the boundary with the specimen surface at positions 1, 2 and 3 respectively. Thus, spot 3 is the zeroth-order spot for the periodicities associated with boundary structure. Each pattern of spots in figures 4.12(*b*), (*c*) and (*d*) indicates the presence of several different periodicities, but only some of these correlate with the resolved fringe systems in the corresponding images. These correlations are: fringe system z with the periodicity 6–4, y with 3–8, x with 3–6, w with 4–5 and, within the limits of experimental error, v with 3–4 or 3–5. This set of periodicities is common to all diffraction patterns for different beam directions and diffracting vectors over the full range of tilt, correlates with resolved fringe systems in corresponding images and will be used to identify the basis periodicities associated with boundary structure.

The pattern of spots 3, 4, 5, 6 and 8 (figure 4.12(*e*)) requires a combination of three independent periodicities for its generation. However, there are several possible combinations which can be used. For example, different combinations of three independent periodicities chosen from the set given by the periodic repeat distances along the directions 3–4, 3–5, 4–5, 3–6, 3–8 and 6–4 would all generate the same overall pattern, but each combination would correspond to a different assignment of three basis periodicities to the grain boundary structure. The periodicities 3–4, 3–5, 4–5, 3–6, 3–8 and 6–4 are designated v, v', w, x, y and z respectively in accordance with their correlation with the fringe systems. These periodicities are analysed in the same manner described for the 6° boundary to determine whether there exists in this set an appropriate combination of three that can be identified with three periodic arrays of grain boundary dislocations which accommodate the observed misorientation between the grains. The analysis, similar to that in figure 4.9, of each periodicity in the above set shows, as for the 6° boundary, that each of the periodicities in the diffraction patterns arises from the intersection of the Ewald sphere with a set of rods in reciprocal space, parallel to the local boundary normal, corresponding to a single periodicity in the boundary. Following the nomenclature used for the 6° boundary, the directions of the lines in real space corresponding to these periodicities are designated $\boldsymbol{V}_{i\perp}$, $\boldsymbol{V}'_{i\perp}$, $\boldsymbol{W}_{i\perp}$, $\boldsymbol{X}_{i\perp}$, $\boldsymbol{Y}_{i\perp}$ and $\boldsymbol{Z}'_{i\perp}$ ($i = 1, 2, 3$ corresponding to the three positions on the boundary), and these directions are shown in figure 4.13. No distinction could be made, within the limits of experimental error, between directions $V_{i\perp}$ and $V'_{i\perp}$, and in figure 4.13 these are shown as the single direction $V_{i\perp}$. Clearly, the directions of each of these lines for the three positions on the boundary are coplanar and the planes are consistent with having a common line of intersection which coincides, within the limits of experimental error, with the determined rotation axis $\boldsymbol{u}$. This result shows that all the lines satisfy property (a), so that it is possible, at this stage, for any combination of three independent sets of these periodicities to be identified with grain boundary dislocations that accommodate the entire misorientation between the two grains. Therefore, this 11° boundary is more complex than the 6° boundary in that the three basis periodicities are not immediately obvious, but have to be determined. The determination of the basis periodicities is made by using properties (a) and (b) to find the set of three non-coplanar Burgers vectors which, in expressions (4.1) and (4.2), predict for the three positions on the boundary the spacings, as well as the directions, of three basis periodicities, which will then give all the other periodicities that are observed.

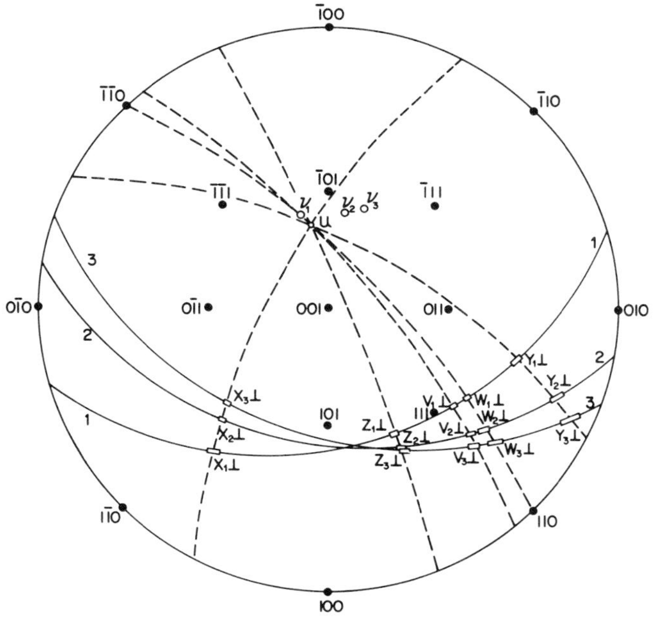

Figure 4.13 Stereographic projection indexed with respect to grain L showing the three boundary normals $\boldsymbol{v}_1 = [\cos 126.2^\circ, \cos 99.8^\circ, \cos 37.9^\circ]$, $\boldsymbol{v}_2 = [\cos 127.5^\circ, \cos 84.3^\circ, \cos 38.0^\circ]$, $\boldsymbol{v}_3 = [\cos 128.4^\circ, \cos 77.5^\circ, \cos 41.1^\circ]$ to an accuracy of $\pm 2^\circ$, and their associated planes 1, 2 and 3, together with the determination of the common line of intersection (identifiable as $\boldsymbol{u}$) of the zones defined by $\boldsymbol{V}_{i\perp}$, $\boldsymbol{W}_{i\perp}$, $\boldsymbol{X}_{i\perp}$, $\boldsymbol{Y}_{i\perp}$ and $\boldsymbol{Z}_{i\perp}$.

When simple $(1/2)\langle 110\rangle$ Burgers vectors from either grain are used in expressions (4.1) and (4.2) they do not predict spacings or directions which can be identified with any of the possible combinations of three independent periodicities. However, when Burgers vectors involving simple sums and differences of $(1/2)\langle 110\rangle$ lattice vectors from each grain (of the type $(1/2)\langle 110\rangle_L + (1/2)\langle 110\rangle_U$) are used in expressions (4.1) and (4.2), computation shows that one of the possible combinations of three independent periodicities, namely that corresponding to $\boldsymbol{X}_{i\perp}$, $\boldsymbol{Y}_{i\perp}$ and $\boldsymbol{Z}_{i\perp}$, does satisfy the experimental results. In fact the measured spacings and directions of these three periodicities are predicted by four sets of three non-coplanar Burgers vectors of the above type, namely:

$$\boldsymbol{b}_x = (1/2)[\bar{1}10]_L + (1/2)[011]_U \text{ or } (1/2)[011]_L + (1/2)[\bar{1}10]_U$$

$$\boldsymbol{b}_y = (1/2)[\bar{1}10]_L + (1/2)[01\bar{1}]_U \text{ or } (1/2)[01\bar{1}]_L + (1/2)[\bar{1}10]_U$$

$$\boldsymbol{b}_z = (1/2)[101]_L + (1/2)[101]_U.$$

Furthermore, all four sets of Burgers vectors fulfil the more restrictive condition of predicting the spacings and directions of the other periodicities corresponding to $\boldsymbol{V}_{i\perp}$, $\boldsymbol{V}'_{i\perp}$ and $\boldsymbol{W}_{i\perp}$ from appropriate combinations of the basis periodicities $\boldsymbol{X}_{i\perp}$, $\boldsymbol{Y}_{i\perp}$ and $\boldsymbol{Z}_{i\perp}$. The agreement between the calculated and experimental results is equally good (within the limits of experimental error) for all four sets of Burgers vectors and is shown in table 4.4 for the case of $\boldsymbol{b}_x = (1/2)[\bar{1}10]_L + (1/2)[011]_U$, $\boldsymbol{b}_y = (1/2)[01\bar{1}]_L + (1/2)[110]_U$ and $\boldsymbol{b}_z = (1/2)[101]_L + (1/2)[101]_U$.

The sense of the Burgers vectors listed satisfies the determined sense of misorientation between the grains when the line directions of the dislocations are along $-\boldsymbol{X}_{i\perp}$, $\boldsymbol{Y}_{i\perp}$ and $\boldsymbol{Z}_{i\perp}$ of figure 4.13. This relation between Burgers vector, line direction and misorientation is obtained from expression (4.1) in the same way as already discussed for the 6° boundary.

Although the analysis describes the 11° boundary as being made up of single dislocations with Burgers vectors which are non-primitive vectors, involving sums of $(1/2)\langle 110\rangle$ vectors from each grain, it is more likely that the actual dislocation structure of the boundary consists of pairs of the component $(1/2)\langle 110\rangle$ dislocations.

The lack of uniqueness in the results of the present analysis, which is indicated by the four sets of fitting Burgers vectors, arises simply because it has not been possible to distinguish the grain, U or L, to which each $(1/2)\langle 110\rangle$ dislocation in a pair should be referred.

The results for the 6° and 11° boundaries obtained using the geometric theory given in section 1.2, although internally consistent, cannot be taken as a strict confirmation of the theory. To test the validity of the theory experimentally a grain boundary is required in which the spacing of the grain boundary dislocations is such that the Burgers vectors found by geometric analysis can be compared with those determined independently from the

Table 4.4 Comparison of observed and calculated periodicities for the low-angle grain boundary of figure 4.11†.

Position on boundary	$V_\perp$ Spacing (Å)	$V_\perp$ Angle ($\rho°$)	$V'_\perp$ Spacing (Å)	$V'_\perp$ Angle ($\rho°$)	$W_\perp$ Spacing (Å)	$W_\perp$ Angle ($\rho°$)	$X_\perp$ Spacing (Å)	$X_\perp$ Angle ($\rho°$)	$Y_\perp$ Spacing (Å)	$Y_\perp$ Angle ($\rho°$)	$Z_\perp$ Spacing (Å)	$Z_\perp$ Angle ($\rho°$)
Observed periodicities												
1	25 ± 1	119.5 ± 0.8	22 ± 1	119.5 ± 0.8	188 ± 19	124.0 ± 1.0	57 ± 2	42.5 ± 1.6	22 ± 1	141.8 ± 1.8	21 ± 1	98.5 ± 1.1
2	26 ± 1	140.6 ± 1.3	23 ± 1	140.6 ± 1.3	192 ± 19	144.2 ± 1.5	58 ± 2	62.6 ± 1.0	23 ± 1	164.3 ± 1.7	22 ± 1	120.0 ± 1.4
3	27 ± 1	148.5 ± 1.5	23 ± 1	148.5 ± 1.5	193 ± 19	154.0 ± 2.0	58 ± 2	69.1 ± 1.0	23 ± 1	172.5 ± 2.3	23 ± 1	128.5 ± 1.6
Calculated periodicities for $\boldsymbol{b}_x = (1/2)[\bar{1}10]_L + (1/2)[011]_U$, $\boldsymbol{b}_y = (1/2)[01\bar{1}]_L + (1/2)[\bar{1}10]_U$, $\boldsymbol{b}_z = (1/2)[101]_L + (1/2)[101]_U$												
1	24.6 ± 0.4	119.3 ± 0.2	21.7 ± 0.4	119.7 ± 0.2	188 ± 19	123 ± 6	56.3 ± 1.0	43.6 ± 0.9	22.1 ± 0.4	142.3 ± 0.3	20.8 ± 0.4	98.5 ± 0.2
2	25.2 ± 0.5	140.3 ± 0.2	22.3 ± 0.4	140.8 ± 0.2	192 ± 19	144 ± 6	56.6 ± 1.0	63.2 ± 0.8	22.4 ± 0.4	163.8 ± 0.3	21.3 ± 0.4	119.0 ± 0.2
3	25.9 ± 0.5	149.5 ± 0.2	22.9 ± 0.4	149.7 ± 0.2	197 ± 19	154 ± 6	57.4 ± 1.0	69.9 ± 0.9	22.7 ± 0.4	172.5 ± 0.3	22.1 ± 0.4	127.2 ± 0.2

† The angle $\rho°$ is measured from $([001] \wedge \boldsymbol{v}_i)_L$, $i = 1, 2, 3$. The values in the table are for the values of $\boldsymbol{v}_i$ given in the caption to figure 4.13. The error limits given for the calculated periodicities arise from the uncertainty in the determined values of $\boldsymbol{u}$ and θ. Uncertainties, not included in the table, due to error limits of $\pm 2°$ in $\boldsymbol{v}_i$ affect observed and calculated periodicities in a coupled way, causing a relative change in agreement of $\pm 2.5\%$ in spacings and 0.5° in the angle ρ.

contrast of the dislocations in electron microscope images. A demonstration of such a validation will be given in section 5.3.2 (i) for a high-angle grain boundary close to a Σ9 CSL orientation, where it will be shown that the same Burgers vectors are obtained for grain boundary dislocations from their geometry and by image matching.

4.4 LOW-ANGLE BOUNDARIES IN FABRICATED THIN-FILM BICRYSTALS

A landmark in experimental investigation of the structure of grain boundaries using transmission electron microscopy was the technique developed originally by Schober and Balluffi (1969) and used extensively by Balluffi and his colleagues (see, for example, Balluffi *et al* 1972a,c), which enabled grain boundaries with selected misorientations to be fabricated in thin-film specimens of gold. The technique involved growing epitaxially from the vapour phase two single-crystal films of gold of the desired orientation, e.g. (001), (110) or (111) films, on appropriate substrates. The two films, while still on their substrates, were lightly clamped together at the misorientation required and then annealed for five minutes at 400 °C in air. This annealing procedure welded the two films together to produce a sandwich containing a bicrystal of the required misorientation. After annealing, the sandwich was cleaved into sections and the substrates dissolved away. Portions of the bicrystal were then mounted on electron microscope grids and given an extra annealing treatment of three minutes at 300 °C to cause further welding at the interface prior to examination in the electron microscope. With grain boundaries produced by this technique it was possible to distinguish contrast and diffraction effects associated with grain boundary dislocations from other contrast and diffraction effects which can arise solely from the superposition of the two crystals, such as moiré fringes and double-diffraction. This was done by comparing the contrast and diffraction behaviour of the grain boundary in welded regions with that in small regions where welding failed to occur and the two grains were merely superposed.

Using this technique, twist and tilt boundaries were studied in a systematic way in thin-film bicrystals of gold over a wide range of misorientation. The Burgers vectors of the grain boundary dislocations were determined, using $\boldsymbol{g} \cdot \boldsymbol{b}$ invisibility criteria, for misorientations where the dislocations were coarsely spaced, and the spacings of the dislocations were measured over the full range of misorientation studied. Only the results for misorientations less than or approximately equal to 15° will be discussed here and those for larger misorientations will be treated in Chapter 5.

The first of these studies was by Schober and Balluffi (1969, 1970) on (001) and (111) pure twist low-angle boundaries. The (001) boundaries were studied for different misorientations θ in the range 1–9°, and in all cases the boundaries were shown to consist of two orthogonal arrays of screw dislocations in agreement with theoretical predictions for the structure of low-angle twist boundaries. Figure 4.14, taken from Schober and Balluffi (1970), shows a cross-grid of screw dislocations in a (001) twist boundary in gold with a misorientation $\theta = 1°$. They established the dislocation structure of the boundaries by showing that the Burgers vectors of the dislocations were $(1/2)[110]$ and $(1/2)[1\bar{1}0]$, and that the spacing d of the dislocations in each of the arrays decreased with θ in the way expected from the dislocation model of low-angle boundaries (section 1.2.2), i.e. in accordance with the simple relation $d = |\boldsymbol{b}|/\theta$. Their investigation of low-angle (111) twist boundaries was not as detailed as that for the (001) boundaries, but they did confirm that these boundaries were made up of hexagonal arrays of dislocations consistent with the interaction of two orthogonal arrays of screw dislocations in the (111) plane. Later, Tan *et al* (1975) made a more detailed investigation of low-angle (001) twist boundaries, extended the number of discrete misorientations studied and confirmed the earlier results. Goodhew *et al* (1976) adapted the same techniques to low-angle (110) twist boundaries in the range of misorientation 1–5° and found that the boundary structure consisted of an hexagonal array of dislocations which they interpreted as having arisen from a rectangular cross-grid of screw dislocations. Figure 4.15 summarises the work of Balluffi and his colleagues on (001) twist boundaries in gold and shows a comparison of experimental and

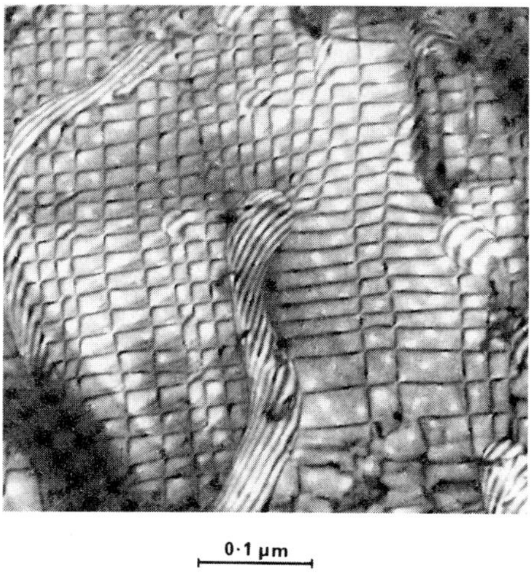

Figure 4.14 Electron micrograph showing a crossed grid of screw dislocations in a low-angle twist boundary in gold (Schober and Balluffi 1970).

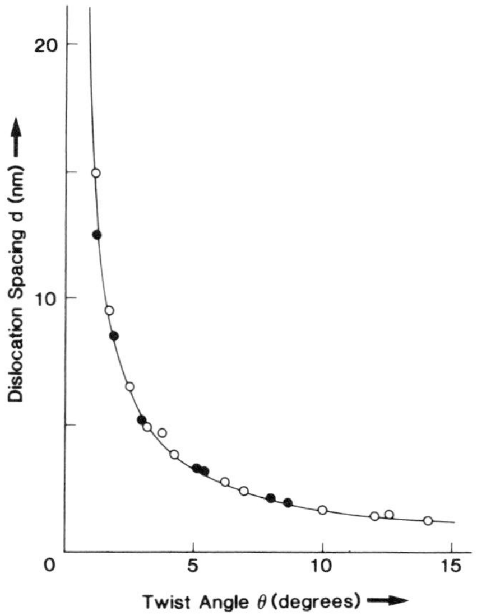

Figure 4.15 Comparison of experimental and calculated spacings of dislocations for (001) twist boundaries in gold. The points ● are the results of Schober and Balluffi (1970), the points ○ are the results of Tan *et al* (1975) and the curve is calculated from the relation $d = |\boldsymbol{b}|/\theta$ (after Babcock and Balluffi 1987).

calculated dislocation spacings as a function of angle of misorientation θ for values of θ up to 15°.

Schober and Balluffi (1971a) adapted their technique of specimen preparation to enable the investigation of low-angle tilt boundaries in which the boundaries were, as for their twist boundaries, parallel to the surfaces of the fabricated thin-film gold bicrystals. They carried out a systematic study of symmetric-tilt boundaries on (100) and (110) planes, with both types of boundary having a tilt axis parallel to [001]. The ranges of misorientation studied were 3–13° for the (100) boundaries and $2\text{–}9\frac{1}{2}°$ for the (110) boundaries. They showed that the structure of both types of boundary consisted of regularly spaced edge dislocations parallel to the [001] tilt axis, as expected on a simple dislocation model. As in their experiments on twist boundaries, they again confirmed the dislocation model by showing that the variation in spacing of the dislocations with misorientation θ obeyed the simple relation $d = |\boldsymbol{b}|/\theta$ for Burgers vectors of (1/2)[110] for the (110) boundaries and [100] for the (100) boundaries. They pointed out that the dislocations with [100] Burgers vectors are likely to have dissociated according to the reaction $[100] \rightarrow (1/2)[110] + (1/2)[1\bar{1}0]$, but this dissociation could not be detected in the diffraction contrast in images of the dislocations in the (100) boundaries.

In describing these early results on tilt boundaries, Schober and Balluffi reported some instances where the grain boundary dislocations, although on average parallel to the tilt axis, appeared to be serrated. This aspect of the structure of low-angle tilt boundaries in gold was investigated in more detail by Darby and Balluffi (1977) for a range of symmetric and asymmetric boundaries involving different tilt axes and boundary planes. They found, in general, that the grain boundary dislocations were not straight, but were serrated into a zig-zag form. They explained this as arising from the alignment of lengths of dislocation along those directions in the boundary plane that enable dislocation dissociation reactions to occur, with the generation of stacking fault on inclined (111) planes in each grain, to give an overall reduction in energy. This interpretation was supported by observations that the grain boundary dislocations were straight in boundaries with tilt axes along $\langle 110 \rangle$ or $\langle 112 \rangle$, where dissociation on $\{111\}$ planes could occur directly without the need of serration. Further support for this interpretation was given later by Liu and Balluffi (1985) who found that grain boundary dislocations were straight in low-angle tilt boundaries in aluminium, where the high stacking fault energy of aluminium precludes dissociation.

A different technique for producing welded boundaries in thin-film bicrystals of gold was developed by Erlings and Schapink (1977) and used by them in studies of (111) twist boundaries. The technique involved overlaying, at a selected misorientation, two single-crystal (111) flakes of gold which had been precipitated from a gel, and then allowing these crystals, mounted on an electron microscope grid, to weld together at room temperature (or at a slightly elevated temperature) in the electron microscope to form a bicrystal. Erlings (1979) studied low-angle (111) twist boundaries in the range of misorientation 0–5° and found that the structure of the boundaries consisted of hexagonal networks of dislocations for which the spacings and Burgers vectors were in agreement with the simple dislocation model (see also Erlings and Schapink 1978).

Pareja and Serna (1979) fabricated low-angle thin-film (001) twist boundaries in silver by superposing evaporated thin films, and Pareja (1980) observed crossed grids of screw dislocations in these boundaries with $(1/2)\langle 110 \rangle$ Burgers vectors and spacings in agreement with calculation for misorientations θ in the range 0–5°.

The work described in this section on fabricated thin-film bicrystals of gold is a striking demonstration of the applicability of the dislocation model to low-angle boundaries.

4.5 n-BEAM LATTICE IMAGES OF LOW-ANGLE BOUNDARIES

High-resolution lattice images of low-angle grain boundaries enable their atomic structure to be studied directly. However, as discussed in section 2.5, quantitative interpretation of lattice images essentially confines the application of n-beam imaging to pure tilt boundaries where the tilt axis is a low-index crystallographic direction and the electron beam direction can be aligned along this axis, i.e. parallel to the line direction of the dislocations. Only recently has the improved resolution of electron microscopes enabled grain boundaries in metals and alloys to be studied using n-beam lattice imaging, so that most of the work that has been done on low-angle boundaries has used two-beam lattice images.

The first observations on the structure of a low-angle tilt boundary in an FCC metal were those of Parsons and Hoekle (1969) on a boundary formed in a deformed single crystal of aluminium. They used two-beam lattice images of the (111) planes spaced at 2.34 Å and found that fringes terminated at the positions of the dislocations. From measurements of the angle between fringes in each grain the misorientation was determined as $2.8^\circ \pm 0.1^\circ$. Using the relation $d = |\boldsymbol{b}|/\theta$, this misorientation correlated with the observed 60 Å spacing of the end-on dislocations in the lattice image on the assumption that their Burgers vector was $(1/2)[110]$. They studied the strains associated with the grain boundary dislocations and concluded that all strains were localised at the boundary over a region comparable with the 60 Å spacing of the dislocations. They measured the strain associated with one of the dislocations in this region and found that, apart from the strain present in the immediate vicinity of the core of the dislocation, the measured strain was consistent with that predicted by linear elasticity for an isolated dislocation on the assumption that aluminium is elastically isotropic.

More extensive investigations of low-angle tilt boundaries in aluminium have been made by Penisson and Bourret (1979) using the techniques of two-beam lattice imaging and weak-beam imaging. The boundaries, in bicrystals grown from the melt, involved $(01\bar{1})$ and (100) boundary planes, and the misorientation corresponded to a rotation in the range 2.5–3° about the [011] tilt axis. Whereas Parsons and Hoekle were restricted to lattice images from one set of $\{111\}$ planes, Penisson and Bourret took advantage of the [011] tilt axis and used $(1\bar{1}1)$ and $(\bar{1}\bar{1}1)$ planes separately to obtain two types of two-beam lattice images of their boundaries.

For the $(01\bar{1})$ boundary plane the dislocation structure was made up of arrays of 10–20 edge dislocations with the expected $(1/2)[01\bar{1}]$ Burgers vector separated by boundary steps consisting of 60° $(1/2)\langle 110\rangle$ dislocations. The 60° dislocations accommodated a small twist component ($\lesssim 0.5^\circ$) in the boundary. Using terminating fringes in lattice images and the application of $\boldsymbol{g}\cdot\boldsymbol{b}$ invisibility criteria to weak-beam images, the Burgers vectors of these 60° glissile dislocations were determined as $(1/2)[\bar{1}10]$ or $(1/2)[\bar{1}0\bar{1}]$ for steps of one sense and $(1/2)[110]$ or $(1/2)[10\bar{1}]$ for steps of the opposite sense. The spacings of the edge dislocations in the arrays were 55 ± 5 Å, which was consistent with the observed misorientation. The displacements of the $\{111\}$ fringes, for the region outside the core of one of these dislocations, were found to be in good agreement with the displacement field calculated for an isolated dislocation by applying linear isotropic elasticity to aluminium and adding to this displacement field that of ten neighbouring dislocations to take the grain boundary into account. These more detailed calculations support the earlier work of Parsons and Hoekle.

In their study of the (100) boundary Penisson and Bourret found that the boundary plane was less well-defined and the dislocation structure in the boundary less periodic. The dislocation structure of this boundary consisted of small groups of 60° $(1/2)\langle 110\rangle$ dislocations with screw components alternating in sign along the [011] tilt axis so that they accommodated tilt across the boundary. These groups of 60° dislocations were separated by non-periodic mixtures of dislocations which contained 60° dislocations accommodating a small twist component between the grains, and $(1/2)[01\bar{1}]$ edge dislocations associated with the deviation of the boundary plane from the (100) plane. Penisson and Bourret pointed out that there was some loss and rearrangement of the 60° dislocations during the thinning of their bicrystals.

Two-beam lattice images of (200) planes in gold were used by Cosandey *et al* (1978) to study a symmetric-tilt boundary, involving a misorientation of $10° \pm 1.2°$ around the [001] axis, in a thin-film bicrystal of gold which was prepared by evaporation on to a bicrystal of sodium chloride. Making the assumption that the Burgers vector of the grain boundary dislocations could be specified solely from the number of lattice fringes terminating at each end-on dislocation, they found that the Burgers vector for these edge dislocations was of the $(1/2)\langle 110\rangle$ type rather than the $\langle 100\rangle$ type. Moreover, for the measured misorientation, a $(1/2)\langle 110\rangle$ Burgers vector predicted a dislocations spacing in agreement with the observed spacing of 16.5 Å. Using boundaries prepared in the same way, Cosandey and Bauer (1981) extended this type of investigation to cover several misorientations θ in the range 5–15° around the [001] tilt axis. They also investigated misorientations for $\theta > 15°$ and these results will be discussed in section 5.4. For low-angle boundaries with values of θ of 5°, 9.5°, 10.0°, 13.9° and 15° they found that the spacing of the edge dislocations in the boundaries agreed with those predicted using a Burgers vector of the $(1/2)\langle 110\rangle$ type. In this range of values of θ the spacing of the dislocations decreased from 31 Å to 11 Å.

The structure of a tilt boundary in molybdenum has been studied by Penisson *et al* (1988) using lattice images of the $\{110\}$ planes in this BCC metal, and this work followed an earlier investigation of a very similar boundary (Penisson *et al* 1982). The misorientation across the boundary in the bicrystal was $14° \pm 0.5°$ around the [001] tilt axis, and the specimen was prepared by electron-beam-welding two cylindrical single crystals with axes along the [001] direction, sectioning the welded cylinders normal to the [001] tilt axis and then thinning electrolytically. In high-resolution images with the electron beam along the [001] tilt axis, the crossing (110) and $(1\bar{1}0)$

fringes defined a grid of black and white contrast in which the black contrast was shown to correspond to the positions of columns of atoms along [001]. A high-resolution lattice image of a portion of the boundary is reproduced in figure 4.16(*a*) and figure 4.16(*b*) is a matching computed image. The boundary is clearly close to being a symmetric-tilt boundary and figures 4.16(*a*) and (*b*) show the detail of the atomic structure in the neighbourhood of a grain boundary dislocation, the core of which is marked by a circle. The Burgers vector of the dislocation in figures 4.16(*a*) and (*b*) was obtained as $\boldsymbol{b} = [100]$, and was associated with two extra half-planes, one in each grain.

The structure of the boundary over an extended length consisted of individual dislocations with [100] Burgers vectors of the type illustrated in

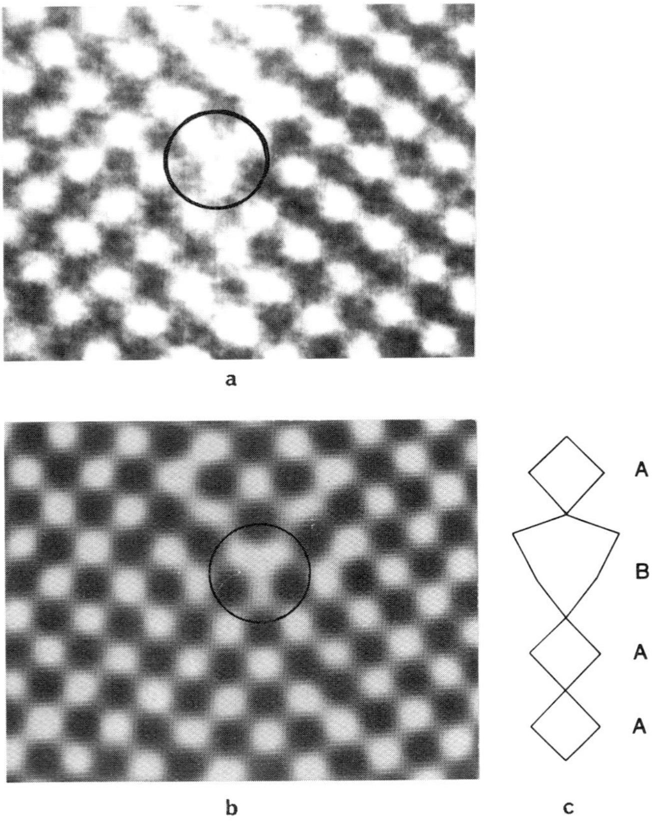

Figure 4.16 Comparison of experimental (*a*) and computed (*b*) *n*-beam lattice images of a low-angle boundary in molybdenum. The sequence of structural units comprising the boundary is illustrated schematically in (*c*) (Penisson *et al* 1988).

figures 4.16(*a*) and (*b*) separating regions of relatively perfect crystal, and Penisson *et al* described this structure in terms of A and B units where the A units corresponded to nearly perfect crystal and the B units to the cores of the dislocations. The sequence of these units for the structure of figures 4.16(*a*) and (*b*) is AABA, as illustrated schematically in figure 4.16(*c*). For the extended boundary it was found that the sequence of units was non-periodic and that the B units were sometimes separated by two A units and sometimes by three. Using computer modelling of boundary structure, they showed that a sequence of the type AAABAAAB corresponded to the atomic structure of a symmetric-tilt boundary with a misorientation of 12.68° around [001], i.e. the exact $\Sigma41$ CSL orientation. Similarly, they found that the sequence AABAAB corresponded to the structure of a symmetric-tilt boundary with a misorientation of 16.25° around [001], i.e. the exact $\Sigma25$ CSL orientation. They concluded that the observed misorientation of $14^\circ \pm 0.5^\circ$ was accommodated by a boundary structure consisting of a mixture of $\Sigma41$ and $\Sigma25$ low-energy CSL structures (see section 5.5). Lattice images computed using the atomic positions obtained for the $\Sigma41$ and $\Sigma25$ models of the boundary were in agreement with the experimental images of the appropriate portions of the boundary and an example of this agreement is shown in figures 4.16(*a*) and (*b*) for the $\Sigma41$ structure AAABAAAB. The spacing of the dislocations in the region of the boundary displaying $\Sigma41$ structure was 14 Å, and that in the region displaying $\Sigma25$ structure was 11 Å. These are the spacings expected from the simple relationship $d = |\boldsymbol{b}|/\theta$ for symmetric-tilt boundaries at the $\Sigma41$ and $\Sigma25$ orientations.

4.6 COMMENT

The electron microscopy of low-angle boundaries described in this chapter has demonstrated that the dislocation structure of grain boundaries, with misorientations $\theta \lesssim 15^\circ$, is consistent with the geometric description given in section 1.2 in that grain boundary dislocations accommodate the entire misorientation between the neighbouring crystals and have Burgers vectors which are lattice vectors.

5

High-angle Grain Boundaries

5.1 INTRODUCTION

It has been seen in Chapter 4 that the entire misorientation across a low-angle grain boundary is accommodated by periodic arrays of dislocations with Burgers vectors which are lattice vectors. A characteristic of this type of boundary structure is that the spacing of the dislocations decreases as the misorientation θ increases, and in an extension of this simple description to high-angle boundaries ($\theta \gtrsim 15°$), the spacing of the dislocations becomes comparable with the interatomic spacing. In such a situation, there will be virtually no elastic displacement field associated with the boundary and the atomic displacements at the boundary will be essentially those associated with the cores of dislocations. Clearly, such a dislocation description can only have formal geometric significance and, for this reason, other descriptions of the structure of high-angle grain boundaries have been proposed such as the CSL model, the plane matching model, and the structural unit model. Each of these models will be introduced here, and the CSL model will be treated in greater detail in section 5.2.

The CSL model was developed to describe the structure of high-angle boundaries in materials with cubic crystal symmetry, and considers the misorientation between the grains to be accommodated in two parts. In the model, the major part of the misorientation corresponds to an exact low-energy CSL orientation which is considered to be accommodated in a formal geometric sense by arrays of closely spaced 'primary' grain boundary dislocations. The remainder of the misorientation, corresponding to a small departure from the CSL orientation, is accommodated by arrays of more coarsely spaced 'secondary' grain boundary dislocations which have elastic displacement fields that can be described in the same way as those for dislocations in low-angle boundaries. These secondary grain boundary dislocations maintain the low-energy structure at the boundary corresponding to the exact CSL orientation by localising the departure from this structure into regions of 'bad fit' in much the same way as the dislocations in low-angle boundaries localise the departure from the perfect crystal structure. The spacing of the secondary grain boundary dislocations decreases with increasing angular separation from the exact CSL orientation in the same way as the spacings of dislocations in low-angle boundaries decrease with increase in misorientation. The original form of this model was put forward by Brandon *et al* (1964) and Brandon (1966) and was based on the earlier application by Kronberg and Wilson (1949) of the concept of the coincident site lattice to special grain boundaries, and on the suggestion by Read and Shockley (1950) that a small departure from a twin orientation could be represented by a dislocation sub-boundary. Later the model was formalised quantitatively by Bollmann (1967a, b, 1970), Warrington and Bollmann (1972), Grimmer *et al* (1974) and Bollmann (1982).

The plane matching model, which is less general than the CSL model, was proposed originally by Pumphrey (1972) to explain the occurrence of single periodic arrays of lines of strain contrast present in electron micrographs of high-angle grain boundaries. The model attributes these arrays to relaxations which occur to minimise the mismatching of the same type of slightly misoriented low-index crystallographic planes in each grain where they terminate at the boundary. Later, in a discussion of the plane matching model Balluffi and Schober (1972) showed that, if these linear relaxations are considered to be associated with dislocations, then the plane matching model reduces to a limiting case of the CSL model.

The structural unit model of high-angle grain boundaries was first introduced by Bishop and Chalmers (1968, 1971). They considered that the boundary structure consisted of repeating units, making up a two-dimensional array, in which each unit consisted of a small group of atoms arranged in a characteristic configuration. Computer calculations were first used to develop this type of model by Weins *et al* (1969, 1970, 1971) and later by Smith *et al* (1977), Ashby *et al* (1978) and Pond *et al* (1978). More recently, the most quantitative expression of the structural unit model has been given in a series of papers by Sutton and Vitek (1983a, b and c). They used computer modelling to determine the structure of both symmetric- and asymmetric-tilt boundaries in aluminium and copper over a wide range of misorientation. The model for tilt boundaries which arises from their computations is one in which for some misorientations the boundary structure consists of only one type of structural unit, and these boundaries are called 'favoured' boundaries and correspond to certain CSL orientations with low values of Σ. However, not all CSLs with low values of Σ correspond to favoured boundaries. Two examples of the structure of favoured $\langle 110 \rangle$ symmetric-tilt boundaries in aluminium are represented in figure 5.1 as projections of the FCC structure along the common $[1\bar{1}0]$ tilt axis (Sutton and Vitek 1983a). Figure 5.1(*a*) illustrates the structure found for a $\Sigma 11$ boundary on $(113)_1/(\bar{1}\bar{1}3)_2$ with

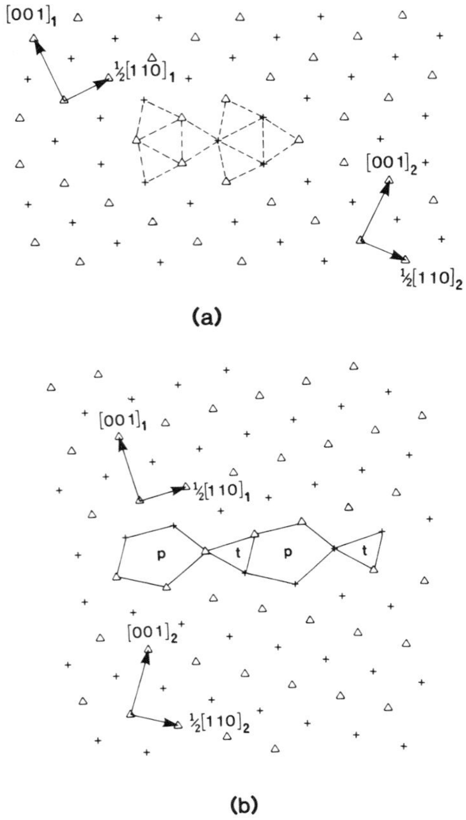

Figure 5.1 Structural units for a $\Sigma 11$ boundary (*a*) and a $\Sigma 27_a$ boundary (*b*) in aluminium (after Sutton and Vitek 1983a).

$\boldsymbol{u} = [1\bar{1}0]$ and $\theta = 50.48°$. The structure of the boundary consists of a contiguous sequence of capped trigonal prisms with axes along $[1\bar{1}0]$, which are delineated in figure 5.1(*a*) by the dashed lines. This structure is a simple one in which the structural unit consists of one of these capped trigonal prisms. Figure 5.1(*b*) illustrates the structure found for a $\Sigma 27_a$ boundary on $(115)_1/(\bar{1}\bar{1}5)_2$ with $\boldsymbol{u} = [1\bar{1}0]$ and $\theta = 31.59°$. The structure of the boundary in this case is more complicated and is based on a structural unit which consists of a nearly regular tetrahedron (labelled t) combined with an irregular pentagonal region (labelled p).

The structures of tilt boundaries, with misorientations intermediate between those of favoured boundaries, consist of mixtures of the structural units from the two nearest favoured boundaries, with the required mixture being given by a simple lever rule. In addition, the sequence of structural units must vary in such a way as to give continuity of structure from one favoured boundary, through the intermediate misorientations, to the next favoured boundary. This enables the sequence of structural units to be predicted for intermediate misorientations.

The boundary structure predicted by the structural unit model can be related to the dislocations in the CSL model. For example, if the misorientation is close to a CSL orientation corresponding to a favoured boundary, so that the structural units are nearly all of one type (say A) with a small number of embedded units of another type (say B) from a distant favoured boundary, then A units would correspond to the cores of the primary dislocations and the B units to the cores of the secondary dislocations in the CSL model.

Similar computer modelling has been done by Schwartz *et al* (1985) for twist boundaries which showed that the structural unit model is more complicated in this case, because it requires three types of structural unit where the third type is not given by a simple mix of the other two.

Sutton (1989) has pointed out that the usefulness of the structural unit model is limited to pure tilt and pure twist boundaries with low-index rotation axes of $\langle 100 \rangle$, $\langle 110 \rangle$, $\langle 111 \rangle$, and possibly $\langle 112 \rangle$, and that it cannot be used in a predictive way for higher index rotation axes. In addition, he emphasises that the structural unit model cannot be used for boundaries with mixed tilt–twist character.

In this chapter, experimental work on the structure of high-angle grain boundaries using transmission electron microscopy will be reviewed for twist and tilt boundaries in fabricated thin-film bicrystals, for general boundaries of mixed tilt–twist character in specimens prepared by thinning bulk polycrystalline metals and alloys, and for tilt boundaries prepared so as to be suitable for n-beam lattice imaging. It will be shown that in all cases the observed structures can be interpreted using the CSL model.

5.2 THE CSL MODEL OF HIGH-ANGLE GRAIN BOUNDARIES

The discussion here will be concerned principally with the properties of the secondary grain boundary dislocations in the CSL model, i.e. the way in which they accommodate a small angular departure from a CSL orientation, maintain the coincident site lattice and maintain the atomic structure at a boundary. In addition, rigid-body displacements between neighbouring grains, which move atoms away from CSL positions, and their associated partial grain boundary dislocations will be discussed.

5.2.1 Secondary Grain Boundary Dislocations

The CSL or secondary grain boundary dislocation model for the structure of a general high-angle grain boundary can be formulated as follows. If, for a general boundary, the misorientation between the grains is given by a rotation

θ about an axis $\boldsymbol{u}$ (with common indices in both grains) and is close to a CSL orientation corresponding to a rotation ω about an axis $\boldsymbol{p}$ †(common to one of the grains and the CSL), then the small angular departure from the exact CSL will be given by a rotation φ about an axis $\boldsymbol{q}$ (common to the other grain and the CSL), according to the matrix equation

$$(\boldsymbol{u}, \theta) = (\boldsymbol{q}, \varphi)\,(\boldsymbol{p}, \omega). \tag{5.1}$$

The secondary grain boundary dislocations required to accommodate the small misorientation $(\boldsymbol{q}, \varphi)$ can be described by an equation, equivalent to equation (1.1) for low-angle grain boundaries, namely

$$\boldsymbol{B} = 2\sin(\varphi/2)(\boldsymbol{x} \wedge \boldsymbol{q}) \tag{5.2}$$

where here $\boldsymbol{B}$ is the net Burgers vector of the secondary grain boundary dislocations intersected by any vector $\boldsymbol{x}$ lying in the plane of the boundary. An important property of equation (5.2) is that it does not suffer from the limitations on uniqueness discussed in section 1.4 for equation (1.1). Equation (5.2) gives a unique value for the net Burgers vector $\boldsymbol{B}$ of the secondary grain boundary dislocations because any re-indexing of the grains causing changes in $(\boldsymbol{u}, \theta)$ will cause compensating changes in $(\boldsymbol{p}, \omega)$ for the CSL involved, so that the small departure $(\boldsymbol{q}, \varphi)$ from the CSL orientation is specified as a unique physical quantity.

The geometric analysis derived in section 1.2 from equation (1.1) can be applied, through equation (5.2), to the secondary grain boundary dislocations accommodating the small angular departure φ from a CSL orientation, by simply replacing $\boldsymbol{u}$ and θ in the equations of section 1.2 with $\boldsymbol{q}$ and φ. This allows the Burgers vectors of secondary grain boundary dislocations in a high-angle boundary to be determined in a similar way to that described in section 4.3.2 for the determination of the Burgers vectors of primary grain boundary dislocations in low-angle boundaries and this procedure will be described in section 5.3.2.

In the CSL model, the secondary grain boundary dislocations, accommodating the departure $(\boldsymbol{q}, \varphi)$ from the exact CSL orientation, maintain the CSL structure and, in order to do this, they need to have a particular type of Burgers vector. The type of Burgers vector required to maintain the CSL structure for two given interpenetrating lattices can be appreciated by considering the effect that relative translations of the interpenetrating lattices have on the CSL. The CSL will be maintained, after a translation of one of the interpenetrating lattices, when this translation is a lattice vector of either of the lattices, or more generally, when it is a sum or difference of lattice vectors

† In section 3.4.2 the CSL orientation is the total misorientation between the lattices and is represented by the symbols $(\boldsymbol{u}, \theta)$. However, since here the CSL orientation is only part of the total misorientation, it is represented by the symbols $(\boldsymbol{p}, \omega)$ so that $(\boldsymbol{u}, \theta)$ can be consistently used to represent the total misorientation between the grains.

of the two lattices. The only effect that such translations can have is a shift in the origin of the CSL. It therefore follows that, in order for secondary grain boundary dislocations to maintain the CSL structure, they must have Burgers vectors which are lattice vectors or sums or differences of lattice vectors. This class of vectors belongs to a lattice, known as the DSC lattice, which is the coarsest lattice that contains all the lattice points of both interpenetrating lattices when they are at the exact CSL orientation. The DSC lattice is a finer lattice than either of the interpenetrating lattices and only some of the DSC lattice sites are occupied by lattice sites of the two interpenetrating lattices. The concept of the DSC lattice or 'complete pattern shift lattice' was first introduced by Bollmann (1967a, b) in his 0-lattice model of interfaces from a consideration of the effects of translation on the maintenance of patterns formed by two interpenetrating lattices. In the acronym DSC, D stands for displacement of one lattice relative to the other, S for shift of the pattern and C for complete.

The net Burgers vector $\boldsymbol{B}$ in equation (5.2) for secondary grain boundary dislocations is specified by a Burgers circuit in the same manner as described in section 1.2 for primary grain boundary dislocations in a low-angle boundary, but in this case of secondary grain boundary dislocations the Burgers circuit is constructed in the DSC lattice (Hirth and Balluffi 1973).

The DSC lattice for two FCC lattices 1 and 2 misoriented at the exact $\Sigma 5$ CSL orientation ($\boldsymbol{p} = [001]$, $\omega = 36.87°$, i.e. the $\boldsymbol{u}$, θ values from table 3.5) is illustrated in figures 5.2(*a*) and (*b*). Figure 5.2(*a*), which is a reproduction of the unit cell of the CSL enclosed by the dashed lines in figure 3.9, shows the [001] projection of the two interpenetrating lattices. As in figure 3.9, lattice 1 is represented by the symbols +, in the plane of the page, and □, $(1/2)[00\bar{1}]$ below the plane of the page. Similarly lattice 2 is represented by the symbols × and ◇, and the coincident lattice sites are shown by the coincident symbols. The coarsest lattice which contains the lattice points of both lattices 1 and 2 is specified by basis DSC vectors:

$$\boldsymbol{a} = (1/10)[3\bar{1}0]_1/(1/10)[310]_2$$

$$\boldsymbol{b} = (1/10)[130]_1/(1/10)[\bar{1}30]_2$$

both of which are indicated in the plane of figure 5.2(*a*), and

$$\boldsymbol{c} = (1/10)[21\bar{5}]_1/(1/10)[12\bar{5}]_2$$

which is inclined to and below the plane of figure 5.2(*a*). The three-dimensional configuration of these basis DSC vectors is illustrated schematically in figure 5.2(*b*). The DSC lattice sites in the plane of the page in figure 5.2(*a*) correspond to the intersections of the square grid of continuous lines, and the DSC lattice sites in the plane $(1/2)[00\bar{1}]$ below the plane of the page correspond to the intersections of the dashed lines. It is clear from figure 5.2(*a*) that only some of the DSC sites are occupied by sites of lattices 1 and 2.

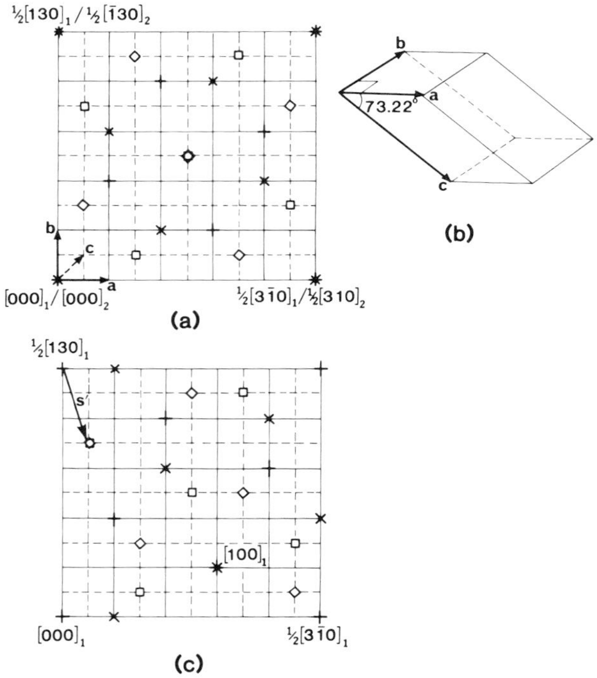

Figure 5.2 The CSL and DSC lattices for two FCC lattices 1 and 2 misoriented at the $\Sigma5$ CSL orientation (a). In (b) basis DSC vectors corresponding to the $\Sigma5$ CSL in (a) are shown and (c) illustrates new coincident sites created by the displacement of lattice 2 relative to lattice 1 by the DSC vector $\boldsymbol{a} = (1/10)[3\bar{1}0]_1$.

If there is a relative displacement of the two interpenetrating lattices by any of these DSC vectors, then the coincident sites of figure 5.2(a) will be destroyed and new coincident sites created. For example, if lattice 1 is considered to be fixed and lattice 2 is moved by the DSC vector $\boldsymbol{a} = (1/10)[3\bar{1}0]_1$, then the new coincident sites of figure 5.2(c) will be created and a new origin of the CSL may be chosen from any of these new coincident sites, e.g. the site $[100]_1$ in figure 5.2(c).

Basis DSC vectors can be determined for a particular CSL orientation without graphical construction and a procedure for calculating DSC vectors for lattices with cubic symmetry has been given by Grimmer *et al* (1974), but a simpler procedure will be described here. This is based on the theorem by Grimmer (1974) that the DSC lattice for a particular CSL is reciprocal to the coincident lattice formed by the interpenetration of the reciprocal lattices of the two crystal lattices involved. The coincident lattice of these two reciprocal lattices is the lattice defined by the three smallest non-coplanar same-$\boldsymbol{g}_c$ diffracting vectors associated with the CSL orientation (see sections 2.4.5 and 3.4.2). Therefore the DSC lattice is the reciprocal lattice of the same-$\boldsymbol{g}_c$ lattice. This type of calculation of DSC vectors will be illustrated using the $\Sigma 5$ boundary of figure 5.2. The rotation matrix $(\boldsymbol{p}, \omega)$ for this $\Sigma 5$ CSL, which re-indexes a vector indexed with respect to lattice 2 into indices with respect to lattice 1, is obtained from equation (3.8), using the rotation axis and angle corresponding to the $\Sigma 5$ CSL of $\boldsymbol{p} = [001]$ and $\omega = -36.87°$, as

$$(\boldsymbol{p}, \omega) = (1/5)\begin{pmatrix} 4 & 3 & 0 \\ -3 & 4 & 0 \\ 0 & 0 & 5 \end{pmatrix}.$$

The three smallest non-coplanar same-$\boldsymbol{g}_c$ vectors are obtained from this matrix by inspection, or by using table 3.6 as discussed in section 3.4.2. These vectors, indexed with respect to lattice 1, are $00\bar{2}_1$, 131_1, $3\bar{1}1_1$, and can be written as rows in a 3×3 matrix to give

$$\boldsymbol{g}_c = \begin{pmatrix} 0 & 0 & -2 \\ 1 & 3 & 1 \\ 3 & -1 & 1 \end{pmatrix}_1.$$

The inverse of this $\boldsymbol{g}_c$ matrix is

$$\boldsymbol{g}_c^{-1} = (1/10)\begin{pmatrix} 2 & 1 & 3 \\ 1 & 3 & -1 \\ -5 & 0 & 1 \end{pmatrix}_1$$

and the DSC vectors, indexed with respect to lattice 1, are given by the column vectors of this matrix. Thus, for an FCC lattice a set of basis DSC vectors for a $\Sigma 5$ CSL is $(1/10)[21\bar{5}]_1$, $(1/10)[130]_1$ and $(1/10)[3\bar{1}0]_1$ and these are the vectors indicated in figure 5.2.

For different CSLs, as the value of Σ increases, the magnitudes of the same-$\boldsymbol{g}_c$ vectors increase and there is an accompanying decrease in the magnitudes of the basis DSC vectors. Thus in the CSL model of the structure of high-angle boundaries, as the value of Σ increases the secondary grain boundary dislocations become more closely spaced in order to accommodate a given angular departure from each CSL orientation.

5.2.2 Secondary Grain Boundary Dislocations and Steps in Grain Boundaries

In the CSL model of a high-angle grain boundary in a bicrystal, secondary grain boundary dislocations with DSC Burgers vectors, in addition to maintaining the coincident site lattice, are required to maintain the local arrangement of atoms at the boundary. In general, the local arrangement of atoms at the boundary is maintained, in the presence of the secondary grain boundary dislocations, by each of these dislocations being associated with a step in the boundary, i.e. the level of the boundary plane in the bicrystal changes on crossing a secondary grain boundary dislocation. Steps associated with secondary grain boundary dislocations have been discussed by Hirth and Balluffi (1973), King and Smith (1980), Brokman (1981) and King (1982).

The way in which secondary grain boundary dislocations with DSC Burgers vectors introduce steps into a boundary can be visualised in the way described by King and Smith (1980). This first involves placing a notional boundary plane in a DSC lattice corresponding to a particular CSL, and then considering how the boundary plane has to be relocated, so as to maintain the same pattern of occupied DSC lattice sites across it, following a relative DSC displacement between the two interpenetrating lattices of the CSL. The DSC lattice of a $\Sigma 5$ CSL will be used to demonstrate this procedure. Figure 5.3(*a*) shows the DSC lattice extending over four of the unit cells of the $\Sigma 5$ CSL of figure 5.2(*a*) with a $(130)_1/(\bar{1}30)_2$ boundary plane located at AA. For this particular choice of boundary plane, those DSC lattice sites in the boundary which are occupied are coincident sites. Figure 5.3(*b*) shows the effect of keeping lattice 1 fixed and moving lattice 2 by the DSC vector $(1/10)[3\bar{1}0]_1$. This displacement, which destroys the original set of coincident sites and generates a new set, rearranges the pattern of occupied DSC lattice sites at the boundary plane AA. In order to maintain the same arrangement of occupied DSC lattice sites at the boundary, it can be seen that the plane of the boundary must be relocated to positions CC or EE, or equivalently to BB or DD, where the pattern is the same but merely displaced in the plane of the boundary by $(1/2)[001]_1$.

This DSC lattice construction can be related in an idealised way to the grains of a bicrystal if the atom sites in grain 1 are taken as the sites of lattice 1 on one side of AA in figure 5.3(*a*), and the atom sites in grain 2 are taken as the sites of lattice 2 on the other side of AA. This representation is shown in the DSC lattice of figure 5.3(*c*) where the grains labelled 1 and 2 are above and below the plane AA respectively, and the coincident sites in the boundary are shown as being occupied alternately by atoms from each grain. Figure 5.3(*d*) shows the change in boundary plane in the bicrystal (corresponding to a relocation of the plane to the position CC of figure 5.3(*b*)) following a displacement of grain 2 relative to grain 1 by the DSC vector $(1/10)[3\bar{1}0]_1$. It can be seen that the arrangement of atoms at the relocated boundary of figure 5.3(*d*) is the same as that in figure 5.3(*c*). Clearly, if a secondary grain boundary dislocation with a DSC Burgers vector of $(1/10)[3\bar{1}0]_1$ was present in a $(130)_1/(\bar{1}30)_2$ $\Sigma 5$ boundary then, for the particular case represented in figures 5.3(*c*) and (*d*), the boundary would step at the dislocation line from the level AA to the level CC. Figure 5.4 shows the step formed in the boundary when this $(1/10)[3\bar{1}0]_1$ secondary grain boundary dislocation has a line direction along the common [001] tilt axis. This geometry is established by using the FS/RH convention and by following the procedure of Thompson (1953)† in the DSC lattice. The FS/RH convention shows that the extra half-plane in the DSC lattice is above the plane of the boundary, and the procedure of Thompson shows that, on crossing the secondary grain boundary dislocation from left to right, the right hand side of grain 2 is displaced relative to its left hand side, and relative to grain 1, by the DSC Burgers vector $(1/10)[3\bar{1}0]_1$. Therefore on crossing the dislocation from left to right, the boundary steps from its original level AA to its new level CC.

For the general case of a secondary grain boundary dislocation with a DSC Burgers vector $\boldsymbol{b}$ in a high-angle grain boundary, there will be a relative displacement $\boldsymbol{b}$ between the two grains on crossing the dislocation. This displacement, which causes a shift in coincident sites on crossing the dislocation from 'original' sites on one side to 'new' sites on the other, requires the introduction of a step in order to maintain the same arrangement of atoms at the boundary on both sides of the dislocation. The height of the step on crossing the dislocation is given by the component normal to the boundary plane of a vector $\boldsymbol{s}$ joining an 'original' coincident site to a 'new' coincident site. If the shortest of these vectors is selected and designated $\boldsymbol{s}'$, then the step height h is given by

$$h = (\boldsymbol{s}' + \boldsymbol{l}) \cdot \boldsymbol{v} \tag{5.3}$$

where $\boldsymbol{l}$ is a lattice vector of the CSL and $\boldsymbol{v}$, as defined previously, is a unit vector normal to the boundary plane. Different step heights will be obtained for different choices of $\boldsymbol{l}$ in equation (5.3), and the minimum step height will be given by the value of $\boldsymbol{l}$ which minimises $(\boldsymbol{s}' + \boldsymbol{l}) \cdot \boldsymbol{v}$. For example, in figure 5.2(c) the shortest vector $\boldsymbol{s}'$ joining an 'original' coincident site to a 'new' coincident site is indicated, but the vector $(\boldsymbol{s}' + \boldsymbol{l})$ which gives the minimum step height is the $\boldsymbol{s}$ vector from $[000]_1$ to $[100]_1$. This corresponds to the shift in the level of the boundary from AA to CC in figures 5.3 and 5.4. This analysis gives the step height in interplanar spacings of planes parallel to the boundary plane and uses the convention of keeping lattice 1 (or grain 1) fixed and displacing lattice 2 (or grain 2) by the Burgers vector

† The procedure of Thompson (1953), adapted for establishing the displacement of one grain relative to the other in a bicrystal on crossing a grain boundary dislocation, is as follows. If an observer is situated in one of the grains, looking towards the boundary, and makes a 90° rotation in a clockwise sense from the positive direction of the line of the grain boundary dislocation, the direction in which the observer is now facing defines the positive direction of dislocation motion in the boundary. If the dislocation is moved relative to the observer in this positive sense, the grain into which the observer is looking will move relative to the one in which the observer is situated by an amount equal to and in the direction of the Burgers vector of the grain boundary dislocation.

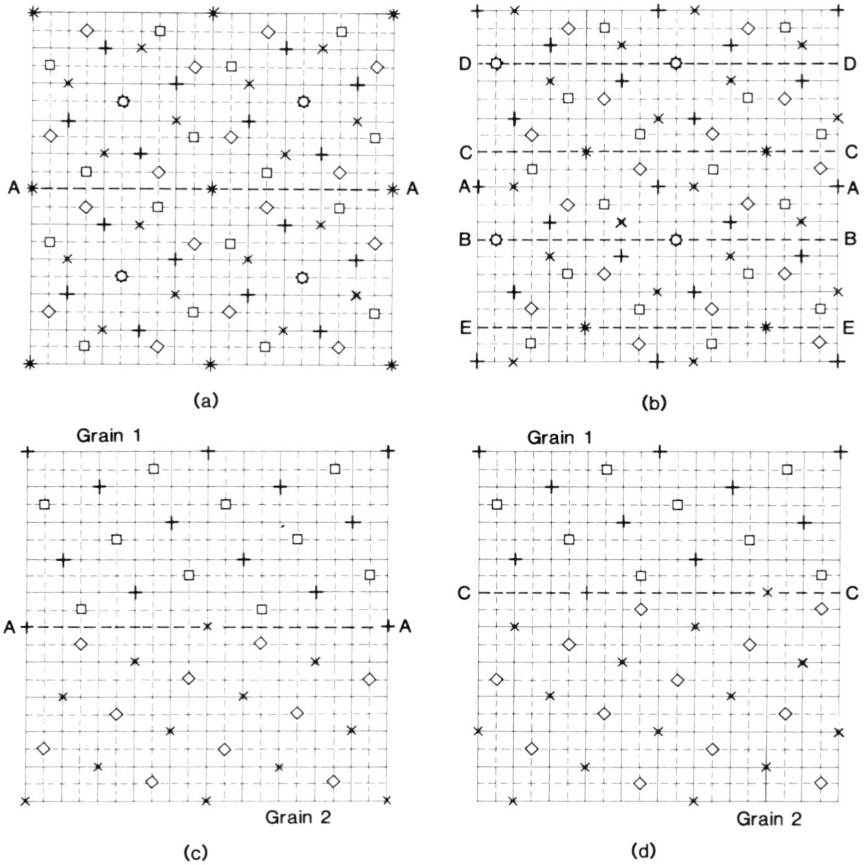

Figure 5.3 Illustration of how the level of a boundary needs to change in order to maintain the same arrangement of lattice sites after a displacement of one lattice relative to the other by a DSC vector. The geometry and displacement are the same as those discussed for the Σ5 CSL orientation of figure 5.2.

$\boldsymbol{b}$, so that the step height is referred to lattice 1 (or grain 1). If the opposite convention were used and the step height was referred to lattice 2 (or grain 2), then the step height obtained would not necessarily be the same. This is the case because $\boldsymbol{s}$ values in lattice 2 (or grain 2) obtained by using an exact reversal of the initial convention and procedure will always differ from equivalent $\boldsymbol{s}$ values in lattice 1 (or grain 1) by the vector $-\boldsymbol{b}$. Therefore, if $\boldsymbol{b}$ lies in the boundary plane the step height will be the same in both grains, but if $\boldsymbol{b}$ is inclined to the boundary plane the step height will differ in the two grains. For the case of the boundary in figure 5.4, the $(1/10)[3\bar{1}0]_1$

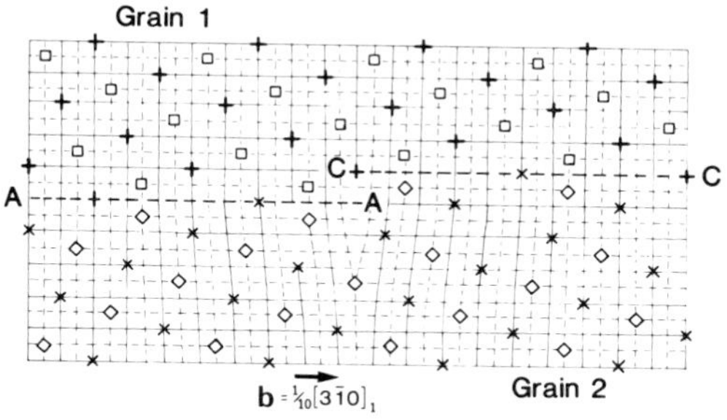

Figure 5.4 Illustration of the step in a $(130)_1/(\bar{1}30)_2$ $\Sigma5$ [001] tilt boundary associated with a grain boundary dislocation with $\boldsymbol{b} = (1/10)[3\bar{1}0]_1$.

Burgers vector lies in the $(130)_1$ boundary plane so that the step height is the same in both grains.

It follows from the type of analysis described here that there will be certain situations in which secondary grain boundary dislocations will not need steps associated with them to maintain the same arrangement of atoms at the boundary. These situations are ones in which the $(\boldsymbol{s}' + \boldsymbol{l})$ vector lies in the boundary plane so that $(\boldsymbol{s}' + \boldsymbol{l})\cdot\boldsymbol{v}$ is zero in equation (5.3).

5.2.3 Rigid-body Displacements and Partial Grain Boundary Dislocations

In the diagrams of figures 5.3(*c*), (*d*) and 5.4 the atoms in the grain boundary occupy coincident sites. However, the occupation of coincident sites in a grain boundary by atoms does not necessarily lead to the lowest-energy atomic configuration for the boundary. This was first pointed out by Chalmers and Gleiter (1971) who suggested that better atomic fit at a boundary, and therefore lower boundary energy, could result if atoms were moved away from coincident sites by a rigid-body displacement of one grain relative to the other by a constant displacement vector. That displacements of this type may occur to lower boundary energy was supported by the computer modelling of Weins *et al* (1971). Subsequently, Pond and Smith (1974) observed a rigid-body displacement at an incoherent $\Sigma3$ twin boundary in aluminium from fringe contrast in same-$\boldsymbol{g}_c$ images. Since then rigid-body displacements have been studied experimentally and theoretically and shown to be a general feature of grain boundary structure (see section 5.4). Different experimental techniques have been used to study rigid-body displacements. For example, in addition to the analysis of fringe contrast in same-$\boldsymbol{g}_c$ images, methods involving the use of high-resolution n-beam lattice images (see e.g. Wood *et al* 1984, moiré fringes (Matthews and Stobbs 1977), convergent beam diffraction (see e.g. Schapink *et al* 1983) and Fresnel fringes (see e.g. Boothroyd *et al* 1986) have been used. The determination of rigid-body displacements from lattice images of grain boundaries is an integral part of the determination of the atomic structure of boundaries using this method and will be discussed in section 5.5. The methods using moiré fringes, convergent beam diffraction and Fresnel fringes have found very little application, e.g. the convergent beam and Fresnel methods have only been applied to coherent $\Sigma 3$ boundaries on $\{111\}$ planes in FCC metals. In the discussions of rigid-body displacements in this chapter, the major emphasis will be on the application of the method of fringe contrast in same-$\boldsymbol{g}_c$ images to obtain solutions to a range of problems involving rigid-body displacements and partial grain boundary dislocations.

Rigid-body displacements between neighbouring grains, which move atoms away from coincident sites, must have displacement vectors which are non-DSC vectors, since DSC displacements always maintain a CSL. The presence of a non-DSC rigid-body displacement at a grain boundary has no effect on the orientation relationship between the grains. Therefore the quantitative description in equations (5.1) and (5.2) of a misorientation across a high-angle boundary, in terms of a departure from an exact CSL orientation accommodated by secondary grain boundary dislocations with DSC Burgers vectors, still applies in the presence of a rigid-body displacement. Moreover, the conclusions concerning steps in the boundary associated with secondary grain boundary dislocations with DSC Burgers vectors also still apply in the presence of a rigid-body displacement. This follows, because the properties of secondary grain boundary dislocations with DSC Burgers vectors are properties of interpenetrating lattices and are independent of the way atoms are located on the lattices in representing a bicrystal.

A rigid-body displacement must either extend over the entire boundary or be terminated by a grain boundary dislocation with a non-DSC Burgers vector. Such dislocations which separate regions of different rigid-body displacements in a boundary are known as partial grain boundary dislocations. A partial grain boundary dislocation bordering a rigid-body displacement between two grains is thus analogous to a partial dislocation bordering a stacking fault in a single crystal, when the DSC lattice of a bicrystal is taken as being analogous to the crystal lattice of a single crystal. Unlike secondary grain boundary dislocations which are perfect dislocations in the DSC lattice with Burgers vectors quantised to that lattice, a partial grain boundary dislocation can have any fractional vector of the DSC lattice, since the rigid-body displacement associated with it is one which minimises the boundary energy. Partial grain boundary dislocations separating regions of different rigid-body displacement occur not only at facet intersections, but also in planar boundaries. A source of partial dislocations in planar grain boundaries is the dissociation of secondary grain boundary dislocations with DSC Burgers vectors (see section 5.4.2).

accommodating the 0.3° departure from the Σ3 CSL orientation. For other near-CSL (111) twist orientations, they used diffraction rather than imaging to study secondary grain boundary dislocation structure. They found that for near-Σ7 and near-Σ31 orientations new diffraction spots appeared in diffraction patterns after welding of the superposed films. When the separations of these diffraction spots were associated with spacings of secondary grain boundary dislocations with DSC Burgers vectors the angular departures from the exact CSL orientations were accounted for. They also found that such interspot spacings in diffraction patterns could be associated with secondary grain boundary dislocations accommodating the angular departure from near-Σ13 and near-Σ19 boundaries.

The technique of using fabricated thin-film bicrystals has been applied to studies of high-angle tilt boundaries as well as to high-angle twist boundaries. For example, Schober and Balluffi (1971b) extended their studies of low-angle symmetric-tilt boundaries in gold (Schober and Balluffi 1971a) to high-angle symmetric-tilt boundaries, with [001] tilt axes, close to Σ5, Σ13, Σ17 and Σ25 CSL orientations. In all these boundaries they observed the expected arrays of secondary boundary dislocations consisting of evenly spaced edge dislocations parallel to the tilt axis. However, because of experimental difficulties, this work was not as definitive as that on high-angle twist boundaries. More recently high-angle symmetric-tilt boundaries in thin-film bicrystals of aluminium have been studied by Liu and Balluffi (1985) and Kvam and Balluffi (1987). The thin-film bicrystals of aluminium were prepared by evaporation on to bicrystals of sodium chloride with preselected orientations and, with this method of preparation, the tilt boundaries were normal to the surfaces of the thin-film bicrystals rather than parallel to the surfaces, as was the case for the tilt boundaries in gold. Liu and Balluffi studied near-Σ5 and Σ13 boundaries with [001] tilt axes, and with (310) and (510) boundary planes respectively. They found arrays of secondary grain boundary dislocations parallel to the tilt axis, and were able to relate the dislocation spacings and angular departures from the exact-CSL orientations to identify the secondary grain boundary dislocations as edge dislocations with basis DSC Burgers vectors of (1/10)[310] for the near-Σ5 boundaries and (1/26)[510] for the near-Σ13 boundaries. Kvam and Balluffi confirmed the results of Liu and Balluffi and extended them to the boundary planes (210) for near-Σ5, (320) for near-Σ13 and (410) and (530) for near-Σ17 symmetric-tilt boundaries. Again they found arrays of secondary grain boundary dislocations parallel to the tilt axis. However, they found that the Burgers vectors that satisfied the relation $d = |\boldsymbol{b}|/\varphi$ were not always basis DSC vectors, but were always the shortest DSC vectors normal to the boundary plane defining edge dislocations, i.e. (1/5)[210] for Σ5, (1/13)[320] for Σ13, and (1/17)[410] and (1/34)[530] for Σ17. In addition, Kvam and Bulluffi studied symmetric-tilt boundaries with [1$\bar{1}$0] tilt axes for a near-Σ9 orientation with a boundary plane of (114) and for near-Σ11 orientations with boundary planes of (113) and (332). As for the cases involving [001] tilt axes, they found that the secondary grain boundary dislocations were parallel to the tilt axis and that the Burgers vectors, which agreed with the observed spacings and angular departures from the exact-CSL orientations, were the shortest DSC vectors normal to the appropriate boundary plane, i.e. the dislocations were edge dislocations with Burgers vectors of (1/18)[114] for the near-$\Sigma 9$ boundary, and (1/11)[113] and (1/22)[332] for the near-$\Sigma 11$ boundaries.

In summary, the experimental investigations on thin-film bicrystals show that the CSL model for the structure of high-angle grain boundaries applies to twist and tilt boundaries in a variety of materials for a range of CSL orientations up to $\Sigma 29$.

5.3.2 High-angle Grain Boundaries in Polycrystalline Metals and Alloys

In addition to the work on high-angle grain boundaries in fabricated thin-film bicrystals, considerable research effort has been concentrated on direct studies of the structure of high-angle boundaries present in practical polycrystalline metals and alloys. While the work described in section 5.3.1 emphasises the idealised cases of planar twist and tilt boundaries at preselected misorientations, the work on boundaries in bulk polycrystalline metals and alloys is concerned with general boundaries between misoriented grains as they occur in polycrystalline aggregates, and these boundaries are not usually planar. Although, at first sight, such boundaries may seem to be more complicated, investigation of their structure is in fact simplified to a degree by advantages which are not associated with thin-film bicrystals. For example, major problems which occur in studies of boundaries in fabricated thin-film bicrystals are associated with the fact that, in most cases, the boundary plane is parallel to the surface of the specimen. This means that it is difficult to distinguish diffraction effects that may be associated with periodic spacings in secondary grain boundary dislocation arrays from effects of double-diffraction. In addition, the image contrast of the dislocations can be confused by moiré effects; further the dislocations lie at a constant depth in the thin-film bicrystal so that their contrast is not ideal for determining Burgers vectors. Neither of these problems are serious ones for boundaries in specimens prepared by thinning bulk polycrystalline material, because such boundaries are usually curved and inclined to the surface of the electron microscope specimen. As pointed out in section 4.3.1, diffraction effects associated with the dislocation structure in these types of boundary can be distinguished from the effects of double-diffraction by the way they are separately influenced by changes in boundary plane; further, in images of dislocations inclined to the surface of the specimen, the full variation of contrast with depth in the specimen is sampled and this enables a more reliable identification of Burgers vectors.

In discussing work on the secondary grain boundary dislocation structure of high-angle boundaries in polycrystals, results for boundaries which are close to exact-CSL orientations ($\varphi \lesssim 1^\circ$) will be treated first, and then the work on boundaries further removed from exact-CSL orientations will be reviewed.

5.3.2 *(i) Grain boundaries with misorientations close to CSL orientations*

For high-angle grain boundaries with misorientations close to CSL orientations ($\varphi \lesssim 1^\circ$), the spacings of the secondary grain boundary dislocations will usually be sufficiently coarse for their contrast to be characteristic of that of individual dislocations rather than of the dislocation array as a whole. Under these conditions the Burgers vectors of the secondary grain boundary dislocations can be determined from their diffraction contrast. In the following, examples of the determination of secondary grain boundary dislocation structure in high-angle grain boundaries will be given for a range of misorientations, all of which are near-CSL orientations.

Near-Σ3 boundary in a Cu–Si alloy

In a study of the structure of high-angle grain boundaries in an FCC polycrystalline Cu − 6 at% Si alloy, many examples of near-Σ3 coherent twin-boundaries were observed to contain coarse hexagonal networks of secondary grain boundary dislocations (Forwood and Clarebrough 1986a). An example of such a boundary is shown in figure 5.7(a) in which the secondary grain boundary dislocations forming the network have spacings of approximately 2000 Å. The boundary plane is (111) and within the limits of experimental error ($\pm$ 0.03°) the misorientation ($\boldsymbol{u}, \theta$) corresponds to a rotation of the upper grain U in a right-handed sense with respect to the lower grain L by the angle $\theta = 60^\circ$ around an axis $\boldsymbol{u}$ parallel to the [111] direction common to both grains, i.e. the misorientation is very close to a Σ3 CSL orientation. In applying the procedure of section 5.2.1 to determine the basis DSC vectors for this case the appropriate Σ3 CSL matrix is taken to describe the misorientation of the two grains, and this matrix, which re-indexes a vector in grain L into its indices in grain U is

$$(1/3)\begin{pmatrix} 2 & 2 & -1 \\ -1 & 2 & 2 \\ 2 & -1 & 2 \end{pmatrix}.$$

From this Σ3 CSL matrix the three smallest non-coplanar same-$\boldsymbol{g}_c$ vectors indexed in grain U give the $\boldsymbol{g}_c$ matrix as

$$\begin{pmatrix} 1 & 1 & 1 \\ 2 & 0 & -2 \\ 0 & 2 & -2 \end{pmatrix}_{\mathrm{U}}$$

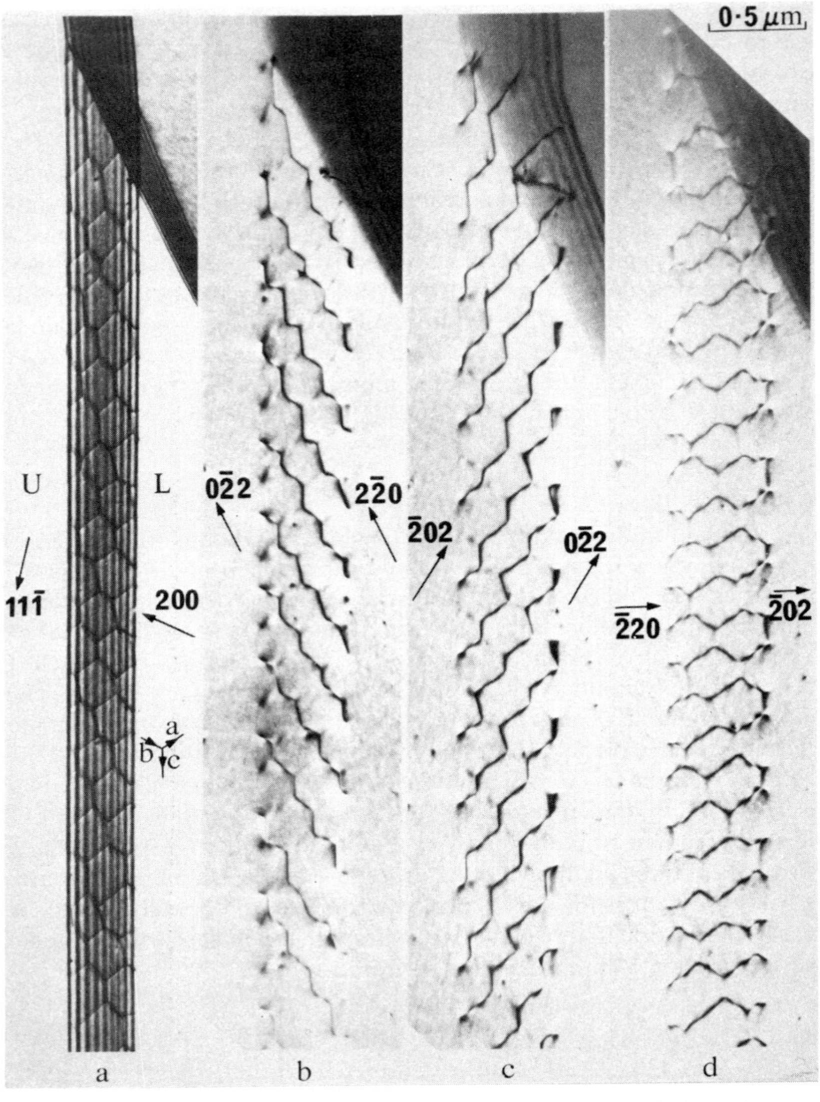

Figure 5.7 Double two-beam images of a near-$\Sigma3$ coherent twin-boundary on $(111)_{U/L}$ in a Cu – 6 at% Si alloy; the diffracting vectors are indicated. In (*a*), (*b*), (*c*) and (*d*) the beam directions are close to $[\bar{1}43]_U/[015]_L$, $[001]_U/[114]_L$, $[111]_U/[111]_L$ and $[111]_U/[111]_L$ respectively.

and its inverse as

$$(1/6)\begin{pmatrix} 2 & 2 & -1 \\ 2 & -1 & 2 \\ 2 & -1 & -1 \end{pmatrix}_{U}$$

so that a set of basis DSC vectors indexed in grain U is $\boldsymbol{x} = (1/3)[111]_U$, $\boldsymbol{y} = (1/6)[2\bar{1}\bar{1}]_U$ and $\boldsymbol{z} = (1/6)[\bar{1}2\bar{1}]_U$.

Figure 5.7(*a*), which is a simultaneous double two-beam electron micrograph with different diffracting vectors $\boldsymbol{g}$ operating in each grain, has three arrays of secondary grain boundary dislocations (indicated by the labelling *a*, *b*, and *c*) in good contrast, showing that they form a well-defined hexagonal network. The images in figures 5.7(*b*), (*c*) and (*d*) are simultaneous double two-beam electron micrographs with the same $\boldsymbol{g}_c$ operating in each grain as indicated on the micrographs. Since there is no fringe contrast due to rigid-body displacements in these images of the boundary, $\boldsymbol{g} \cdot \boldsymbol{b}$ criteria can be used as a guide to determining the Burgers vectors of the secondary grain boundary dislocations (see section 2.4.5). The absence of contrast corresponding to $\boldsymbol{g}_c \cdot \boldsymbol{b} = 0$, for dislocation segments *a* in figure 5.7(*b*), segments *b* in figure 5.7(*c*) and segments *c* in figure 5.7(*d*), suggests that the Burgers vectors involved in the network are the DSC vectors $\pm\boldsymbol{y} = \pm(1/6)[2\bar{1}\bar{1}]_U$ for $\boldsymbol{b}_a$, $\pm\boldsymbol{z} = \pm(1/6)[\bar{1}2\bar{1}]_U$ for $\boldsymbol{b}_b$ and $\pm(\boldsymbol{y} + \boldsymbol{z}) = \pm(1/6)[\bar{1}\bar{1}2]_U$ for $\boldsymbol{b}_c$, where $\boldsymbol{b}_a$, $\boldsymbol{b}_b$ and $\boldsymbol{b}_c$ are the Burgers vectors of segments *a*, *b* and *c* respectively. These Burgers vectors were confirmed by other absences of contrast in same-$\boldsymbol{g}_c$ images with $31\bar{1}_U/13\bar{1}_L$ and $3\bar{1}1_U/31\bar{1}_L$ diffracting vectors. However, to obtain positive identification of the Burgers vectors and to determine their sign the technique of image matching for simultaneous double two-beam images was used. For a positive sense of the line direction $\boldsymbol{r}$ of each grain boundary dislocation segment taken from the bottom to the top of the foil (namely $\boldsymbol{r}_a = [\bar{5}23]_U$, $\boldsymbol{r}_b = [\bar{3}7\bar{4}]_U$ and $\boldsymbol{r}_c = [13\bar{4}]_U$), it was found that $\boldsymbol{b}_a = (1/6)[2\bar{1}\bar{1}]_U$, $\boldsymbol{b}_b = (1/6)[1\bar{2}1]_U$ and $\boldsymbol{b}_c = (1/6)[\bar{1}\bar{1}2]_U$, giving the nodal balance in the network $\boldsymbol{b}_b = \boldsymbol{b}_a + \boldsymbol{b}_c$. Figure 5.8 shows a portion of the experimental image of figure 5.7(*a*) together with matching and mismatching theoretical images for the three segments *a*, *b* and *c* of the hexagonal network, indicating the distinction that can be made by the image matching process between Burgers vectors of opposite sign.

Using the determined Burgers vectors and the measured spacings of the dislocations, and applying equations (1.20) and (1.21) shows that the hexagonal network would accommodate a small twist departure of $\varphi \approx 0.02°$ around the common [111] axis from the exact-$\Sigma 3$ CSL orientation. This angular departure φ could not have been determined directly from equation (5.1) as it lies within the experimental error in $(\boldsymbol{u}, \theta)$.

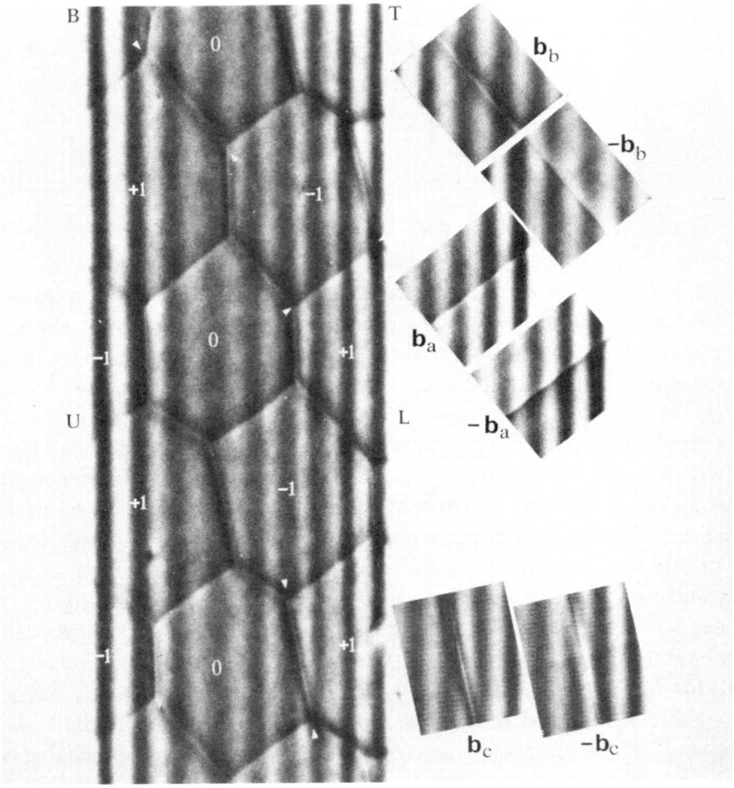

Figure 5.8 Higher magnification image of portion of figure 5.7(a) together with matching computed images for $\boldsymbol{b}_a = (1/6)[2\bar{1}\bar{1}]_{\mathrm{U}}$, $\boldsymbol{b}_b = (1/6)[1\bar{2}1]_{\mathrm{U}}$ and $\boldsymbol{b}_c = (1/6)[\bar{1}\bar{1}2]_{\mathrm{U}}$ and mismatching computed images for $-\boldsymbol{b}_a$, $-\boldsymbol{b}_b$ and $-\boldsymbol{b}_c$ for the grain boundary dislocation segments indicated by the pairs of white arrowheads in the experimental image. $\boldsymbol{B} = [\bar{1}43]_{\mathrm{U}}$, $w_{\mathrm{U}} = 0.3$ and $w_{\mathrm{L}} = 0.2$. The intersections of the boundary with the top and bottom of the specimen are marked T and B respectively.

The dislocations in the hexagonal network will have steps associated with them (see section 5.2.1), but such steps need not change the mean boundary plane from (111). For example, no change in the mean boundary plane will occur if the level of the boundary in cells marked $+1$ in figure 5.8 is displaced from the level in cells marked 0 by one (111) interplanar spacing into grain U, nor if the level of the boundary in cells marked -1 is displaced from the level in cells marked 0 by one (111) interplanar spacing into grain L. With this arrangement of levels for the boundary plane, two different step heights are associated with dislocations of the same Burgers vector, namely one (111) interplanar spacing in one sense or two (111) interplanar spacings in the opposite sense. As discussed in section 5.2.1 these two different step heights maintain an identical atomic structure in the boundary since they are related by a translation vector of the Σ3 CSL lattice.

Near-Σ3 boundary in iron

Image matching has been used to determine the Burgers vectors of the secondary grain boundary dislocations in three independent arrays in a near-Σ3 symmetric-tilt boundary on a {112} plane in BCC α-iron (99.96%Fe) (Clarebrough and Forwood 1988). In contrast to the previous example, $\boldsymbol{g}\cdot\boldsymbol{b}$ invisibility criteria could not be used as a guide to possible Burgers vectors in this case because a rigid-body displacement between the grains, corresponding to an expansion normal to the {112} plane of the boundary, gave strong fringe contrast in same-$\boldsymbol{g}_c$ images (see section 5.4).

Figures 5.9(*a*)–(*d*) show four different simultaneous double two-beam electron micrographs of the {112} Σ3 boundary studied. In all micrographs grain 2 is on the left and grain 1 is on the right, but the beam directions are such that in figures 5.9(*a*) and (*b*) grain 1 is the upper grain and in figures 5.9(*c*) and (*d*) grain 2 is the upper grain with respect to the electron source. This switch in the grain nearer the electron source is brought about in this example, because the boundary plane has been tilted through the optic axis of the microscope during the process of establishing the different simultaneous double two-beam conditions. The boundary plane is $(1\bar{2}1)_1/(2\bar{1}\bar{1})_2$ and, within the limits of experimental error, the measured misorientation between the two grains corresponds to a rotation of grain 1 with respect to grain 2 by 60° in a right-handed sense around an axis with common indices [111] in both grains. The Σ3 CSL matrix which describes this misorientation and re-indexes a vector in grain 1 into its indices in grain 2 is

$$(1/3)\begin{pmatrix} 2 & 2 & -1 \\ -1 & 2 & 2 \\ 2 & -1 & 2 \end{pmatrix}.$$

The boundary contains three independent arrays of secondary grain boundary dislocations and these are illustrated schematically in figure 5.10 which corresponds to the projection in figures 5.9(*a*) and (*b*). Dislocation elements in the arrays of figure 5.10 are labelled *A*, *B*, *C* and *D* with the sense of the line directions from the bottom to the top of the specimen as indicated by the arrows. The elements *B*, *C* and *D* form an hexagonal network in which there can be only two independent elements (say *B* and *C*) with the third element (*D*) formed as a reaction product of the other two. There is no obvious reaction product associated with the intersections of the third independent element *A* with the other elements of the network. The micrograph of figure 5.9(*d*) is a double two-beam electron micrograph with the same diffracting vector $\boldsymbol{g}_c = \bar{1}10_1/\bar{1}01_2$ operating in each grain and the

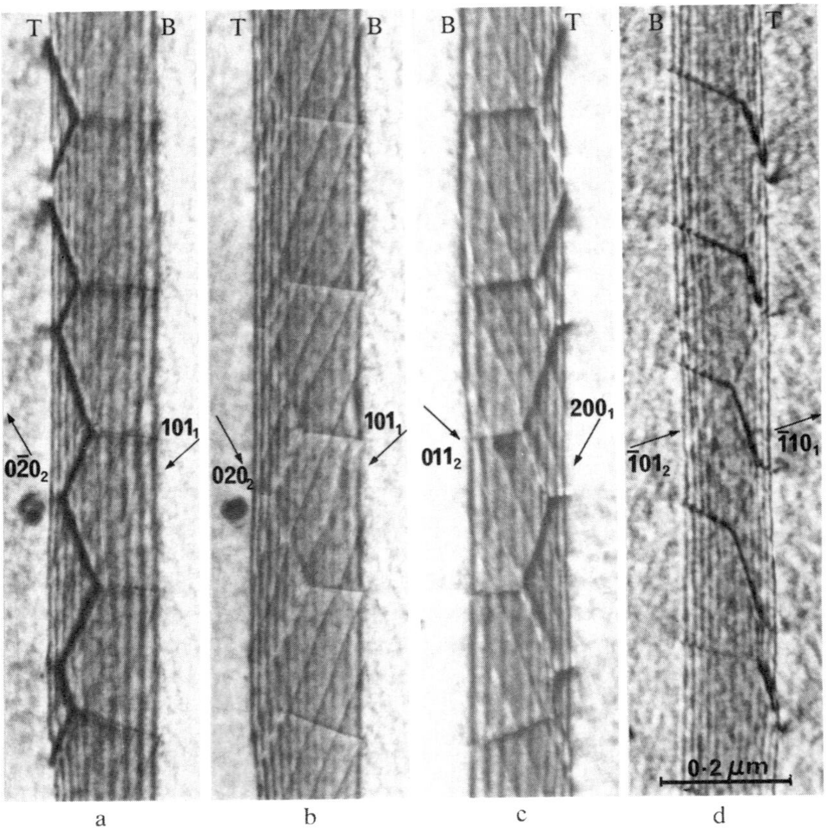

Figure 5.9 Double two-beam images of a $(1\bar{2}1)_1/(2\bar{1}\bar{1})_2$ $\Sigma 3$ boundary in α-iron showing three independent arrays of grain boundary dislocations. The intersections of the boundary with the top and bottom surfaces of the specimen are indicated T and B respectively. The diffracting vectors are indicated and the beam directions are $[\bar{2}32]_1$ in (*a*) and (*b*), $[3\bar{2}2]_2$ in (*c*) and $[4\bar{7}4]_2$ in (*d*).

fringe contrast, arising from the rigid-body displacement between the grains, is continuous across all the elements of the secondary grain boundary dislocation arrays. As will be pointed out for this boundary in section 5.4 this continuity of fringe contrast is also obtained for other same-$\boldsymbol{g}_c$ images in a set of non-coplanar $\boldsymbol{g}_c$, indicating that the vector dot product of each $\boldsymbol{g}_c$ with the change in displacement across each dislocation element is an integer. Thus, all the secondary grain boundary dislocations must have Burgers vectors of the DSC lattice, since a set of basis DSC vectors is always reciprocal to the corresponding set of basis same-$\boldsymbol{g}_c$ vectors. The basis vectors of the DSC lattice for this case can be determined as described previously.

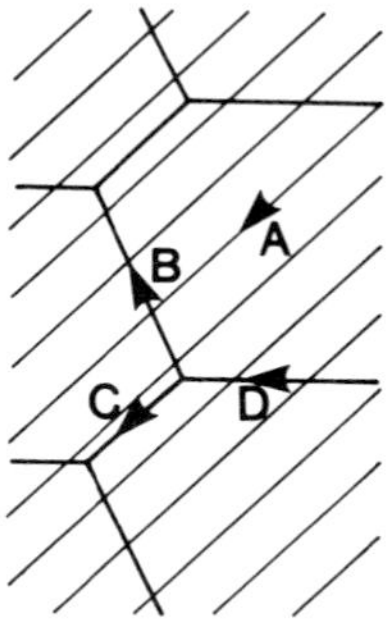

Figure 5.10 Schematic illustration of the secondary grain boundary dislocation arrays with elements *A*, *B*, *C* and *D* corresponding to the projection in the beam direction of figures 5.9(*a*) and (*b*). The arrows indicate the positive sense of line direction of the dislocations.

Thus from the Σ3 CSL matrix, the three smallest non-coplanar same-$\boldsymbol{g}_c$ vectors indexed with respect to grain 1 are, 222_1, $\bar{1}01_1$, and $\bar{1}10_1$, giving the $\boldsymbol{g}_c$ matrix and its inverse as

$$\begin{pmatrix} 2 & 2 & 2 \\ -1 & 0 & 1 \\ -1 & 1 & 0 \end{pmatrix}_1 \quad \text{and} \quad (1/6)\begin{pmatrix} 1 & -2 & -2 \\ 1 & -2 & 4 \\ 1 & 4 & -2 \end{pmatrix}_1$$

respectively, so that three basis DSC vectors are $\boldsymbol{x} = (1/6)[111]_1$, $\boldsymbol{y} = (1/3)[\bar{1}\bar{1}2]_1$ and $\boldsymbol{z} = (1/3)[\bar{1}2\bar{1}]_1$. The basis vectors $\boldsymbol{x}$, $\boldsymbol{y}$ and $\boldsymbol{z}$ and their linear combinations, up to a magnitude of $\langle 001 \rangle$, were tested by image matching as possible Burgers vectors for each of the elements in the secondary grain boundary dislocation network. The elements *A* extend completely through the foil and have a line direction parallel to $[111]_1$. A set of experimental images (i) together with a set of theoretical images (ii) computed for a Burgers vector of $(1/6)[111]_1$ are shown in figure 5.11. There is good agreement between the experimental and theoretical images and this is the only set of theoretical images which matched the detail of the contrast in the experimental images. This set of images therefore identifies the Burgers vector of the secondary grain boundary dislocations *A* as $\boldsymbol{b}_A = (1/6)[111]_1$, i.e. these dislocations are right-handed screw dislocations and would be glissile in the boundary. The elements *B*, which form part of the hexagonal network, have a line direction parallel to $[\bar{1}\bar{3}\bar{4}5]_1$, and do not extend completely through the specimen. Figure 5.12 shows a comparison between experimental images (i) and theoretical images (ii) and (iii) for two *B* elements located at different depths in the foil. The agreement between the experimental and theoretical images identifies the Burgers vector of the secondary grain boundary dislocations *B* as $\boldsymbol{b}_B = (1/3)[\bar{1}\bar{1}2]_1$. The elements *C* also form

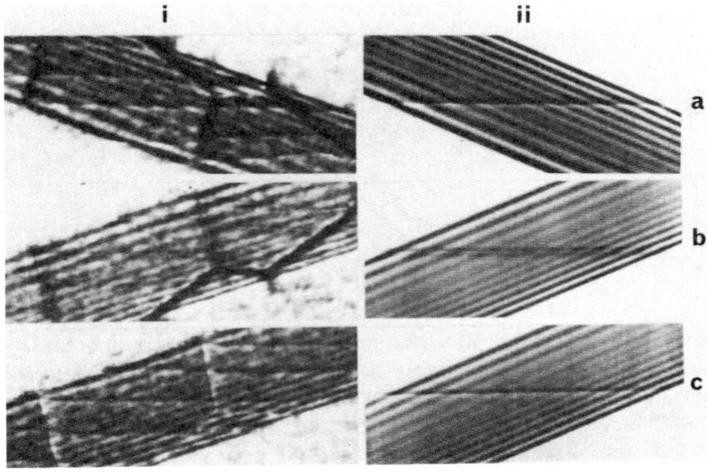

Figure 5.11 Comparison of experimental (i) and matching computed (ii) images for a secondary grain boundary dislocation A with $\boldsymbol{b}_A = (1/6)[111]_1$. The intersection of the boundary with the top of the specimen is on the right in all the images, grain 1 is on the right in (a) and on the left in (b) and (c). The diffracting conditions are:

	$\boldsymbol{B}$	g	w
(a)	$[3\bar{2}2]_2$	$011_2, 200_1$	$0.29_2, 0.29_1$
(b)	$[\bar{2}32]_1$	$101_1, 0\bar{2}0_2$	$0.4_1, 0.5_2$
(c)	$[\bar{2}32]_1$	$101_1, 020_2$	$0.43_1, 0.65_2$

part of the hexagonal network and have a line direction parallel to that of elements A, i.e. parallel to $[111]_1$. Figure 5.13 shows a comparison between experimental images (i) and their matching computed images (ii), which identifies the Burgers vector of the secondary grain boundary dislocations C as $\boldsymbol{b}_C = (1/3)[\bar{1}2\bar{1}]_1$, so that these dislocations are edge dislocations. The secondary grain boundary dislocations A, B and C constitute three independent arrays, since they have Burgers vectors which are basis DSC vectors, namely $\boldsymbol{b}_A = \boldsymbol{x}$, $\boldsymbol{b}_B = \boldsymbol{y}$ and $\boldsymbol{b}_C = \boldsymbol{z}$. Since there must be a balance of the Burgers vectors of the elements B, C and D at the nodal points in the hexagonal network, the Burgers vector of the elements D must be

$$\boldsymbol{b}_D = \boldsymbol{b}_B + \boldsymbol{b}_C = (1/3)[\bar{2}11]_1.$$

Figure 5.14 shows a comparison of experimental images (i) with theoretical images (ii) computed for $\boldsymbol{b}_D = (1/3)[\bar{2}11]_1$ for an element D in which the line direction is parallel $[\bar{4}\,3\,10]_1$ and the agreement shown confirms that the Burgers vector is that expected from the nodal balance.

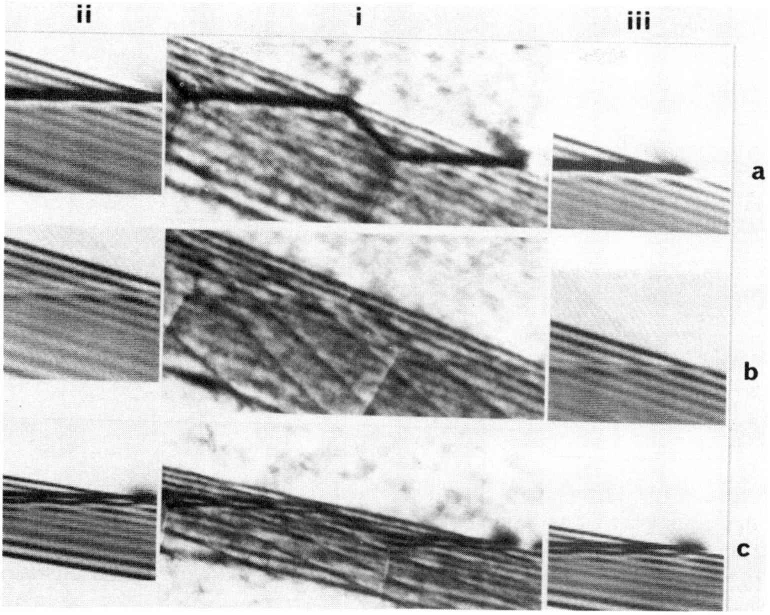

Figure 5.12 Comparison of experimental (i) and matching computed (ii) and (iii) images for a secondary grain boundary dislocation B with $\boldsymbol{b}_B = (1/3)[\bar{1}\bar{1}2]_1$. The intersection of the boundary with the top of the specimen is on the right and grain 1 is on the left in all images. The diffracting conditions are:

	B	**g**	w
(a)	$[\bar{2}32]_2$	$101_1, 0\bar{2}0_2$	$0.5_1, 0.3_2$
(b)	$[\bar{2}32]_1$	$101_1, 020_2$	$0.43_1, 0.65_2$
(c)	$[\bar{2}23]_1$	$110_1, 200_2$	$0.25_1, 0.25_2$

The observed interaction between different elements of the secondary grain boundary dislocation arrays is consistent with the determined Burgers vectors in that, on a $|\boldsymbol{b}|^2$ criterion, the reaction involved in the formation of the elements D in the hexagonal network is an energy lowering reaction; whereas the lack of interaction of elements A with other elements of the network is to be expected since all such interactions would be energetically neutral.

All the secondary grain boundary dislocations A, B, C and D in this example have steps associated with them. On the assumption that the average boundary plane is maintained as $(1\bar{2}1)_1/(2\bar{1}\bar{1})_2$, then, in accordance with the procedure of section 5.2.1, these steps will have the form illustrated in figure 5.15. For the hexagonal network in figure 5.15 the levels are marked 0, +1, −1, where 0 indicates some reference level, +1 indicates a step of grain 2

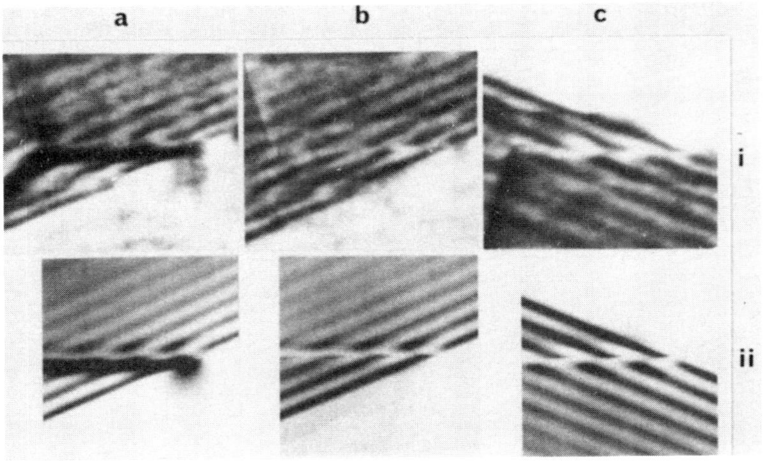

Figure 5.13 Comparison of experimental (i) and matching computed (ii) images for a secondary grain boundary dislocation *C* with $\boldsymbol{b}_C = (1/3)[\bar{1}2\bar{1}]_1$. The intersection of the boundary with the top of the specimen is on the right in all images, grain 1 is on the left in (*a*) and (*b*) and on the right in (*c*). The diffracting conditions are:

	$\boldsymbol{B}$	g	w
(*a*)	$[\bar{2}32]_1$	$101_1, 0\bar{2}0_2$	$0.5_1, 0.3_2$
(*b*)	$[\bar{2}32]_1$	$101_1, 020_2$	$0.43_1, 0.65_2$
(*c*)	$[3\bar{2}2]_2$	$011_2, 200_1$	$0.29_2, 0.29_1$

into grain 1 and -1 a step of grain 1 into grain 2, with all steps having a magnitude of one {112} interplanar spacing. Superimposed on the step pattern associated with the hexagonal network there is a step pattern associated with the secondary grain boundary dislocations *A*. In figure 5.15 this step pattern is illustrated schematically as separate from the hexagonal network and the levels are specified in the same way as for the hexagonal network.

Near-Σ9 boundary in copper

In this example of the analysis of the structure of a near-Σ9 boundary in an FCC metal, the Burgers vectors of the dislocations are determined by two independent methods (Clarebrough and Forwood 1980a, b). First the geometric method of section 1.2, which has already been applied to low-angle boundaries in section 4.3.2, will be used via its extension through equation (5.2) to identify the secondary grain boundary dislocations. Second, the

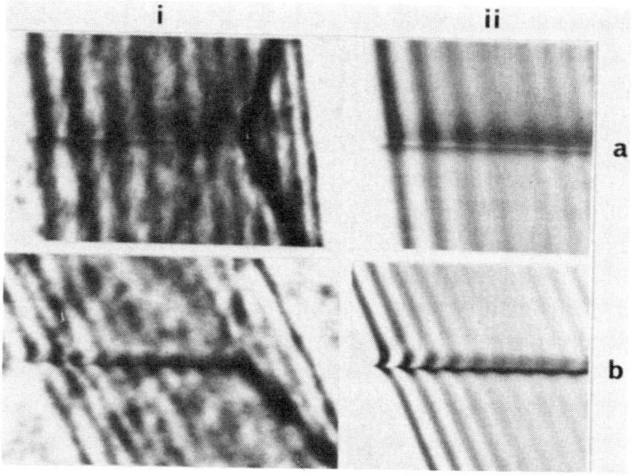

Figure 5.14 Comparison of experimental (i) and matching computed (ii) images for a secondary grain boundary dislocation D with $\boldsymbol{b}_D = (1/3)[\bar{2}11]_1$. The intersection of the boundary with the top of the specimen is on the right, grain 1 is on the left in (a) and on the right in (b). The diffracting conditions are:

	$\boldsymbol{B}$	$\boldsymbol{g}$	w
(a)	$[\bar{2}32]_1$	$101_1, 0\bar{2}0_2$	$0.4_1, 0.5_2$
(b)	$[4\bar{7}4]_2$	$\bar{1}01_2, \bar{1}10_1$	$0.15_2, 0.15_1$

The value of $\boldsymbol{g}_c \cdot \boldsymbol{R}$ used in the computed image (ii) (b) is 0.14, which is the displacement of grain 1 relative to grain 2 (see table 5.5 in section 5.4).

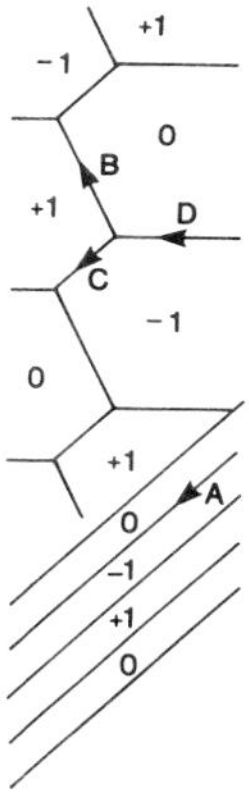

Figure 5.15 Schematic diagram showing the levels of the boundary plane associated with the different elements in the arrays of secondary grain boundary dislocations.

method of image matching will be used independently to identify the Burgers vectors of the same secondary grain boundary dislocations.

As already indicated in section 5.2, the geometric method for analysing boundary structure as applied to secondary grain boundary dislocations follows the same principles as those described for the analysis of the dislocation structure in low-angle boundaries. Thus for a high-angle grain boundary with boundary normal $\boldsymbol{v}$, if three independent arrays of secondary grain boundary dislocations with non-coplanar Burgers vectors $\boldsymbol{b}_1$, $\boldsymbol{b}_2$, $\boldsymbol{b}_3$, spacings d_1, d_2, d_3 and line directions $\boldsymbol{r}_1, \boldsymbol{r}_2, \boldsymbol{r}_3$ accommodate a small angular departure $(\boldsymbol{q}, \varphi)$ from a CSL orientation $(\boldsymbol{p}, \omega)$ as defined by equation (5.1), then it follows from equation (5.2) and expressions (1.9) and (1.10) that,

$$\boldsymbol{r}_1 \parallel [\boldsymbol{q} \wedge (\boldsymbol{b}_2 \wedge \boldsymbol{b}_3)] \wedge \boldsymbol{v} \tag{5.4}$$

with similar expressions for $\boldsymbol{r}_2$ and $\boldsymbol{r}_3$, and

$$d_1 = \left(2 \sin(\varphi/2) \left| \frac{[\boldsymbol{q} \wedge (\boldsymbol{b}_2 \wedge \boldsymbol{b}_3)] \wedge \boldsymbol{v}}{\boldsymbol{b}_1 \cdot (\boldsymbol{b}_2 \wedge \boldsymbol{b}_3)} \right| \right)^{-1} \tag{5.5}$$

with similar expressions for d_2 and d_3. As for the case of grain boundary dislocations in a low-angle boundary, the Burgers vectors of secondary grain boundary dislocations can be determined from changes in the direction and spacing of the dislocations with variation in boundary normal by using the two properties of expression (5.4) that:

(a) the different directions of $\boldsymbol{r}_1$ for different boundary plane normals $\boldsymbol{v}$ must be coplanar with $\boldsymbol{q}$, and so also for $\boldsymbol{r}_2$ and $\boldsymbol{r}_3$;
(b) the different directions of $\boldsymbol{r}_1$ for different boundary plane normals $\boldsymbol{v}$ must also be coplanar with the direction $(\boldsymbol{b}_2 \wedge \boldsymbol{b}_3)$, and so also for $\boldsymbol{r}_2$ with $(\boldsymbol{b}_3 \wedge \boldsymbol{b}_1)$ and $\boldsymbol{r}_3$ with $(\boldsymbol{b}_1 \wedge \boldsymbol{b}_2)$.

If measurements of the directions of the secondary grain boundary dislocations show that property (a) holds, then these changes in direction can be associated with dislocations having Burgers vectors which remain constant with change in boundary plane. These Burgers vectors can then be determined from property (b) as those which give spacings from equation (5.5) that are in agreement with the measured spacings.

Figure 5.16 shows two electron micrographs of the near-$\Sigma 9$ boundary in 99.999% Cu, where the grains are labelled L and R and the intersection of the boundary with the top surface of the specimen indicates that grain R is the upper grain. The misorientation between the two grains was determined as a rotation of grain R with respect to grain L by the angle $\theta = 38.74°$ in a right-handed sense about an axis $\boldsymbol{u} = [\cos 45.13°, \cos 89.82°, \cos 44.87°]$ common to both grains to an accuracy of $\pm 0.05°$, and the corresponding

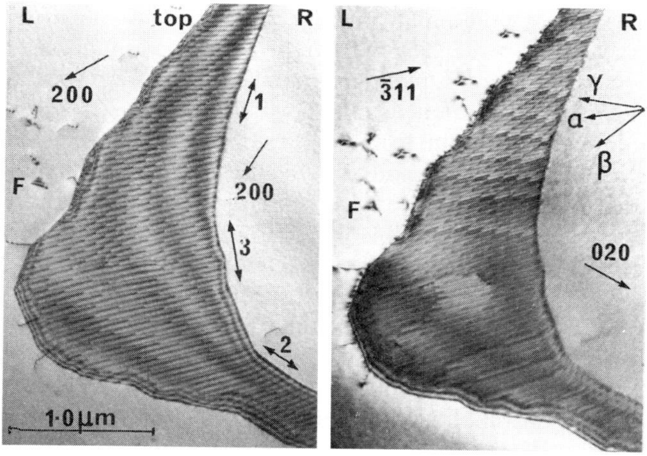

a b

Figure 5.16 Double two-beam images of a curved near-Σ9 boundary in copper. ***B*** is close to $[0\bar{1}5]_L$ and $[015]_R$ in (*a*) and close to $[1\bar{3}6]_L$ and $[001]_R$ in (*b*). The diffracting vectors are indicated.

matrix which re-indexes a vector in grain R into its indices in grain L is

$$(\boldsymbol{u}, \theta) = \begin{pmatrix} 0.889\,512\,1 & -0.442\,978\,8 & 0.111\,973\,5 \\ 0.443\,965\,8 & 0.780\,030\,8 & -0.440\,960\,5 \\ 0.107\,993\,4 & 0.441\,952\,1 & 0.890\,514\,3 \end{pmatrix}. \qquad (5.6)\dagger$$

The closest CSL orientation to this measured misorientation is the Σ9 CSL orientation which corresponds to a rotation of grain R with respect to a crystal lattice at the exact-Σ9 CSL orientation by the angle $\omega = 38.94°$ in a right-handed sense around an axis ***p*** of [101] common to grain R and the CSL, and the matrix which re-indexes a vector in grain R into its indices in a crystal lattice at the exact-Σ9 CSL orientation is

$$(\boldsymbol{p}, \omega) = (1/9)\begin{pmatrix} 8 & -4 & 1 \\ 4 & 7 & -4 \\ 1 & 4 & 8 \end{pmatrix}. \qquad (5.7)$$

† These matrix elements are quoted to seven significant figures and although this exceeds the determined accuracy in $(\boldsymbol{u}, \theta)$, it ensures that the matrix retains its property of being a pure rotation matrix (i.e. being orthonormal) within the experimental error of $(\boldsymbol{u}, \theta)$. In all further examples of this type the same criterion is applied in quoting matrix elements.

From equation (5.1) the small angular departure of the measured misorientation $(\boldsymbol{u}, \theta)$ from the Σ9 orientation $(\boldsymbol{p}, \omega)$ corresponds to a rotation $\varphi = 0.25° \pm 0.05°$ of grain L in a right-handed sense with respect to a crystal lattice in the CSL orientation around an axis $\boldsymbol{q} = [\cos 24.21°, \cos 109.6°, \cos 76.5°]$ common to grain L and the CSL. The relationship between the misorientations $(\boldsymbol{u}, \theta)$, $(\boldsymbol{q}, \varphi)$ and the Σ9 CSL orientation is illustrated in schematic form in figure 5.17. For the sense of the misorientations in figure 5.17, expression (5.4) is compatible with the FS/RH convention (see section 1.2) when the boundary normal $\boldsymbol{v}$ points from the CSL into grain L, i.e. in the sense from grain L into grain R. The error limits for the rotation axis $\boldsymbol{q}$ are comparatively large because the departure from the Σ9 orientation is small, so that small uncertainties in $\boldsymbol{u}$ and θ become magnified in $\boldsymbol{q}$. These error limits for $\boldsymbol{q}$ are specified by a solid angle with semicone angle of 5°, which is represented by the small circle around $\boldsymbol{q}$ on the stereographic projection of figure 5.19.

From the Σ9 CSL matrix of equation (5.7) the three smallest non-coplanar same-$\boldsymbol{g}_c$ vectors indexed with respect to grain R are 202_R, $\bar{3}11_R$ and $04\bar{2}_R$, or equivalently 202_L, $\bar{3}\bar{1}1_L$ and $\bar{2}40_L$ in grain L, giving the $\boldsymbol{g}_c$ matrix and its inverse as

$$\begin{pmatrix} 2 & 0 & 2 \\ -3 & -1 & 1 \\ -2 & 4 & 0 \end{pmatrix}_L \quad \text{and} \quad (1/18)\begin{pmatrix} 2 & -4 & -1 \\ 1 & -2 & 4 \\ 7 & 4 & 1 \end{pmatrix}_L$$

respectively, so that a set of three basis DSC vectors in grain L is $\boldsymbol{x} = (1/18)[217]_L$, $\boldsymbol{y} = (1/9)[\bar{2}\bar{1}2]_L$ and $\boldsymbol{z} = (1/18)[\bar{1}41]_L$.

The boundary in figure 5.16 shows a variation in boundary plane along its length and three different boundary planes are marked 1, 2 and 3. Only the planes marked 1 and 2 will be considered here for the geometric analysis, because in plane 3 the secondary grain boundary dislocation structure has been extensively modified by interactions with slip dislocations from the

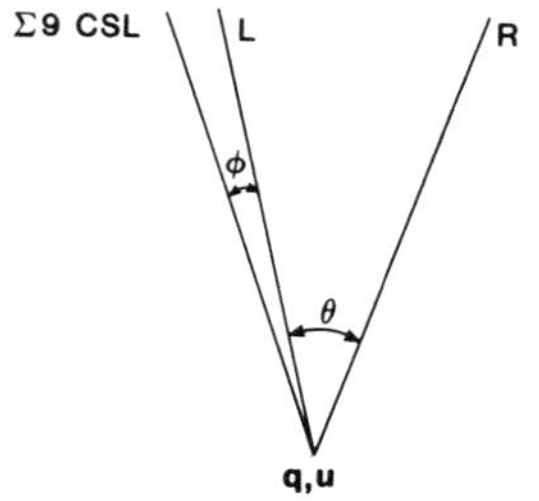

Figure 5.17 Schematic illustration of relationships between the Σ9 CSL orientation, $(\boldsymbol{u}, \theta)$ and $(\boldsymbol{q}, \varphi)$ for the boundary of figure 5.16, where $\boldsymbol{q}$ and $\boldsymbol{u}$ are directed into the page.

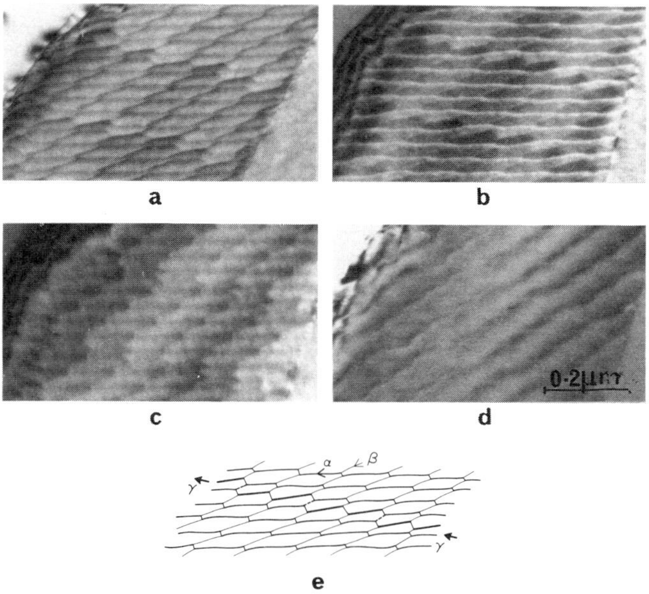

Figure 5.20 Electron micrographs (*a*)–(*d*) of position 1 on the boundary for different diffracting conditions: (*a*) as for figure 5.16(*b*); (*b*) grain L non-diffracting, $\boldsymbol{B}$ is close to $[\bar{1}34]_R$ and $\boldsymbol{g} = \bar{1}1\bar{1}_R$; (*c*) grain L non-diffracting, $\boldsymbol{B}$ is close to $[134]_R$ and $\boldsymbol{g} = 11\bar{1}_R$; (*d*) $\boldsymbol{B}$ is close to $[103]_L$ and $[112]_R$ with a same-$\boldsymbol{g}_c = 31\bar{1}_L, 3\bar{1}\bar{1}_R$; (*e*) schematic illustration of network in position 1 where γ crosses the α, β array.

additional $A + C$ segments. The different segments of the secondary grain boundary dislocation network shown in figure 5.21(*d*) can be identified in the images of figure 5.20(*a*)–(*c*) where they show different diffraction contrast behaviour, and in figure 5.21(*e*) which is a portion of the network in figure 5.20(*a*) at higher magnification with an appropriately magnified schematic drawing. The image in figure 5.20(*d*) is a same-$\boldsymbol{g}_c$ image with diffracting vectors $31\bar{1}_L$ and $3\bar{1}\bar{1}_R$ which does not show any strong fringe contrast due to rigid-body displacement. Apart from those segments containing a component of Burgers vector $\boldsymbol{B}$, all other segments in the network show no contrast in this same-$\boldsymbol{g}_c$ image, implying that $\boldsymbol{g}_c \cdot \boldsymbol{b} = 0$ for these segments. This is consistent with the identification of $\boldsymbol{A} = (1/18)[217]_L$, $\boldsymbol{B} = (1/9)[\bar{2}\bar{1}2]_L$ and $\boldsymbol{C} = (1/6)[\bar{1}1\bar{2}]_L$ by the geometric analysis and with the identification of the segments in the network as represented in figure 5.21(*d*), namely $\boldsymbol{D} = (1/6)[\bar{2}\bar{1}\bar{1}]_L$, $\boldsymbol{A} + \boldsymbol{C} = (1/18)[141]_L$, and $\boldsymbol{A} + \boldsymbol{B} + \boldsymbol{C} = (1/18)[\bar{5}25]_L$.

The spacings of the secondary grain boundary dislocations in this near-$\Sigma 9$ boundary are sufficiently coarse for image matching to be used to give an

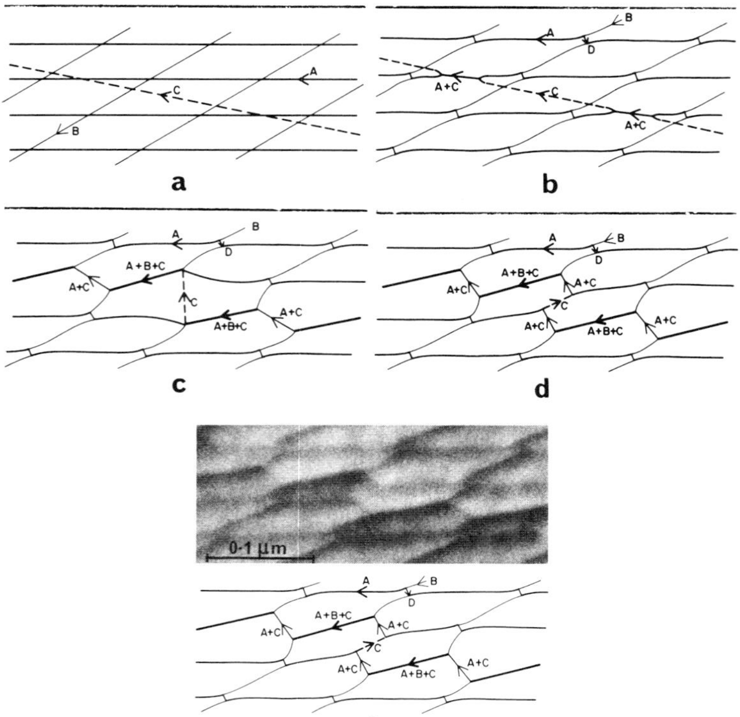

Figure 5.21 Schematic illustration of the stages in the formation of the secondary grain boundary dislocation network at position 1 in figure 5.16(*a*).

independent identification of the Burgers vectors. Images of segments of the network labelled *A*, *B* and $A + B + C$ in figure 5.21(*d*) were computed with 16 different DSC vectors as possible Burgers vectors. These 16 vectors were the basis DSC vectors $\boldsymbol{x}$, $\boldsymbol{y}$, $\boldsymbol{z}$ and their linear combinations up to the magnitude of $(1/6)\langle 112 \rangle$. For each segment, the computed images were compared with a set of simultaneous double two-beam experimental images which contained three non-coplanar diffracting vectors. The only theoretical images which matched the experimental images for dislocation segments *A* were those computed for the Burgers vector $\boldsymbol{A} = (1/18)[217]_L$, and an example of matching experimental and computed images is shown in figure 5.22. In the experimental images in figure 5.22 three parallel lines of *A* are intersected by two lines of *B*, whereas the theoretical images are computed for only one line *A* (arrowed in figure 5.22) and take no account of the intersecting dislocation lines *B*. For dislocation segment *B* the only theoretical images which matched the experimental images were those computed for the Burgers vector $\boldsymbol{B} = (1/9)[\bar{2}\bar{1}2]_L$, and a matching set of experimental and

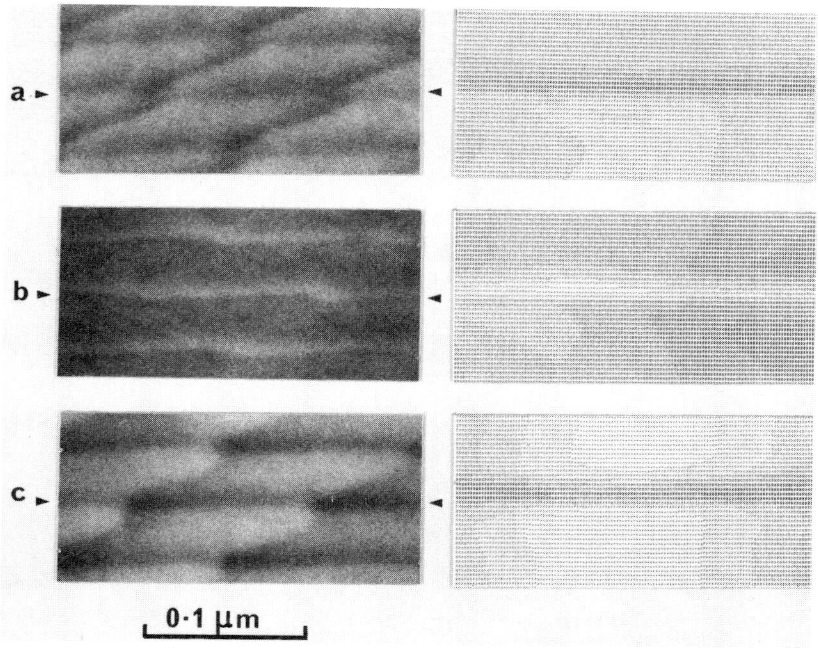

Figure 5.22 Three double two-beam images of a secondary grain boundary dislocation A with a matching set of computed images computed for the Burgers vector $\boldsymbol{A} = (1/18)[217]_L$. For ($a$) $\boldsymbol{B}$ is close to $[001]_R$ and $[1\bar{3}6]_L$; $\boldsymbol{g}_R = 020_R$, $\boldsymbol{g}_L = \bar{3}11_L$. For ($b$) $\boldsymbol{B}$ is close to $[\bar{1}34]_R$ and $[\bar{1}03]_L$; $\boldsymbol{g}_R = \bar{1}1\bar{1}_R$, $\boldsymbol{g}_L = 020_L$. For (c) $\boldsymbol{B}$ is close to $[015]_R$ and $[0\bar{1}5]_L$; $\boldsymbol{g}_R = 200_R$, $\boldsymbol{g}_L = 200_L$.

computed images is shown in figure 5.23. In this case, the line B (arrowed in figure 5.23) is intersected by several lines of A in the experimental images and again the intersections are not taken into account in the theoretical images. The only theoretical images which matched the experimental images for dislocation segments $A + B + C$ were those computed for $\boldsymbol{A} + \boldsymbol{B} + \boldsymbol{C} = (1/18)[\bar{5}25]_L$ and a matching set of experimental and computed images is shown in figure 5.24. The dislocation line $A + B + C$ is arrowed in figure 5.24 and the other dislocation lines present in the experimental images are the A and B components of the network. Despite the complications introduced by the intersections of crossing arrays of secondary grain boundary dislocations in the experimental images, each of the dislocation segments A, B and $A + B + C$ showed sufficiently distinctive contrast characteristics to enable a positive identification of the Burgers vectors to be made. In practice each dislocation image could be characterised as being either nominally in or out of contrast, as being either a dark or light line, or as being either light above and dark below or dark above and light below, and these characteristics proved adequate to decide between matching and non-matching images.

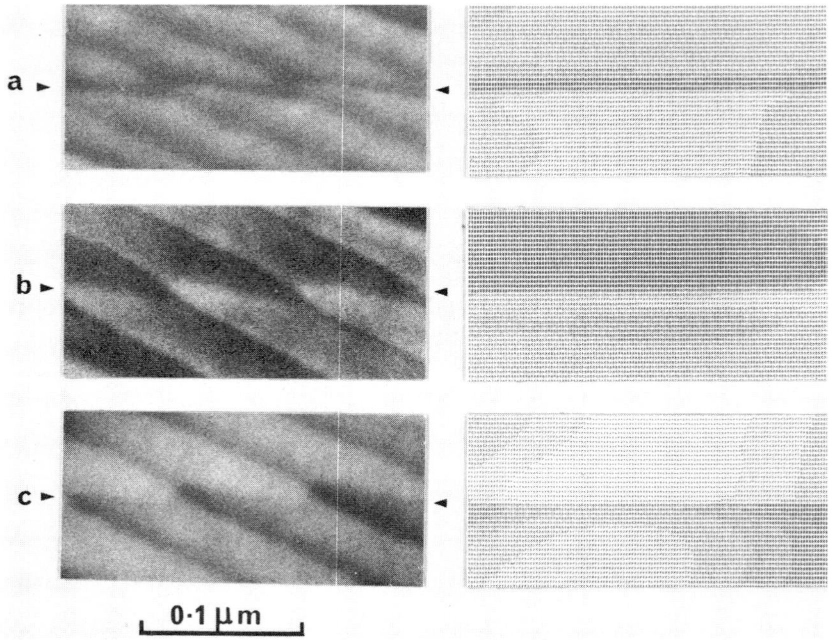

Figure 5.23 Three double two-beam images of a secondary grain boundary dislocation B with the matching set of computed images computed for the Burgers vector $\boldsymbol{B} = (1/9)[\bar{2}\bar{1}2]_L$. For ($a$) $\boldsymbol{B}$ is close to $[001]_R$ and $[1\bar{3}6]_L$; $\boldsymbol{g}_R = 020_R$, $\boldsymbol{g}_L = \bar{3}11_L$. For ($b$) $\boldsymbol{B}$ is close to $[165]_R$ and $[\bar{1}25]_L$; $\boldsymbol{g}_R = \bar{1}1\bar{1}_R$, $\boldsymbol{g}_L = \bar{1}\bar{3}1_L$. For ($c$) $\boldsymbol{B}$ is close to $[015]_R$ and $[0\bar{1}5]_L$; $\boldsymbol{g}_R = 200_R$, $\boldsymbol{g}_L = 200_L$.

In summary, for this near-Σ9 boundary the Burgers vectors of the secondary grain boundary dislocation structure have been determined independently by the geometric method and by the method of image matching with the same results being obtained by both procedures, namely $\boldsymbol{A} = (1/18)[217]_L$, $\boldsymbol{B} = (1/9)[\bar{2}12]_L$ and $\boldsymbol{A} + \boldsymbol{B} + \boldsymbol{C} = (1/18)[\bar{5}25]_L$. This agreement demonstrated the soundness of the geometric method for determining the DSC Burgers vectors of secondary grain boundary dislocations. Therefore this method is a valid way for determining the Burgers vectors of such dislocations in situations where their spacings are too close for the image matching procedure to be applied.

Near-Σ9 boundary in a Cu–Si alloy

Many near-Σ9 boundaries containing coarse arrays of secondary grain boundary dislocations have been observed by the authors in investigations of the structure of high-angle boundaries in a Cu – 6 at% Si alloy. Figure

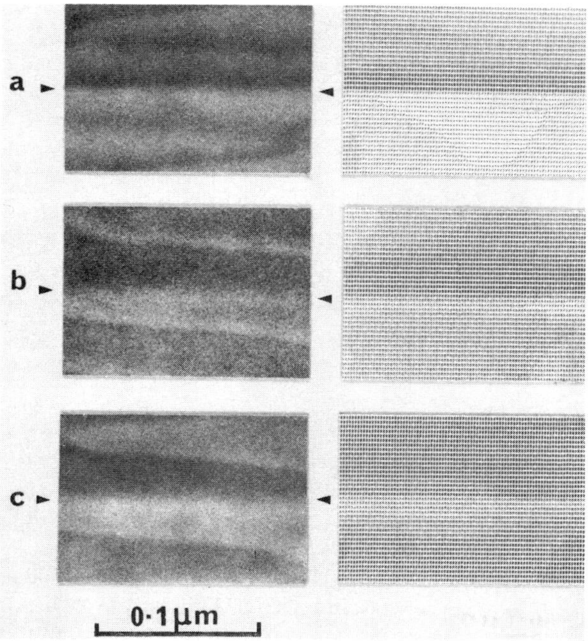

Figure 5.24 Three double two-beam images of a secondary grain boundary dislocation $A + B + C$ with the matching set of computed images computed for the Burgers vector $[\boldsymbol{A} + \boldsymbol{B} + \boldsymbol{C}] = (1/18)[\bar{5}25]_L$. For (*a*) $\boldsymbol{B}$ is close to $[001]_R$ and $[1\bar{3}6]_L$; $\boldsymbol{g}_R = 020_R$, $\boldsymbol{g}_L = \bar{3}11_L$. For (*b*) $\boldsymbol{B}$ is close to $[165]_R$ and $[\bar{1}25]_L$; $\boldsymbol{g}_R = \bar{1}1\bar{1}_R$, $\boldsymbol{g}_L = \bar{1}\bar{3}1_L$. For (*c*) $\boldsymbol{B}$ is close to $[\bar{1}34]_R$ and $[\bar{1}03]_L$; $\boldsymbol{g}_R = \bar{1}1\bar{1}_R$, $\boldsymbol{g}_L = 020_L$.

5.25(*a*) is a simultaneous double two-beam electron micrograph showing a portion of such a near-Σ9 boundary which terminates at the bottom right of the micrograph at a triple junction, where it joins a coherent-Σ3 boundary and a near-Σ27 boundary (Forwood and Clarebrough 1985a). The coarse secondary grain boundary dislocation network of figure 5.25(*a*) is illustrated schematically in figure 5.25(*b*) and it consists of four arrays of dislocations in which the individual dislocations are labelled *A*, *B*, *C* and *D*. The arrays of dislocations *A*, *B* and *D* are three independent arrays and the dislocations *A* and *B* have interacted to form dislocations *C* resulting in a hexagonal network of dislocations *A*, *B*, *C* which is crossed by dislocations *D*. Although the dislocations *D* change direction where they cross *A*, there is no indication in images of the formation of reaction products.

If the misorientation between the two grains is taken as an exact-Σ9 CSL orientation, then the appropriate Σ9 orientation is one which gives a right-handed rotation of grain 2 with respect to grain 1 by 38.94° around

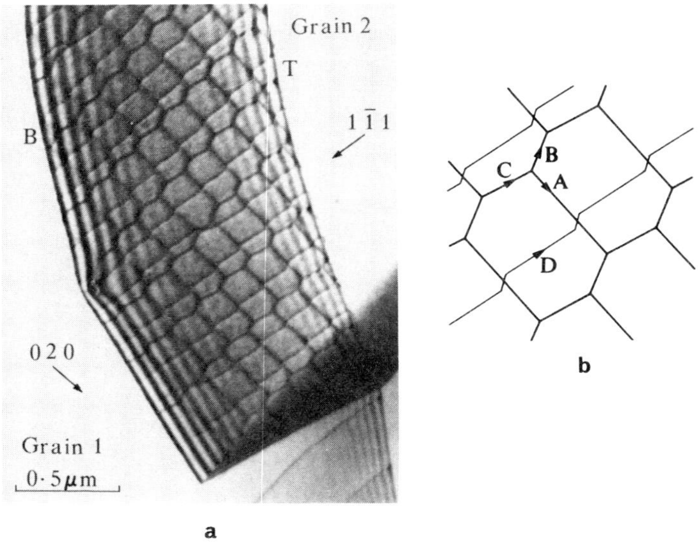

Figure 5.25 Double two-beam image (*a*) showing a secondary grain boundary dislocation network in a near-Σ9 boundary in a Cu − 6 at% Si alloy. The intersections of the boundary with the top and bottom of the specimen are marked T and B respectively, so that grain 1 is the upper grain. The diffracting vectors are indicated and $\boldsymbol{B}$ is close to $[\bar{1}05]_1$ and $[\bar{3}14]_2$. A schematic representation of the secondary grain boundary dislocation network is shown in (*b*) in which the arrows indicate the sense of the dislocation line directions from the bottom to the top of the specimen.

an axis parallel to [011] in both grains and is defined by the matrix

$$(1/9)\begin{pmatrix} 7 & -4 & 4 \\ 4 & 8 & 1 \\ -4 & 1 & 8 \end{pmatrix}$$

which re-indexes a vector indexed with respect to grain 2 into indices with respect to grain 1. The $\boldsymbol{g}_c$ matrix, indexed with respect to grain 1, is

$$\begin{pmatrix} 0 & 2 & 2 \\ 4 & 2 & 0 \\ 1 & -3 & 1 \end{pmatrix}_1$$

and its inverse gives the basis DSC vectors $\boldsymbol{x} = (1/18)[41\bar{1}]_1$, $\boldsymbol{y} = (1/9)[1\bar{2}2]_1$ and $\boldsymbol{z} = (1/18)[1\bar{2}\bar{7}]_1$.

Figure 5.26 shows matching theoretical and experimental images for dislocation segments *A*, *B* and *D* and the reaction product *C* of the network, where the theoretical images were computed for $\boldsymbol{b}_A = (1/18)[1\bar{2}2]_1$, $\boldsymbol{b}_B = (1/18)[12\bar{7}]_1$, $\boldsymbol{b}_C = (1/6)[1\bar{2}\bar{1}]_1$ and $\boldsymbol{b}_D = (1/18)[41\bar{1}]_1$. For dislocations *A* (figures 5.26(*a*) and (*b*)), *B* (figures 5.26(*c*) and (*d*)) and *D* (figures 5.26(*h*) and (*i*)), the theoretical images were computed for two segments at different depths in the specimen and in these figures the images of the dislocations are arranged to be horizontal. The experimental image figure 5.26(*e*) shows the dislocations *A* (centre left), *B* (lower) and *C* (upper) meeting at a node. The matching theoretical image for the reaction product *C* is shown in figure 5.26(*f*) and the matching theoretical image for segment *B*, which under these diffracting conditions is a double line of dark contrast, is shown in figure 5.26(*g*). The character of the experimental images in figures 5.26(*a*) and (*h*) is typical of that found for secondary grain boundary dislocations and involves a change from black to white contrast at the line of the dislocation. It was found, as in the previous example, that a great many possibilities for the Burgers vectors of secondary grain boundary dislocations could be eliminated simply on the basis of whether, in computed images, black and white contrast appeared on the appropriate sides of the dislocation. In some cases special features of contrast involving fine detail helped in the identification, and an example of this is shown in figures 5.26(*c*) and (*d*) where the contrast of dislocation *B* consists of a narrow black line (approximately 30 Å wide) bordered by two diffuse white bands.

Image matching of the type illustrated in figure 5.26 identified the Burgers vectors of the secondary grain boundary dislocations as $\boldsymbol{b}_A = \boldsymbol{y} = (1/9)[1\bar{2}2]_1, \boldsymbol{b}_B = \boldsymbol{z} = (1/18)[12\bar{7}]_1, \boldsymbol{b}_C = (\boldsymbol{y} + \boldsymbol{z}) = (1/6)[1\bar{2}\bar{1}]_1$ and $\boldsymbol{b}_D = \boldsymbol{x} = (1/18)[41\bar{1}]_1$. Thus the three independent arrays of secondary grain boundary dislocations *A*, *B* and *D* have basis Σ9 DSC Burgers vectors and dislocation *C* is the reaction product formed by the reaction

$$\boldsymbol{b}_C = \boldsymbol{b}_A + \boldsymbol{b}_B$$

that is $(1/6)[1\bar{2}\bar{1}]_1 = (1/9)[1\bar{2}2]_1 + (1/18)[12\bar{7}]_1$, which is an energy-lowering reaction on the simple $|\boldsymbol{b}|^2$ criterion.

Near-$\Sigma 27_a$ boundary in a Cu–Si alloy

Figure 5.27 shows a small portion of a near-$\Sigma 27_a$ boundary in a Cu – 6 at% Si alloy which extended over a length of approximately 15 μm in the thin-foil specimen (Forwood and Clarebrough 1985a). The coarse secondary grain boundary dislocation network, which extended over the full length of the boundary, accommodates the small departure of the misorientation between grains 1 and 2 from the exact-$\Sigma 27_a$ CSL orientation. The network contains four arrays of dislocations and individual dislocations in these arrays are

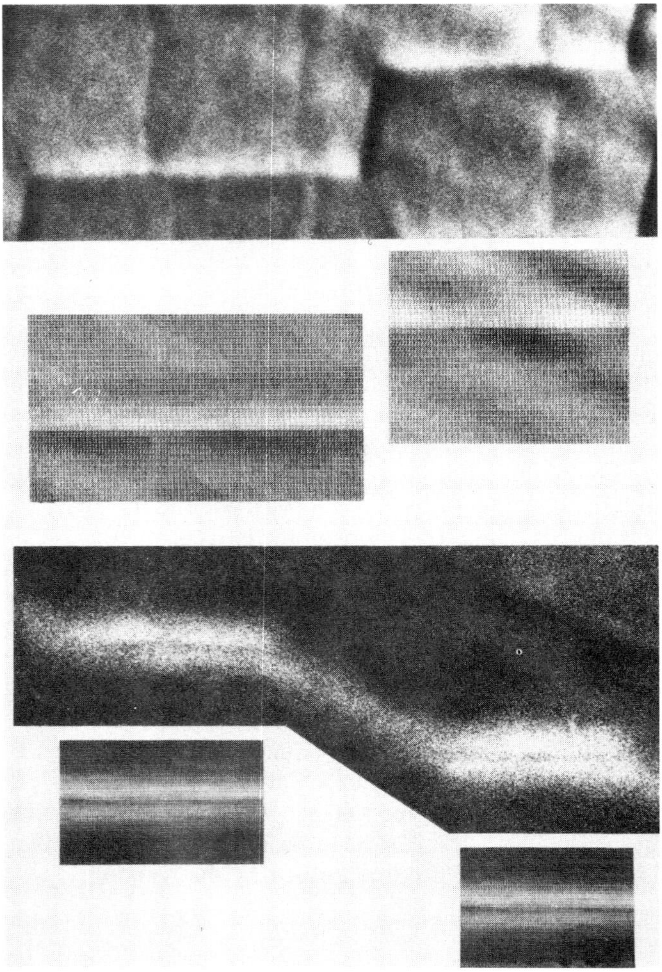

a

b

c

d

Figure 5.26 Experimental and matching computed images of secondary grain boundary dislocations (GBD) in the near-$\Sigma 9$ boundary of figure 5.25. The experimental images (*a*), (*c*), (*e*) and (*h*) have the $\boldsymbol{g}$, $\boldsymbol{B}$ and w values listed below. The matching computed images are for $\boldsymbol{b}_A = (1/9)[1\bar{2}2]_1$ in (*b*), for $\boldsymbol{b}_B = (1/18)[1\bar{2}\bar{7}]_1$ in (*d*), for $\boldsymbol{b}_C = (1/6)[1\bar{2}\bar{1}]_1$ in (*f*), for $\boldsymbol{b}_B = (1/18)[1\bar{2}\bar{7}]_1$ in (*g*) and for $\boldsymbol{b}_D = (1/18)[41\bar{1}]_1$ in (*i*).

	GBD	$\boldsymbol{B}$	$\boldsymbol{g}$	w
(*a*), (*b*)	*A*	$[\bar{1}05]_1, [\bar{3}14]_2$	$020_1, \bar{1}1\bar{1}_2$	$0.35_1, 0.29_2$
(*c*), (*d*)	*B*	$[105]_1, [\bar{1}03]_2$	$020_1, 020_2$	$0.20_1, 0.20_2$
(*e*), (*f*)	*C*	$[\bar{1}05]_1, [\bar{3}14]_2$	$020_1, \bar{1}1\bar{1}_2$	$0.35_1, 0.29_2$
(*e*), (*g*)	*B*			
(*h*), (*i*)	*D*	$[\bar{1}05]_1, [\bar{3}14]_2$	$020_1, 1\bar{1}\bar{1}_2$	$0.40_1, 0.40_2$

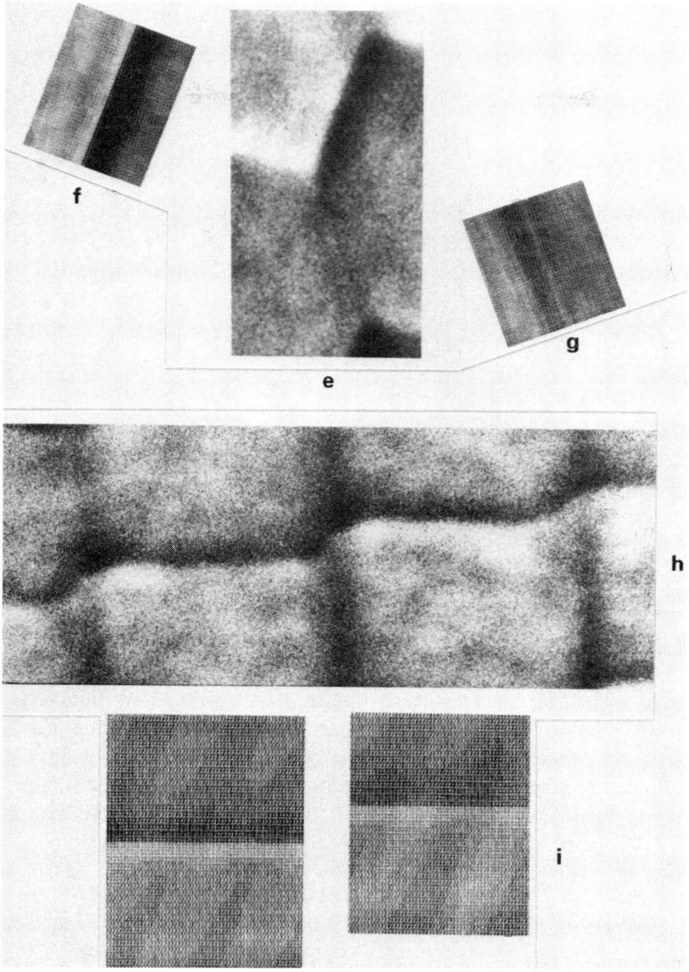

Figure 5.26 *continued.*

labelled A, B, C and D in the schematic diagram of figure 5.27(b). The dislocations A, B and C, with Burgers vectors $\boldsymbol{b}_A$, $\boldsymbol{b}_B$ and $\boldsymbol{b}_C$, form an hexagonal network in which C is the reaction product formed by the interaction of A and B. The dislocations D with Burgers vector $\boldsymbol{b}_D$ cross the elements of the hexagonal network without reacting. Thus the three independent arrays of dislocations in the network are composed of dislocations A, B and D.

If this near-$\Sigma 27_a$ boundary is approximated by an exact CSL orientation then the appropriate $\Sigma 27_a$ orientation is one which gives a right-handed rotation of grain 2 with respect to grain 1 by 31.58° around an axis parallel

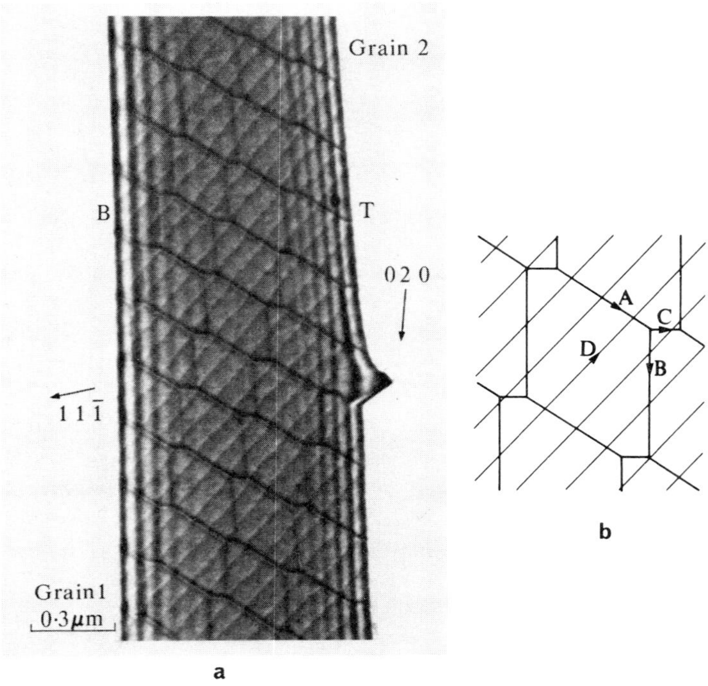

Figure 5.27 Double two-beam image (a) showing a secondary grain boundary dislocation network in a near-$\Sigma 27_a$ boundary in a Cu − 6 at% Si alloy. The intersections of the boundary with the top and bottom of the specimen are marked T and B respectively and grain 1 is the upper grain. The diffracting vectors are indicated and the beam directions are close to $[314]_1$ and $[103]_2$. A schematic representation of the secondary grain boundary dislocation network is shown in (b).

to [011] in both grains, and is defined by the matrix

$$(1/27)\begin{pmatrix} 23 & -10 & 10 \\ 10 & 25 & 2 \\ -10 & 2 & 25 \end{pmatrix}$$

which re-indexes a vector indexed with respect to grain 2 into indices with respect to grain 1. The same-$\boldsymbol{g}_c$ matrix indexed with respect to grain 1 is therefore

$$\begin{pmatrix} 0 & 2 & 2 \\ -5 & -1 & 1 \\ 2 & -6 & 4 \end{pmatrix}_1$$

which gives the basis DSC vectors $\boldsymbol{x} = (1/54)[1\ 11\ 16]_1$, $\boldsymbol{y} = (1/27)[\bar{5}\bar{1}1]_1$ and $\boldsymbol{z} = (1/54)[2\bar{5}5]_1$.

Figure 5.28 shows an example in which matching and mismatching theoretical images are compared with experimental images for a dislocation *A*. The comparison is made for one segment of a dislocation *A* lying between two dislocations *B*, and the segment of *A* is arranged to be horizontal in all the experimental and computed images. The experimental images involving three non-coplanar diffracting vectors are in column (i) and theoretical images computed for Burgers vectors of $(\boldsymbol{x} + \boldsymbol{y})$, $\boldsymbol{x}$, $\boldsymbol{y}$ and $\boldsymbol{z}$ are in columns (ii), (iii), (iv) and (v) respectively. The matching computed images in column (ii) identify the Burgers vector of dislocation *A* as $\boldsymbol{b}_A = (\boldsymbol{x} + \boldsymbol{y}) = (1/6)[\bar{1}21]_1$ and this agreement between computed and experimental images is shown in figure 5.29 for two segments of dislocation *A* at different depths in the specimen. These experimental and theoretical images, which cover a wide range of image character (black–white contrast in figure 5.29(*b*) and double images with different character in figures 5.29(*a*) and (*c*)), illustrate the degree of agreement obtained in the image matching technique which leads to the positive identification of Burgers vectors.

Examples of matching experimental and computed images which identified the Burgers vectors of dislocations *B* and *D* as $\boldsymbol{b}_B = \boldsymbol{y} = (1/27)[\bar{5}\bar{1}1]_1$ and $\boldsymbol{b}_D = \boldsymbol{z} = (1/54)[255]_1$ are shown in figure 5.30. The experimental and matching theoretical images for dislocation *B* in figures 5.30(*a*) and (*b*) illustrate a common feature of images of secondary grain boundary dislocations taken under simultaneous double two-beam conditions with different diffracting vectors in each grain, in that there is a reversal of contrast from black to white on reversing the sign of the diffracting vectors in both grains. The Burgers vector of the short segment *C* was also identified by image matching as $\boldsymbol{b}_C = \boldsymbol{x} = (1/54)[1\ 11\ 16]_1$ which is the reaction product formed by the energy-lowering reaction $\boldsymbol{b}_C = \boldsymbol{b}_A - \boldsymbol{b}_B$. In this boundary the dislocations *B*, *C* and *D* have basis DSC vectors, but the dislocations *A* which form a major part of the network have a Burgers vector which is the sum of two basis vectors.

Near-$\Sigma 27_b$ boundary in a Cu–Si alloy

Figure 5.31(*a*) is a simultaneous double two-beam image of the secondary grain boundary dislocation structure in a near-$\Sigma 27_b$ boundary in the same Cu–Si alloy (Forwood and Clarebrough 1985a). The network is composed of three independent arrays of dislocations and these are illustrated schematically in figure 5.31(*b*) where the dislocations are labelled *A*, *B* and *C*. In the network, dislocations *B* and *C* interact where they cross to form very short lengths of reaction product, whereas dislocation *A* does not form a reaction product with *C*.

The appropriate $\Sigma 27_b$ CSL which is a close approximation to the observed misorientation is one which gives a right-handed rotation of grain 2 with

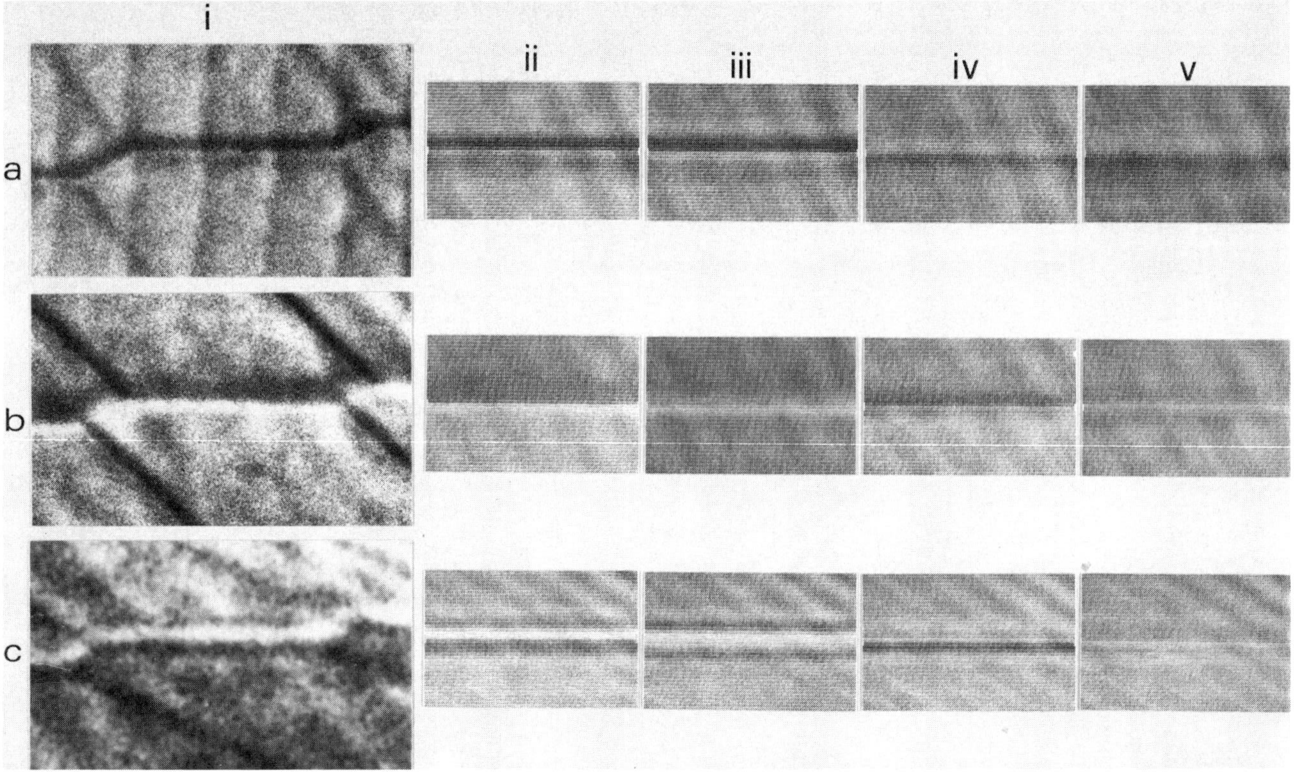

Figure 5.28 Double two-beam images (i) of a dislocation A in the network of figure 5.27. Images (ii)–(v) are computed for the Burgers vectors $(1/6)[\bar{1}21]_1$ (ii), $(1/54)[1\ 11\ 16]_1$ (iii), $(1/27)[\bar{5}\bar{1}1]_1$ (iv) and $(1/54)[2\bar{5}\bar{5}]_1$ (v). A comparison of the experimental and computed images shows that the only set of computed images which matches the set of experimental images is that in column (ii) for $\boldsymbol{b}_A = (1/6)[\bar{1}21]_1$. The values of $\boldsymbol{g}$, $\boldsymbol{B}$ and w are:

	$\boldsymbol{B}$	$\boldsymbol{g}$	w
(*a*)	$[314]_1, [103]_2$	$11\bar{1}_1, 020_2$	$0.30_1, 0.22_2$
(*b*)	$[409]_1, [0\ \bar{1}\ 11]_2$	$020_1, \bar{2}00_2$	$0.40_1, 0.02_2$

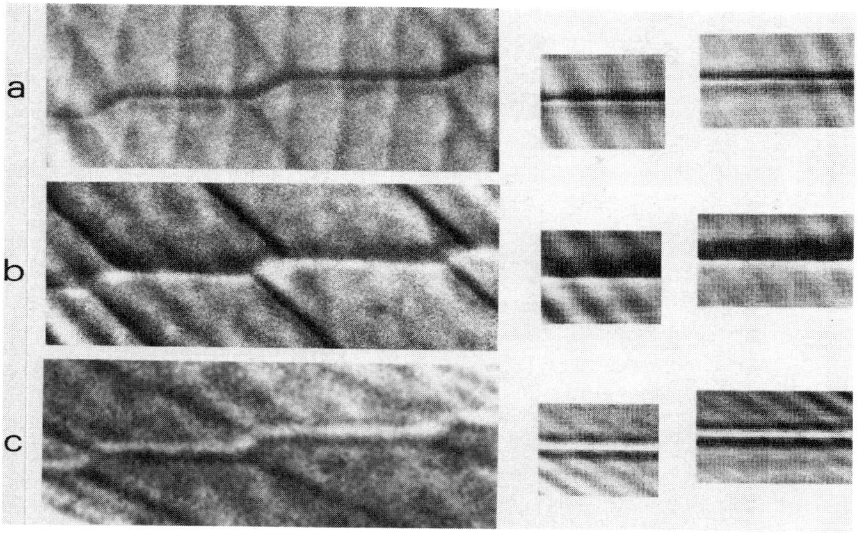

Figure 5.29 Comparison of experimental and matching computed images for two segments of a dislocation A in the network of figure 5.27 with $\boldsymbol{b}_A = (1/6)[\bar{1}21]_1$. The diffraction conditions are those of figure 5.28.

respect to grain 1 by 79.33° around an axis parallel to $[1\bar{3}1]$ in both grains, and is defined by the matrix

$$(1/27)\begin{pmatrix} 7 & -14 & -22 \\ 2 & 23 & -14 \\ 26 & 2 & 7 \end{pmatrix}$$

which re-indexes a vector indexed in grain 2 into indices in grain 1. The $\boldsymbol{g}_c$ matrix indexed with respect to grain 1 is therefore

$$\begin{pmatrix} 1 & -3 & 1 \\ 0 & 2 & 4 \\ 7 & 3 & 1 \end{pmatrix}_1$$

which gives the basis DSC vectors $\boldsymbol{x} = (1/54)[5\ \overline{14}\ 7]_1$, $\boldsymbol{y} = (1/18)[\bar{1}14]_1$ and $\boldsymbol{z} = (1/54)[\bar{7}\bar{2}1]_1$.

Figures 5.32(a)–(c) show matching experimental and theoretical images for dislocations A, B and C respectively, where the theoretical images are computed for $\boldsymbol{b}_A = (1/54)[1\ \overline{19}\ \bar{4}]_1$, $\boldsymbol{b}_B = (1/18)[\bar{1}14]_1$ and $\boldsymbol{b}_C = (1/54)[\bar{7}\bar{2}1]_1$. In all cases the relevant dislocation segment is arranged to be horizontal in the figure, and dislocations A and B are computed throughout the full thickness of the specimen, while dislocation C is computed

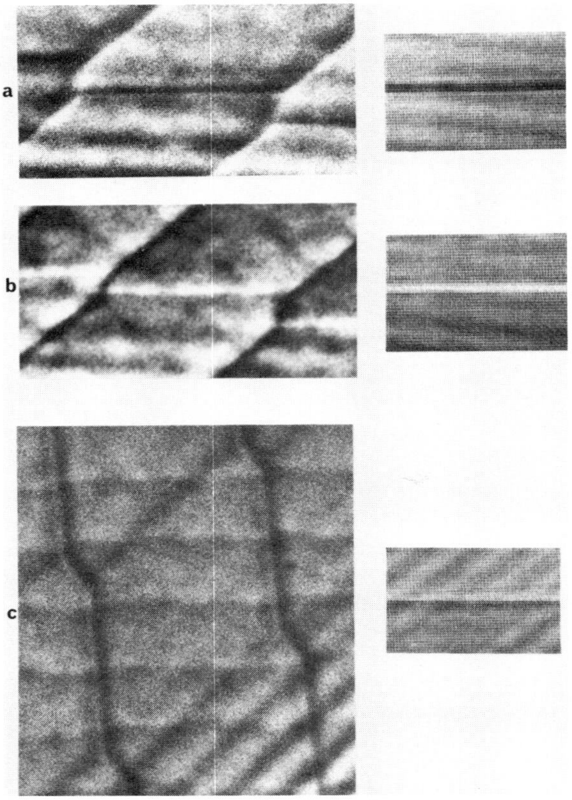

Figure 5.30 Double two-beam experimental and theoretical images of the secondary grain boundary dislocations (GBD) B (in (a) and (b)) and D (in (c)) of the network in figure 5.27. The diffraction parameters for the experimental images and the images computed for $\boldsymbol{b}_B = (1/57)[\bar{5}\bar{1}1]_1$ and $\boldsymbol{b}_D = (1/54)[2\bar{5}5]_1$ are:

	GBD	$\boldsymbol{B}$	$\boldsymbol{g}$	w
(a)	B	$[409]_1, [0\ \bar{1}\ 11]_2$	$020_1, \bar{2}00_2$	$0.40_1, 0.02_2$
(b)	B	$[409]_1, [0\ \bar{1}\ 11]_2$	$0\bar{2}0_1, 200_2$	$0.40_1, 0.02_2$
(c)	D	$[314]_1, [103]_2$	$11\bar{1}_1, 020_2$	$0.30_1, 0.22_2$

between nodal points in the dislocation network. The full image matching procedure identified the Burgers vectors of the dislocations A, B and C as $\boldsymbol{b}_A = (\boldsymbol{x} - \boldsymbol{y} + \boldsymbol{z}) = (1/54)[1\ \overline{19}\ \bar{4}]_1$, $\boldsymbol{b}_B = \boldsymbol{y} = (1/18)[\bar{1}14]_1$ and $\boldsymbol{b}_C = \boldsymbol{z} = (1/54)[\bar{7}\bar{2}1]_1$. The very short lengths of dislocation in the network, formed by the interaction of B and C, were not sufficiently extensive to be reliably identified by image matching, but their Burgers vector would correspond to $\boldsymbol{b}_B - \boldsymbol{b}_C = (1/54)[4\ \ 5\ \ 11]_1$.

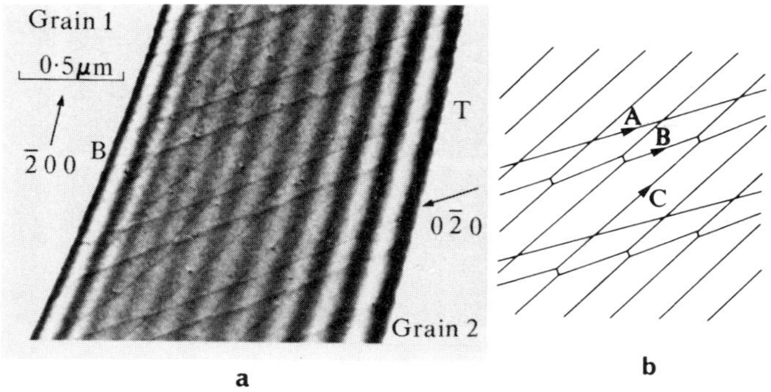

Figure 5.31 Double two-beam image (*a*) showing a secondary grain boundary dislocation network in a near-$\Sigma 27_b$ boundary in a Cu − 6 at% Si alloy. The intersections of the boundary with the top and bottom of the specimen are marked T and B respectively and grain 1 is the upper grain. The diffracting vectors are indicated and the beam directions are close to $[0\ \bar{1}\ 12]_1$ and $[301]_2$. A schematic representation of the secondary grain boundary dislocation network is shown in (*b*).

Near-$\Sigma 81_a$ boundary in a Cu–Si alloy

High-angle grain boundaries with near-Σ81 CSL orientations are frequently observed at triple junctions with Σ27 and Σ3 boundaries. Figure 5.33(*a*) shows an example of a near-$\Sigma 81_a$ boundary in the Cu–Si alloy where the secondary grain boundary network contains three independent arrays of dislocations *A*, *B* and *C* as illustrated schematically in figure 5.33(*b*) (Forwood and Clarebrough 1985a).

The exact CSL orientation corresponding to this near-$\Sigma 81_a$ boundary involves a right-handed rotation of grain 2 with respect to grain 1 by 38.38° around an axis parallel to $[\bar{5}\bar{1}3]$ in both grains and is defined by the matrix

$$(1/81)\begin{pmatrix} 76 & -23 & -16 \\ 28 & 64 & 41 \\ 1 & -44 & 68 \end{pmatrix}$$

which re-indexes a vector indexed in grain 2 into indices in grain 1. The $\boldsymbol{g}_c$ matrix indexed with respect to grain 1 is therefore

$$\begin{pmatrix} -5 & -1 & 3 \\ 2 & 6 & 4 \\ -5 & 5 & -3 \end{pmatrix}_1$$

which gives the basis DSC vectors $\boldsymbol{x} = (1/162)[11\ \overline{13}\ 14]_1$, $\boldsymbol{y} = (1/162)[19\ 7\ \overline{20}]_1$ and $\boldsymbol{z} = (1/54)[\bar{2}\bar{5}\bar{5}]_1$.

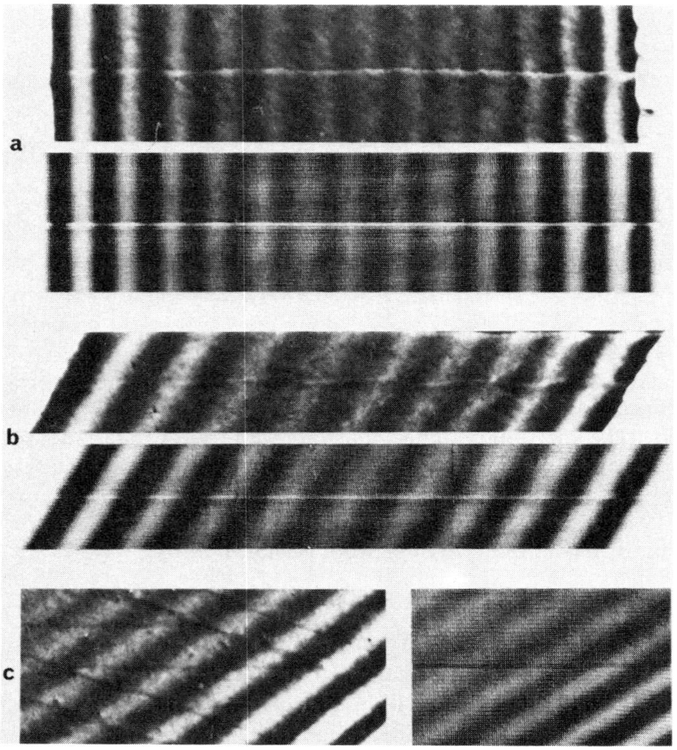

Figure 5.32 Matching experimental and computed images for the secondary grain boundary dislocations *A* in (*a*), *B* in (*b*) and *C* in (*c*) for the network of figure 5.31 with $\boldsymbol{b}_A = (1/54)[1\ \overline{19}\ \bar{4}]_1$, $\boldsymbol{b}_B = (1/18)[\bar{1}14]_1$ and $\boldsymbol{b}_C = (1/54)[\bar{7}\bar{2}1]_1$ in the computed images. The diffraction parameters are:

	GBD	$\boldsymbol{B}$	$\boldsymbol{g}$	w
(*a*)	*A*	$[\bar{9}\bar{1}8]_1, [112]_2$	$1\bar{1}1_1, \bar{1}\bar{1}1_2$	$0.20_1, 0.12_2$
(*b*)	*B*	$[0\ \bar{1}\ 12]_1, [301]_2$	$200_1, 0\bar{2}0_2$	$0.30_1, 0.10_2$
(*c*)	*C*	$[0\ \bar{1}\ 12]_1, [301]_2$	$\bar{2}00_1, 0\bar{2}0_2$	$0.02_1, 0.44_2$

For dislocation *A*, figures 5.34(*a*) and (*b*) show experimental images and matching theoretical images for $\boldsymbol{b}_A = (1/162)[11\ \overline{13}\ 14]_1$. The experimental image in figure 5.34(*b*) proved to be a particularly definitive image for eliminating a large number of possible Burgers vectors for dislocation *A*. It can be seen that in this experimental image the contrast of dislocation *A* is very weak, with weak dark intensity below the line of the dislocation, whereas many of the possible Burgers vectors gave images with stronger contrast with strong light intensity above the line of the dislocation, as shown by the

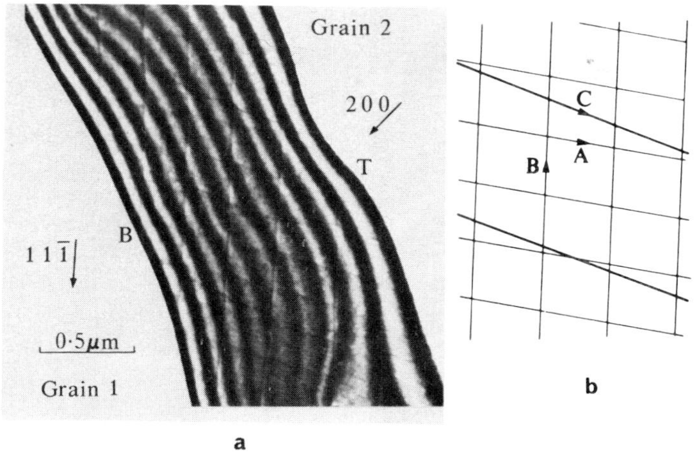

Figure 5.33 Double two-beam image (a) showing a secondary grain boundary dislocation network in a near-$\Sigma 81_a$ boundary in a Cu − 6 at% Si alloy. The intersections of the boundary with the top and bottom of the specimen are marked T and B respectively and grain 1 is the upper grain. The diffracting vectors are indicated and the beam directions are close to $[\bar{3}52]_1$ and $[\bar{1}45]_2$. A schematic representation of the secondary grain boundary dislocation network is shown in (b).

examples of the non-matching theoretical images in figures 5.34(c)–(e). Figures 5.35(a) and (b) show experimental and matching theoretical images for dislocation segments B and C respectively, with $\boldsymbol{b}_B = (1/162)[19\ 7\ \overline{20}]_1$ and $\boldsymbol{b}_C = (1/54)[\bar{2}\bar{5}\bar{5}]_1$. In the experimental image shown in figure 5.36(a), the dislocation segment C, which is arrowed, shows weak contrast throughout the thickness of the specimen and is another example of an image which enabled many of the possible Burgers vectors to be eliminated in the identification procedure. For example, the theoretical image in figure 5.36(b), corresponding to $\boldsymbol{b}_C = (1/54)[\bar{2}\bar{5}\bar{5}]_1$, is a good match in that it shows the type of weak contrast observed at the line of the dislocation in the experimental image, with this contrast virtually disappearing near the bottom surface of the specimen (left hand side of the image). However, the other two theoretical images (c) and (d), computed for the Burgers vectors $(1/162)[\overline{17}\ \bar{2}\ \overline{29}]_1$ and $(1/162)[\overline{23}\ \overline{17}\ \overline{44}]_1$, which had given theoretical images that matched experimental images for other diffracting conditions, are definite mismatches in this case as they show strong contrast throughout the entire thickness of the specimen, and therefore can be eliminated. In summary the image matching technique identified the Burgers vectors of the secondary grain boundary dislocation network in this near-$\Sigma 81_a$ boundary as the basis DSC vectors $\boldsymbol{b}_A = \boldsymbol{x} = (1/162)[11\ \overline{13}\ 14]_1$, $\boldsymbol{b}_B = \boldsymbol{y} = (1/162)[19\ 7\ \overline{20}]_1$ and $\boldsymbol{b}_C = \boldsymbol{z} = (1/54)[\bar{2}\bar{5}\bar{5}]_1$.

Figure 5.34 Matching experimental and computed images (*a*) and (*b*) for dislocation *A* of figure 5.33 for $\boldsymbol{b}_A = (1/162)[11\ 13\ 14]_1$. The diffraction parameters are:

	$\boldsymbol{B}$	$\boldsymbol{g}$	w
(*a*)	$[\bar{1}\ 11\ 11]_1, [115]_2$	$02\bar{2}_1, \bar{2}20_2$	$0.14_1, 0.0_2$
(*b*)	$[\bar{3}52]_1, [\bar{1}34]_2$	$1\bar{1}1_1, \bar{1}1\bar{1}_2$	$0.36_1, 0.21_2$

The additional computed images (*c*)–(*e*) were computed for the diffraction parameters of (*b*) with $\boldsymbol{b}_A = (1/162)[17\ 2\ 29]_1$ in (*c*); $\boldsymbol{b}_A = (1/162)[23\ 17\ 44]_1$ in (*d*) and $\boldsymbol{b}_A = (1/162)[28\ \overline{11}\ 43]_1$ in (*e*).

In conclusion, the studies that have been described on the structure of high-angle grain boundaries in polycrystals with misorientations close to CSL orientations show the following:

(i) The CSL model for the structure of high-angle grain boundaries applies to general boundaries in polycrystalline metals and alloys.

(ii) In general three independent arrays of secondary grain boundary dislocations accommodate the departure from a near-CSL orientation with additional arrays arising from interaction between these independent arrays.

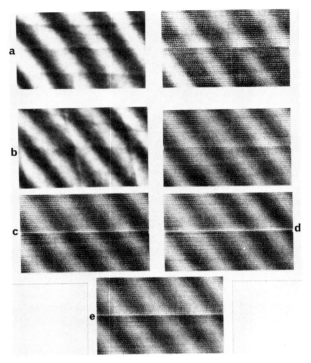

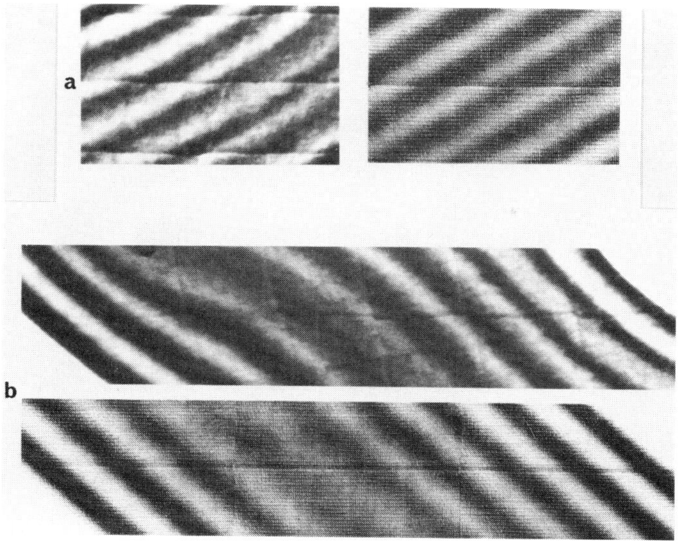

Figure 5.35 Matching experimental and computed images for the secondary grain boundary dislocations B in (*a*) and C in (*b*) for the network of figure 5.33. The matching computed images are for $\boldsymbol{b}_B = (1/162)[19\ 7\ \overline{20}]_1$ and $\boldsymbol{b}_C = (1/54)[\overline{2}5\overline{5}]_1$. The diffraction parameters are:

	$\boldsymbol{B}$	$\boldsymbol{g}$	w
(*a*)	$[\overline{1}32]_1, [012]_2$	$11\overline{1}_1, 200_2$	$0.13_1, 0.10_2$
(*b*)	$[\overline{2}57]_1, [0\ 1\ 12]_2$	$\overline{1}1\overline{1}_1, \overline{2}00_2$	$0.23_1, 0.23_2$

(iii) The geometric analysis of grain boundary structure, which uses changes in direction and spacing of secondary grain boundary dislocations with boundary plane, is a valid method for determining the Burgers vectors of the secondary grain boundary dislocations.

(iv) For different types of near-CSL boundary the Burgers vectors for most of the secondary grain boundary dislocations are the basis vectors of the appropriate DSC lattice. Ignoring the energy of steps in a boundary associated with secondary grain boundary dislocations, the fact that the Burgers vectors are predominantly basis vectors is to be expected, because such vectors minimise the overall self-energy of the arrays of secondary grain boundary dislocations required in a boundary to accommodate the departure of the misorientation from a particular CSL orientation. However, this self-energy argument cannot be the sole factor determining the values of the DSC Burgers vectors because, on occasions, Burgers vectors are found for segments of grain boundary dislocation networks which are not basis DSC vectors, but are simple

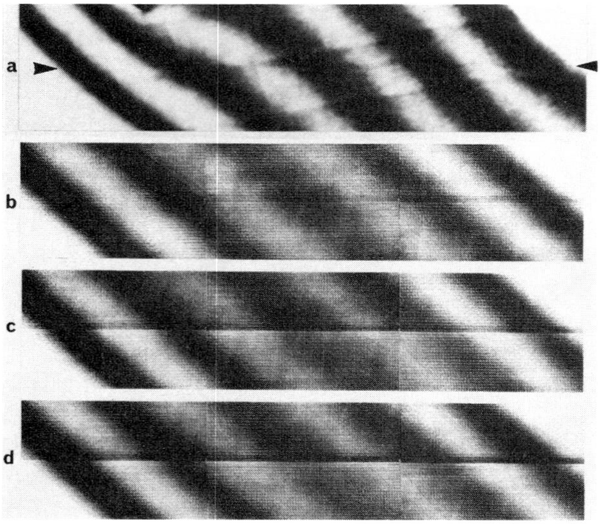

Figure 5.36 Experimental image (*a*) and matching computed image (*b*) for a grain boundary dislocation *C* of figure 5.33 for $\boldsymbol{b}_C = (1/54)[\bar{2}\bar{5}\bar{5}]_1$. The diffraction parameters are:

$\boldsymbol{B}$	$\boldsymbol{g}$	w
$[\bar{1}\ 11\ 11]_1, [115]_2$	$02\bar{2}_1, \bar{2}20_2$	$0.45_1, 0.10_2$

The additional computed images (*c*) and (*d*) were computed for the same diffraction parameters with $\boldsymbol{b}_C = (1/162)[\overline{17}\ \bar{2}\ \overline{29}]_1$ in (*c*) and $\boldsymbol{b}_C = (1/162)[\overline{23}\ \overline{17}\ \overline{44}]_1$ in (*d*).

linear combinations of basis vectors, e.g. dislocations *A* in the near-$\Sigma 27_a$ and near-$\Sigma 27_b$ boundaries which have Burgers vectors $(1/6)[\bar{1}12]_1$ and $(1/54)[1\ \overline{19}\ \bar{4}]_1$ respectively.

(v) With increasing values of Σ, and consequent decreasing magnitudes of the corresponding basis DSC Burgers vectors, there is a decrease in the intensity of the diffraction contrast associated with secondary grain boundary dislocations in simultaneous double two-beam images. However, this decrease in intensity of diffraction contrast does not impair the identification of the Burgers vectors by image matching for Σ values up to $\Sigma 81$, but identification of smaller basis vectors, associated with appreciably higher Σ values, could present difficulties for the image matching technique.

(vi) In some secondary grain boundary dislocation networks, dislocations cross one another without forming reaction products so as to give fourfold nodes. This occurs because in some cases reactions would be energetically neutral on a $|\boldsymbol{b}|^2$ criterion, for example a reaction between dislocations *A* and *D* for the near-$\Sigma 9$ boundary and between dislocations B and D for the near-$\Sigma 27_a$ boundary in the Cu–Si alloy. However, in other cases reactions do not appear to occur, possibly because of the large difference in magnitudes of the Burgers vectors, and examples are dislocations A and D in the near-$\Sigma 27_a$ boundary, and A and C in the near-$\Sigma 27_b$ boundary.

5.3.2 (ii) Grain boundaries with misorientations far from CSL *orientations*

Under this heading results on the structure of high-angle grain boundaries with misorientations that depart from CSL orientations by more than 1° will be discussed. For any near-Σ boundary the spacing of the secondary grain boundary dislocations decreases as φ increases, and for values of $\varphi \gtrsim 1°$ the spacing of the dislocations, in one or more of the three independent arrays, will be too small for their Burgers vectors to be identified by $\boldsymbol{g} \cdot \boldsymbol{b}$ criteria or image matching. For example, for the near-$\Sigma 3$ boundary in figure 5.7, if the departure from the exact-$\Sigma 3$ orientation had been as large as 1°, then the spacing of the hexagonal dislocation network would have been ~85 Å and $\boldsymbol{g} \cdot \boldsymbol{b}$ criteria and image matching could not have been used. For a given angular departure φ the spacings of the secondary grain boundary dislocations will decrease with increase in Σ, because the magnitudes of the Burgers vectors of the secondary grain boundary dislocations decrease with increase in Σ and the spacings of the dislocations are directly proportional to the magnitudes of the Burgers vectors. In polycrystalline metals and alloys high-angle grain boundaries frequently occur with high values of Σ and/or relatively large values of φ, so that the need to determine the Burgers vectors of closely spaced secondary grain boundary dislocations is a common one, and problems associated with such determinations will be illustrated by the following examples.

Near-$\Sigma 57_a$ boundary in a Cu–Si alloy

In this example of the analysis of the structure of a high-angle boundary in a Cu – 8 at% Si alloy (Forwood and Clarebrough 1977), the periodic structure in images associated with secondary grain boundary dislocations was as fine as 80 Å so that image matching could not be used to determine Burgers vectors. However, variations in periodicities present in diffraction patterns, associated with changes in spacings and directions of secondary grain boundary dislocations with change in boundary plane, enabled the geometric method to be used in this case to determine the Burgers vectors of the dislocations involved. Figure 5.37(a) is a low-magnification electron micrograph of the near-$\Sigma 57_a$ boundary between grains 1 and 2 that extends over 40 μm and intersects the top and bottom surfaces of the specimen as indicated by T and B, so that grain 1 is the upper grain. The misorientation between the two grains is a right-handed rotation of grain 2 with respect to grain 1 by the angle $\theta = 133.8°$ about an axis $\boldsymbol{u} = [\cos 125.1° \cos 126.0°$

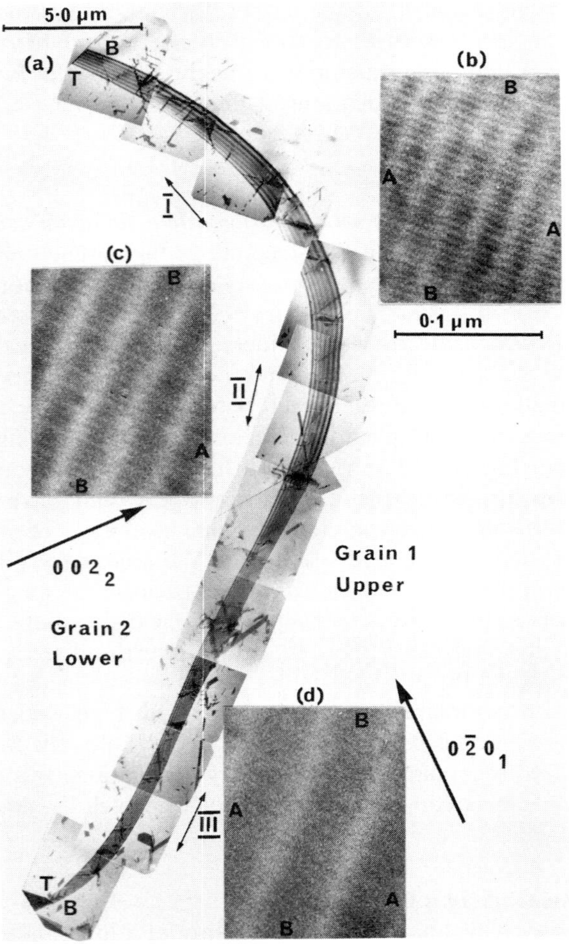

Figure 5.37 Double two-beam images of a curved near-$\Sigma 57_a$ grain boundary in a Cu – 8 at% Si alloy. In the low-magnification image (*a*) the top and bottom of the specimen are indicated by T and B respectively. $\boldsymbol{B}$ is close to $[\bar{1}06]_1$ and $[\bar{1}\bar{8}0]_2$ and the diffracting vectors are indicated. High-magnification images (*b*), (*c*) and (*d*) corresponding to positions I, II and III of (*a*) respectively are shown.

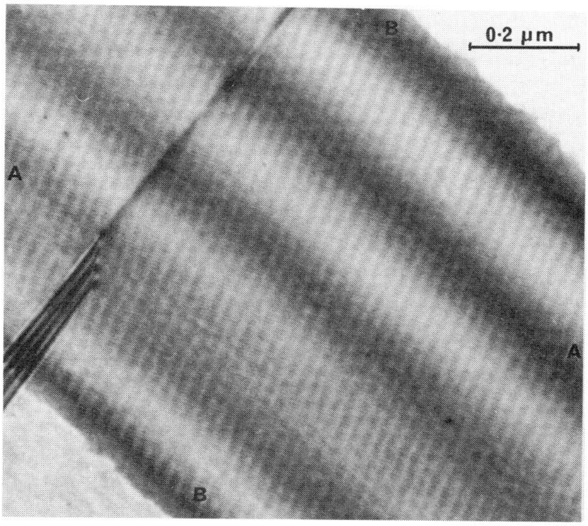

Figure 5.37 (*e*) Image from position I at an intermediate magnification.

cos 55.3°] common to both grains to an accuracy of $\pm 0.2°$†, and the corresponding matrix which re-indexes a vector in grain 2 into its indices in grain 1 is

$$(\boldsymbol{u}, \theta) = \begin{pmatrix} -0.132\,70 & 0.160\,74 & -0.978\,04 \\ 0.982\,74 & -0.106\,99 & -0.150\,93 \\ -0.128\,90 & -0.981\,18 & -0.143\,76 \end{pmatrix}.$$

The closest CSL orientation to this measured misorientation is the $\Sigma 57_a$ orientation which corresponds to a right-handed rotation of grain 2 with respect to a crystal lattice at the exact-$\Sigma 57_a$ orientation by the angle $\omega = 133.17°$ around an axis $\boldsymbol{p}$ of $[\bar{1}\bar{1}1]$ common to grain 2 and the CSL, so that the matrix which re-indexes a vector in grain 2 into its indices in a crystal lattice at the exact-$\Sigma 57_a$ CSL orientation is

$$(\boldsymbol{p}, \omega) = (1/57)\begin{pmatrix} -7 & 8 & -56 \\ 56 & -7 & -8 \\ -8 & -56 & -7 \end{pmatrix}.$$

† The low accuracy in $(\boldsymbol{u}, \theta)$ in this early work is due to the use of the first method for determining orientation relationships described in section 3.4.1 and illustrated in figure 3.8; a higher accuracy would have resulted had the averaging method of Mackenzie (1957) been used.

From equation (5.1), the small angular departure of the measured misorientation ($\boldsymbol{u}$, θ) from the $\Sigma 57_a$ orientation ($\boldsymbol{p}$, ω) corresponds to a rotation $\varphi = -1.6° \pm 0.3°$ of grain 1 in a right-handed sense with respect to a crystal lattice in the CSL orientation around an axis $\boldsymbol{q} = [\cos 58.5° \cos 135.4° \cos 62.0°]$ common to grain 1 and the CSL. The error limits for this rotation axis are large because the errors, which are already large in $\boldsymbol{u}$ and θ, are magnified in $\boldsymbol{q}$, and for $\boldsymbol{q}$ the errors correspond to a solid angle with semicone angle of 15°. The relationship between the misorientations ($\boldsymbol{u}$, θ), ($\boldsymbol{q}$, φ) and the $\Sigma 57_a$ CSL orientation is illustrated schematically in figure 5.38.

From the $\Sigma 57_a$ CSL matrix the three smallest non-coplanar same-$\boldsymbol{g}_c$ vectors indexed with respect to grain 1 are $\bar{1}\bar{1}1_1$, $10\ \bar{4}\ 6_1$ and $\bar{6}\ 10\ 4_1$ giving the $\boldsymbol{g}_c$ matrix and its inverse as

$$\begin{pmatrix} -1 & -1 & 1 \\ 10 & -4 & 6 \\ -6 & 10 & 4 \end{pmatrix}_1 \quad \text{and} \; (1/114)\begin{pmatrix} -38 & 7 & -1 \\ -38 & 1 & 8 \\ 38 & 8 & 7 \end{pmatrix}_1$$

respectively. so that three basis DSC vectors indexed with respect to grain 1 are $\boldsymbol{x} = (1/3)[\bar{1}\bar{1}1]_1$, $\boldsymbol{y} = (1/114)[718]_1$ and $\boldsymbol{z} = (1/114)[\bar{1}87]_1$.

The boundary in figure 5.37(a) has a marked change in boundary plane along its length and the periodic structure present in the boundary can be seen in figures 5.37 (b), (c), (d) and (e), where (b), (c) and (d) are electron micrographs of the regions marked I, II and III at high magnification and (e) is a micrograph of region I at an intermediate magnification. The higher magnification micrographs show two fringe systems, a coarse system marked BB and a fine system marked AA. The spacing and direction of both sets of fringes change with position on the boundary and the changes in spacing are more marked for the coarse fringes, whereas the changes in direction are more marked for the fine. It will be seen that neither of these fringe systems are direct images of secondary grain boundary dislocations and that their significance can be understood from an analysis of the periodic rows of diffraction spots that occur in diffraction patterns from the boundary. These fringe patterns are in fact similar to those observed in the low-angle boundary

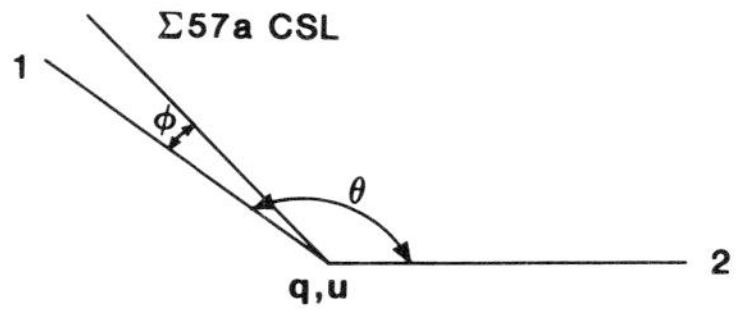

Figure 5.38 Schematic illustration of the relationships between the $\Sigma 57_a$ CSL orientation, ($\boldsymbol{u}$, θ) and ($\boldsymbol{q}$, φ) for the boundary of figure 5.37 where $\boldsymbol{q}$ and $\boldsymbol{u}$ are directed into the page.

of figure 4.11, and the geometric analysis of the secondary grain boundary dislocations, structure from periodicities in the diffraction patterns from this high-angle boundary is similar to that already described for low-angle boundaries in Chapter 4.

Figures 5.39 (*a*) and (*b*) show periodicities in diffraction patterns associated with boundary structure taken from the boundary at position I for two different beam directions. The patterns show, in addition to diffraction spots from both grains, periodic rows of diffraction spots in two directions which make a small angle with one another, and these directions are indicated by the lines *l* and *s*. The periodic rows from the regions marked × in figures 5.39(*a*) and (*b*) are shown at higher magnification in the inserts. Measurements of the spacings and the directions of the periodicities *l* and *s* in diffraction patterns and of the spacings and directions of the fringe systems AA and BB in the corresponding images show that the fine fringe system AA arises from the mean of the two periodicities *l* and *s*, and that the coarse fringe system BB arises from the moiré difference between *l* and *s*. This correlation between diffraction patterns and images shows that the periodic lines of contrast in the images do not result from direct imaging of individual dislocations, but arise from diffraction interference contrast which has its origin in the periodicities associated with the secondary grain boundary dislocations.

The stereographic projection in figure 5.40, with indices referred to grain 1, illustrates, for position I on the boundary, the geometric analysis of diffraction patterns and images. The points marked × are the lines of intersection of the $\{111\}$ planes in the two grains with the boundary plane and define the boundary plane normal $\boldsymbol{v}_{\mathrm{I}}$. The error bars marked *l* and *s* represent the directions of the periodic rows of diffraction spots in the diffraction patterns for four different beam directions, and these different directions for *l* and *s* lie on separate great circles containing $\boldsymbol{v}_{\mathrm{I}}$ which intersect the boundary plane at $\boldsymbol{L}_{\mathrm{I}}$ and $\boldsymbol{S}_{\mathrm{I}}$. This construction indicates that the directions of the rows of periodic diffraction spots *l* and *s* in each diffraction pattern are the projections of fixed direction $\boldsymbol{L}_{\mathrm{I}}$ and $\boldsymbol{S}_{\mathrm{I}}$ in the boundary along the boundary normal $\boldsymbol{v}_{\mathrm{I}}$ on to the plane of each diffraction pattern. Further, when the spacings of the periodicities *l* and *s* in the four different beam directions are projected on to the boundary plane $\boldsymbol{v}_{\mathrm{I}}$, constant values are obtained for the spacings of $\boldsymbol{L}_{\mathrm{I}}$ and $\boldsymbol{S}_{\mathrm{I}}$. In other words, the rows of diffraction spots arise from intersections of the Ewald sphere with rods in reciprocal space parallel to the boundary plane normal and, as discussed in section 2.6, correspond in real space to two periodic arrays of lines in the boundary. For the images to be compatible with this analysis of the diffraction patterns, the normals to the fine and coarse fringe directions, in images taken in different beam directions, must define a great circle through $\boldsymbol{v}_{\mathrm{I}}$ for the fine fringes and a different great circle through $\boldsymbol{v}_{\mathrm{I}}$ for the coarse fringes. It can be seen from figure 5.40 that the normals to the fine fringes, plotted as the

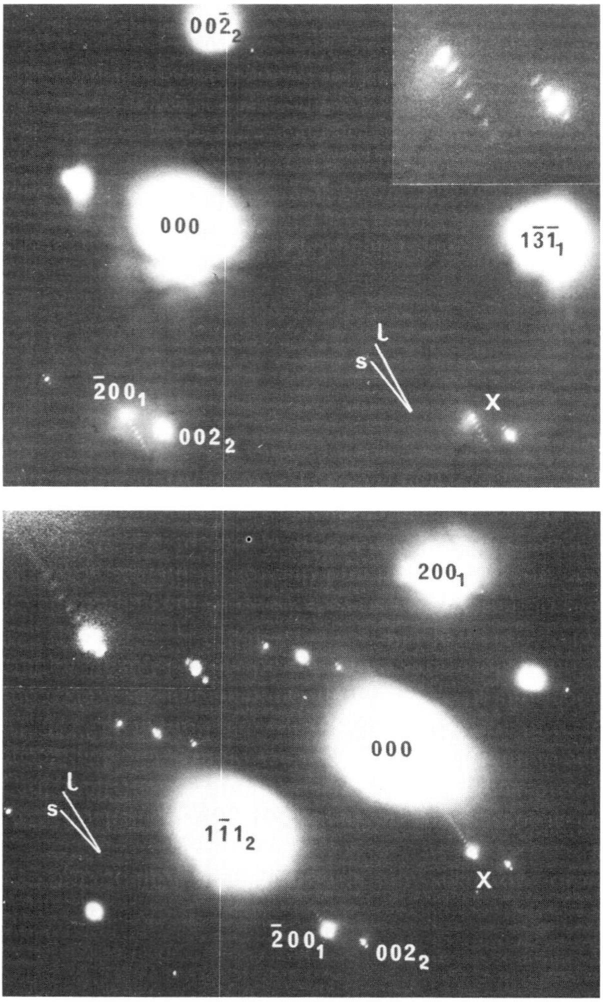

Figure 5.39 Diffraction patterns from the near-$\Sigma 57_a$ boundary at position I of figure 5.37(a) for two different beam directions. For (a) $\boldsymbol{B}$ is $[0\bar{1}3]_1$, 4.3° to $[\bar{1}\bar{1}2]_1$; $[\bar{1}\bar{2}0]_2$, 0.8° to $[\bar{1}\bar{1}0]_2$. For (b) $\boldsymbol{B}$ is $[0\bar{1}1]_1$, 9.3° to $[001]_1$; $[\bar{1}\bar{1}0]_2$, 3.5° to $[\bar{1}2\bar{1}]_2$. The insets are enlargements of the regions marked X.

symbols ▲, and the normals to the coarse fringes, plotted as the symbols ■, do lie on great circles containing $\boldsymbol{v}_{\mathrm{I}}$ and intersect the plane of the boundary at $\boldsymbol{F}_{\mathrm{I}}$ and $\boldsymbol{C}_{\mathrm{I}}$ respectively.

For position II on the boundary, the periodicities $\boldsymbol{L}_{\mathrm{II}}$ and $\boldsymbol{S}_{\mathrm{II}}$ obtained from diffraction patterns and the directions $\boldsymbol{F}_{\mathrm{II}}$ and $\boldsymbol{C}_{\mathrm{II}}$ obtained from images

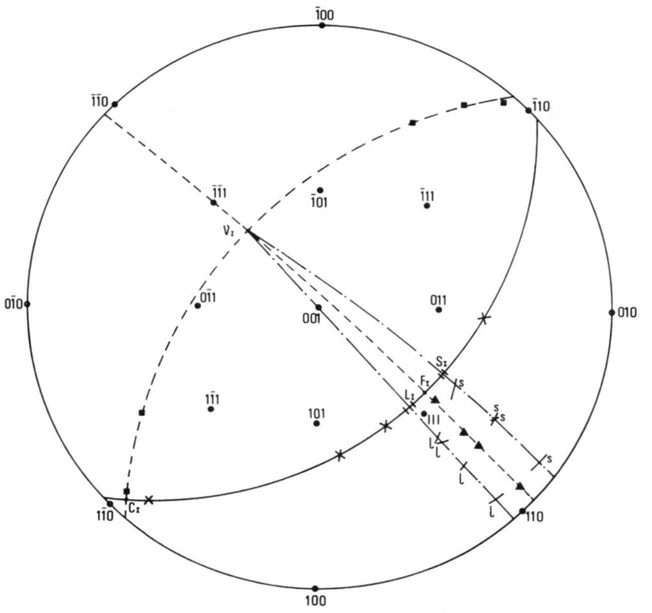

Figure 5.40 Stereographic projection indexed with respect to grain 1 showing analysis of rows of diffraction spots in diffraction patterns and the normals to the coarse and fine fringes in images for position I on the boundary of figure 5.37(*a*).

are determined in exactly the same way. For position III on the boundary, only the directions $\boldsymbol{F}_{\mathrm{III}}$ and $\boldsymbol{C}_{\mathrm{III}}$ could be obtained, as only images were available for this position.

The correlation that the fine fringes in the images represent the mean fringe pattern arising from the two periodicities in the diffraction patterns, and that the coarse set of fringes represents the moiré difference of the two periodicities, is summarised in table 5.2(*a*). This table correlates, for positions I and II, the spacing and direction of the fringes with the spacing and direction of the diffraction spots when both are projected on to the boundary, and includes the data from images for position III. For example, the fine fringe spacing from the images (column (v) of table 5.2(*a*)) agrees within the limits of experimental error with the mean spacings from the diffraction patterns (columns (i) and (iii)) and likewise the direction of the normal to the fine fringes in the images (column (vi)) agrees with the mean direction from the diffraction patterns (columns (ii) and (iv)). Similarly the spacing of the coarse fringes and the direction of their normal in the images (columns (ix) and (x)) agree within the limits of experimental error with those calculated from the diffraction patterns (columns (vii) and (viii)).

Table 5.2 Comparison of experimental and theoretical results for a near-$\Sigma 57_a$ boundary.
(a) Experimental data from diffraction patterns and images.

Position on boundary	Fine fringes						Coarse fringes			
	Data from diffraction patterns				Data from images		Data from diffraction patterns		Data from images	
	Periodicity S		Periodicity L				As calculated from the moiré of S and L			
	Spacing (Å)	Direction† ($\rho°$)	Spacing (Å)	Direction ($\rho°$)	Spacing (Å)	Direction of fringe normal ($\rho°$)	Spacing (Å)	Direction ($\rho°$)	Spacing (Å)	Direction of fringe normal ($\rho°$)
I	79 ± 5	105 ± 2	79 ± 4	90 ± 2	82 ± 5	97 ± 2	302.6	7	299 ± 21	5 ± 2
II	83 ± 4	112 ± 2	83 ± 4	100 ± 2	93 ± 6	107 ± 2	381.2	17	389 ± 28	15 ± 2
III	–	–	–	–	91 ± 6	118 ± 2	–	–	630 ± 70	36 ± 2
	(i)	(ii)	(iii)	(iv)	(v)	(vi)	(vii)	(viii)	(ix)	(x)

(b) Dislocation model based on $\Sigma 57_a$.

Position on boundary	Fine fringes						Coarse fringes	
	Array 1, $\boldsymbol{b}_1 = (1/3)[\bar{1}\bar{1}1]$ (periodicity S)		Array 2, $\boldsymbol{b}_2 = (7/114)[\bar{1}87]$ (periodicity L)		Array 3, $\boldsymbol{b}_3 = (1/114)[13\ 10\ 23]$		Calculated from arrays 1 and 2	
	Spacing (Å)	Direction† ($\rho°$)	Spacing (Å)	Direction ($\rho°$)	Spacing (Å)	Direction ($\rho°$)	Spacing (Å)	Direction ($\rho°$)
I	80 ± 14	105 ± 1	83 ± 14	90 ± 1	40 ± 8	137 ± 1	310 ± 60	3 ± 3
II	83 ± 13	112 ± 1	82 ± 14	100 ± 1	50 ± 8	146 ± 1	380 ± 70	16 ± 5
III	83 ± 14	121 ± 1	82 ± 14	114 ± 1	56 ± 10	138 ± 1	710 ± 150	37 ± 8

† Directions are specified by angles $\rho°$ measured in the plane of the boundary from the intersection of $(001)_1$ with the boundary plane for each position (I, II and III) of the boundary, i.e. from $[(001)_1 \wedge \boldsymbol{v}_i]$ (i = I, II, III). $\boldsymbol{v}_{\mathrm{I}} = [\cos(118.1° \pm 2°) \cos(115.8° \pm 2°) \cos(39.9° \pm 2°)]_1$, $\boldsymbol{v}_{\mathrm{II}} = [\cos(120.4° \pm 2°) \cos(107.1° \pm 2°) \cos(35.8° \pm 2°)]_1$, $\boldsymbol{v}_{\mathrm{III}} = [\cos(131.1° \pm 2°) \cos(98.1° \pm 2°) \cos(42.2° \pm 2°)]_1$.

The stereographic projection of figure 5.41 summarises the geometric analysis of diffraction patterns and images for the three different boundary planes $\boldsymbol{v}_{\mathrm{I}}$, $\boldsymbol{v}_{\mathrm{II}}$ and $\boldsymbol{v}_{\mathrm{III}}$. In this projection the boundary planes are indicated and the directions of the two-dimensional periodic structure in real space, $\boldsymbol{S}_{\perp}$ and $\boldsymbol{L}_{\perp}$, are plotted as normals to the directions of the periodicities S and L respectively for positions I and II on the boundary. The directions $\boldsymbol{F}'$, of the fine fringes in the images, which are the mean directions of $\boldsymbol{S}_{\perp}$ and $\boldsymbol{L}_{\perp}$, are available for positions I, II and III on the boundary, as are the directions $\boldsymbol{C}'$, of the coarse fringes in the images, which are the moiré differences of $\boldsymbol{S}_{\perp}$ and $\boldsymbol{L}_{\perp}$. These, together with the directions $\boldsymbol{S}_{\perp}$ and $\boldsymbol{L}_{\perp}$ for positions I and II, define a rotation axis $\boldsymbol{\eta}$ in figure 5.41 where $\boldsymbol{\eta} = [\cos(71^{\circ} \pm 2^{\circ}) \cos(146^{\circ} \pm 2^{\circ}) \cos(63^{\circ} \pm 2^{\circ})]$. Within the limits of experimental error $\boldsymbol{\eta}$ agrees with the calculated rotation axis $\boldsymbol{q}$ and thus satisfies property (a) of expression (5.4), indicating that the periodicities in the diffraction patterns from the boundary are associated with arrays of secondary grain boundary dislocations.

Since experimental evidence for only two arrays of dislocations is available, a two-dislocation model for the structure of the boundary based on equations (1.16) and (1.17) might, at first, be considered to apply. However, such a model suggests Burgers vectors which do not predict the observed spacings. On the assumption that a third array of dislocations is present, but is not

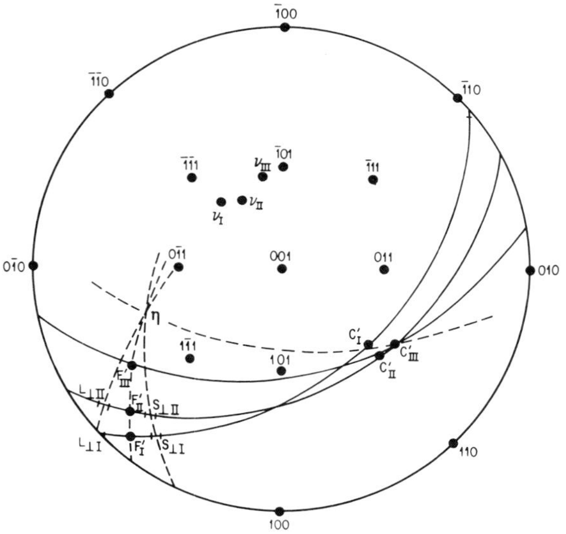

Figure 5.41 Stereographic projection showing the normals $\boldsymbol{S}_{\perp\mathrm{I}}$, $\boldsymbol{S}_{\perp\mathrm{II}}$ and $\boldsymbol{L}_{\perp\mathrm{I}}$, $\boldsymbol{L}_{\perp\mathrm{II}}$ to the periodicities S and L for positions I and II on the boundary, and the directions $\boldsymbol{C}'$ and $\boldsymbol{F}'$ of the coarse and fine fringes respectively for positions I, II and III on the boundary.

detected experimentally due, say, to a small Burgers vector giving weak diffraction effects, then a three-dislocation model for the structure of the boundary can be tested. It is necessary to find three Burgers vectors $\boldsymbol{b}_1$, $\boldsymbol{b}_2$ and $\boldsymbol{b}_3$ of the $\Sigma 57_a$ DSC lattice, for three arrays of dislocations 1, 2 and 3, which not only predict the experimentally observed spacings of the periodicities S and L, but also predict the directions and spacings of the fine and coarse fringes for each position on the boundary. If $\boldsymbol{b}_1$ and $\boldsymbol{b}_2$ are taken to be the Burgers vectors of the dislocations in arrays corresponding to periodicities S and L respectively then it is found, with $\boldsymbol{b}_1$ equal to the basis DSC vector $\pm(1/3)[\bar{1}\bar{1}1]_1$ and with $(\boldsymbol{b}_2 \wedge \boldsymbol{b}_3)$ parallel to $\pm[\bar{1}\bar{1}1]_1$, that, within the limits of experimental error, expressions (5.4) and (5.5) predict directions and spacings of periodicity S for positions I and II on the boundary in agreement with the experimental results. With $\boldsymbol{b}_1 = \pm(1/3)[\bar{1}\bar{1}1]_1$, the direction of $\boldsymbol{b}_3$ is found by altering $\boldsymbol{b}_3$ in the $(\bar{1}\bar{1}1)_1$ plane of the DSC lattice (using different combinations of basis DSC vectors) until a direction for the dislocations with Burgers vector $\boldsymbol{b}_2$ is obtained from expression (5.4) which agrees, within the limits of experimental error, with the direction of periodicity L. This gives a direction for $\boldsymbol{b}_3$ of $\pm[13\ 10\ 23]_1$, and $\boldsymbol{b}_3$ is chosen as $\pm(1/114)[13\ 10\ 23]_1 = \pm[(2/114)[718]_1 + (1/114)[\bar{1}87]_1]$, i.e. $\pm(2y + z)$. With $\boldsymbol{b}_1 = \pm(1/3)[\bar{1}\bar{1}1]_1$ and $\boldsymbol{b}_3 = \pm(1/114)[13\ 10\ 23]_1$ the Burgers vector $\boldsymbol{b}_2$ is selected by testing various combinations of basis DSC vectors in the $(\bar{1}\bar{1}1)_1$ plane until a spacing for periodicity L is obtained which, within the limits of experimental error, is in agreement with the measured spacings for positions I and II on the boundary; the Burgers vector which satisfies these conditions is $\boldsymbol{b}_2 = \pm(7/114)[\bar{1}87]_1$, i.e. $\pm 7z$.

The signs of the Burgers vectors $\boldsymbol{b}_1$, $\boldsymbol{b}_2$ and $\boldsymbol{b}_3$ can be specified for given senses of the line directions of the dislocations from the determined sense of the misorientation between the grains. The misorientation between the grains is such that the $\Sigma 57_a$ CSL is rotated with respect to grain 1 in a right-handed sense by the angle $1.6^\circ \pm 0.3^\circ$ around the axis $\boldsymbol{q}$ (see figure 5.38), so that expression (5.4) satisfies the FS/RH convention when the sense taken for the boundary normals at positions I, II and III points from grain 1 into the CSL, i.e. from the upper grain 1 to the lower grain 2 which is opposite to the upward-drawn sense of $\boldsymbol{v}_{\mathrm{I}}$, $\boldsymbol{v}_{\mathrm{II}}$ and $\boldsymbol{v}_{\mathrm{III}}$ in the stereographic projection of figure 5.41. When the correct sense for the boundary normals is used in expression (5.4) with the sign for the Burgers vectors chosen as $\boldsymbol{b}_1 = (1/3)[\bar{1}\bar{1}1]_1$, $\boldsymbol{b}_2 = (7/114)[\bar{1}87]_1$ and $\boldsymbol{b}_3 = (1/114)[13\ 10\ 23]_1$, the sense obtained for the line directions of the secondary grain boundary dislocations is opposite to those of $\boldsymbol{S}_\perp$ and $\boldsymbol{L}_\perp$ in figure 5.41.

Table 5.2(b) gives a summary of the directions and spacings calculated from expressions (5.4) and (5.5) for the determined values of $\boldsymbol{b}_1$, $\boldsymbol{b}_2$ and $\boldsymbol{b}_3$, where $\boldsymbol{q}$ is taken as the experimentally determined rotation axis $\boldsymbol{\eta}$ and φ is taken as the angle $-1.8^\circ \pm 0.3^\circ$ obtained from equation (5.1) with $\boldsymbol{q}$ equal to $\boldsymbol{\eta}$. The directions in table 5.2(b) are specified as normals to the dislocation lines and as coarse fringe normals to facilitate direct comparison with the experimental data in table 5.2(*a*). The errors quoted in table 5.2(*b*) take account of uncertainties in $\boldsymbol{\eta}$, φ and boundary plane normal $\boldsymbol{v}$, and it can be seen from a comparison of table 5.2(*a*) and table 5.2(*b*) that the computed structure agrees within the limits of experimental error with the observed directions and spacings of the periodicities *S* and *L*, and of the coarse fringes for each position on the boundary. Although the third array of dislocation lines has a periodicity which should be in the range of experimental observation, the magnitude of the Burgers vector involved, $|\boldsymbol{b}_3|$, is less than half $|\boldsymbol{b}_1|$ which is associated with the short row of diffraction spots, *S*, and approximately one-third of $|\boldsymbol{b}_2|$ which is associated with the long row of diffraction spots, *L*. It is likely therefore that the intensity of the diffraction spots with the periodicity of array 3 is below the level of detection.

This example demonstrates that a correlation of information from diffraction patterns and images can be used to obtain Burgers vectors for secondary grain boundary dislocations even when these dislocations are finely spaced and not imaged directly.

Near-Σ9 boundary in stainless steel

Clark and Smith (1978) examined the secondary grain boundary dislocation structure in a near-Σ9 boundary in polycrystalline stainless steel (17.5% Cr, 10% Ni, 2.5% Mo, 2.0% Mn, 0.5% Si, 0.1% C) for which the angular departure from the exact CSL was similar to that for the near-$\Sigma 57_a$ boundary of the previous example. The misorientation between grains 1 and 2 corresponded to a rotation of 37.36° about an axis common to both grains of [1 70 71], and this misorientation corresponded to a departure from the exact-Σ9 CSL orientation (38.94° around [011]) by 1.61° ± 0.22° around an axis indexed with respect to grain 1 of $[\bar{7}\ \overline{40}\ \overline{32}]_1$. A set of basis DSC vectors in grain 1 corresponding to this Σ9 CSL are $\boldsymbol{x} = (1/18)[41\bar{1}]_1$, $\boldsymbol{y} = (1/9)[\bar{1}2\bar{2}]_1$ and $\boldsymbol{z} = (1/6)[112]_1$.

The portion of the boundary, with boundary normal $(232)_1$, selected for detailed examination contained three independent arrays of secondary grain boundary dislocations *A*, *B* and *C*. The arrays *A* and *B*, with spacings of 97 Å and 80 Å respectively, formed a finely spaced orthogonal cross-grid and the array *C*, with a spacing of 460 Å, crossed this grid diagonally. The experimental information consisted of same-$\boldsymbol{g}_c$ images and images taken with only one grain diffracting in two-beam conditions, and $\boldsymbol{g} \cdot \boldsymbol{b}$ criteria were used to determine the Burgers vectors of the secondary grain boundary dislocations. In their paper, Clark and Smith point out the considerable difficulties associated with using image contrast to determine Burgers vectors when the secondary grain boundary dislocations are closely spaced. Nevertheless, they attributed the DSC Burgers vectors $\boldsymbol{b}_1 = \pm(1/18)[41\bar{1}]_1$ to the dislocations in array *A*, $\boldsymbol{b}_2 = \pm(1/9)[\bar{1}2\bar{2}]_1$ to the dislocations in array *B* and $\boldsymbol{b}_3 = \pm(1/6)[112]_1$ to the dislocations in array *C*. However, these Burgers vector determinations were not supported by the geometry of the dislocation arrays because when they were used in expressions similar to expressions (5.4) and (5.5), or in an equivalent 0-lattice analysis (see section 1.5), they gave spacings and directions for the dislocations in the three arrays which were in marked disagreement with the experimental measurements. For example, experimental and calculated directions differed by up to 17.5° and spacings by factors of two to three.

This work demonstrates the difficulties that are encountered in analysing secondary grain boundary dislocation structures, when the misorientation is far from an exact CSL orientation, without detailed geometric information on the changes in spacing and direction of secondary grain boundary dislocations with change in boundary plane.

Near-$\Sigma 13_a$ boundary in an Al–Mg alloy

Mori and Ishida (1978), following an earlier investigation by Ishida *et al* (1977), used the matching of experimental and theoretical simultaneous double two-beam electron micrographs to determine the Burgers vectors of secondary grain boundary dislocations for a single coarsely spaced array in a high-angle boundary in an Al – 4.4 wt% Mg alloy. The misorientation between the two grains A and B was within 2.86° of the exact $\Sigma 13_a$ CSL orientation and the Burgers vectors tested were basis DSC vectors corresponding to this CSL, namely $\boldsymbol{x} = (1/26)[\bar{5}01]_A$, $\boldsymbol{y} = (1/26)[105]_A$, $\boldsymbol{z} = (1/26)[\bar{2}\ \overline{13}\ 3]_A$ and their linear combinations. It was found that the Burgers vectors that gave theoretical images in agreement with the experimental images were the DSC vectors (i) $\boldsymbol{z} = (1/26)[\bar{2}\ \overline{13}\ 3]_A$, (ii) $\boldsymbol{z} - \boldsymbol{x} - \boldsymbol{y} = (1/26)[2\ \overline{13}\ \bar{3}]_A$, (iii) $\boldsymbol{z} - \boldsymbol{y} = (1/26)[\bar{3}\ \overline{13}\ \bar{2}]_A$ and (iv) $\boldsymbol{z} - \boldsymbol{x} = (1/26)[3\ \overline{13}\ 2]_A$, suggesting that one of these vectors was the Burgers vector of the observed secondary grain boundary dislocations. However, with the experimental images available, they could not distinguish between these four possibilities despite the fact that the angular separation of these vectors is either 21.8° or 31.0°.

In order to assess whether images of secondary grain boundary dislocations with the above four Burgers vectors would have sufficiently different character to enable them to be distinguished by the image matching technique, sets of theoretical images were computed by the authors using the program PCGBD for these four Burgers vectors under different diffracting conditions, and the results are shown in figure 5.42. The three columns, (*a*)–(*c*), of theoretical images correspond to different diffracting conditions with non-coplanar diffracting vectors in each grain and the four rows, (i)–(iv), correspond to the four different possible Burgers vectors in the order given above. In column (*a*) the image (i) can be distinguished from images (ii)–(iv) in that it shows a single line of strong contrast at the dislocation, while the others are double images. Similarly in column (*b*), the image (ii) can be distinguished from the other three, so that it only remains to distinguish between images (iii) and

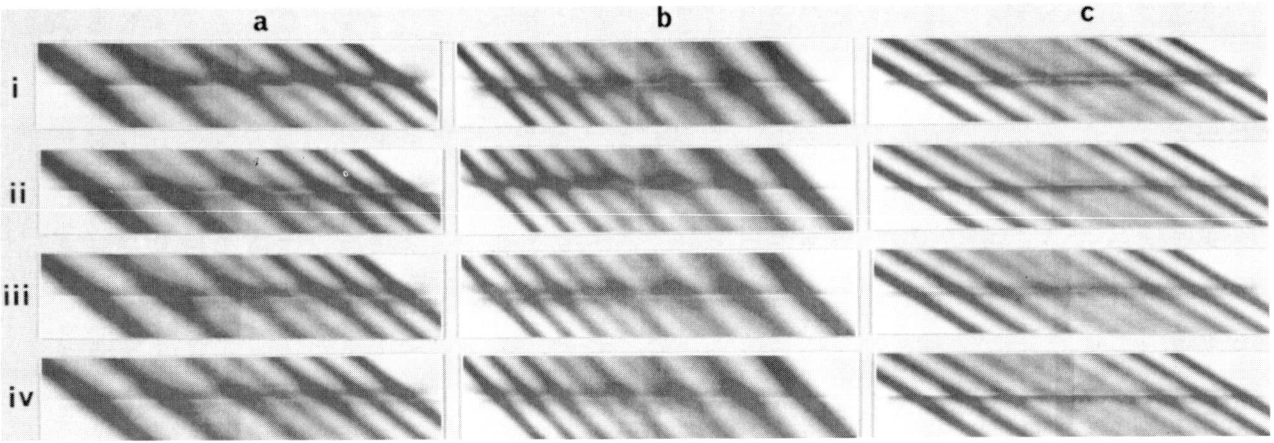

Figure 5.42 Demonstration of the use of computed images to distinguish between the DSC Burgers vectors $(1/26)[\bar{2}\ \overline{13}\ 3]_A$ in (i), $(1/26)[2\ \overline{13}\ \bar{3}]_A$ in (ii), $(1/26)[\bar{3}\ \overline{13}\ \bar{2}]_A$ in (iii) and $(1/26)[3\ \overline{13}\ 2]_A$ in (iv) for a secondary grain boundary dislocation in a $\Sigma 13_a$ boundary in an Al–Mg alloy separating an upper grain *A* from a lower grain *B*. In this computer experiment a range of tilt of $\pm 45°$ was used to give beam directions in each grain which enabled double two-beam images (*a*), (*b*) and (*c*) to be obtained with the following diffracting vectors in each grain: $\bar{1}\bar{1}\bar{1}_A$, $\bar{2}02_B$ in (*a*), $20\bar{2}_A$, $\bar{1}\bar{1}\bar{1}_B$ in (*b*) and $\bar{1}1\bar{1}_A$, $\bar{2}00_B$ in (*c*), with $w_A = w_B = 0.3$ in all cases.

(iv). This distinction is made in column (*c*) where the image (iii) is a definite double image while that in (iv) is not. These characteristic features of contrast should enable the four Burgers vectors to be distinguished if a suitable set of experimental images is available.

Plane matching boundaries

So far consideration has been confined to high-angle grain boundaries in polycrystals for which misorientations only depart from low-Σ CSL orientations by less than 3°. However, it is expected that in a random polycrystalline aggregate misorientations will exist between grains which depart from low-Σ CSL orientations by much larger angles (Warrington and Boon 1975), and, in order to continue to describe these misorientations in terms of small angular departures from CSL orientations, it would be necessary to invoke unreasonably high values of Σ. It was for situations such as these that the plane matching model for the structures of high-angle boundaries was considered to have particular significance.

The plane matching model was developed by Pumphrey (1972) to explain what was a common observation in early studies of grain boundaries using transmission electron microscopy, namely that only one periodic set of lines was observed in electron micrographs of high-angle grain boundaries. The model is concerned with the mismatching of lines of intersection with the boundary of the same type of slightly misoriented low-index crystallographic planes in each grain. Figure 5.43 is a schematic illustration of the plane matching model as envisaged by Pumphrey. It shows two periodic sets of lines of intersection 1 and 2, with spacings S_1 and S_2 inclined at an angle δ in the boundary plane, and their moiré resultant. If these lines represent regions of high atomic density, then their moiré resultant would also

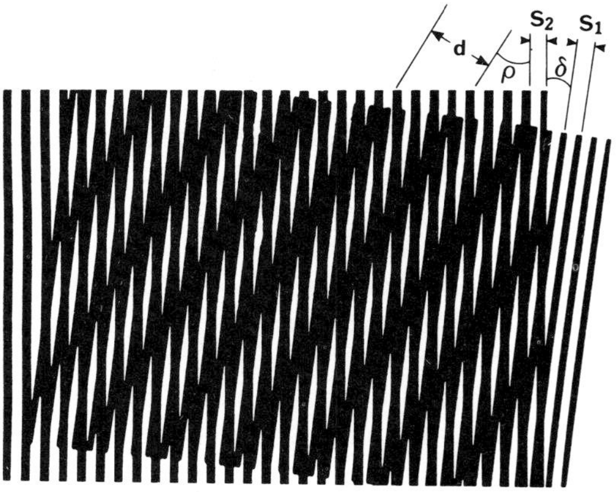

Figure 5.43 Formation of a moiré pattern by two periodic sets of lines inclined at a small angle (after Pashley *et al* 1964).

correspond to regions of high atomic density. The moiré resultant set of lines has a spacing given by

$$d = S_1 S_2/(S_1^2 + S_2^2 - 2S_1 S_2 \cos \delta)^{1/2}$$

and is inclined at an angle to the directions of the original lines with the angle of inclination from set 2 given by

$$\rho = \sin^{-1} [d \sin(\delta/S_2)].$$

Pumphrey considered that in an actual boundary atomic relaxations would occur giving rise to strain fields with a periodicity corresponding to that of the moiré resultant, and that the imaging of these strain fields in the electron microscope would explain the observed linear periodic structure.

Balluffi and Schober (1972) (see also, Schlinder *et al*, 1979 and Warrington 1980) showed that the plane matching model is merely a limiting case of the CSL model. They pointed out, for a high-angle grain boundary in which the only form of matching is the near-matching of lines of intersection with the boundary of a set of planes with common indices (hkl) in each grain, that the misorientation between the grains would be close to a high-Σ CSL orientation with a low-index rotation axis $[hkl]$. In this situation, the two basis DSC vectors $\boldsymbol{x}$ and $\boldsymbol{y}$ perpendicular to the rotation axis would be very small, but the third basis vector $\boldsymbol{z}$ would have a significant magnitude. As Σ tends to infinity, the DSC vectors $\boldsymbol{x}$ and $\boldsymbol{y}$ would tend to zero and in the limit the DSC lattice would degenerate into a lattice defined by the single basis vector $\boldsymbol{z}$, which would be normal to the plane (hkl) with a magnitude equal to the interplanar spacing of the (hkl) planes. In these circumstances, only one array of secondary grain boundary dislocations with Burgers vector $\boldsymbol{z}$ would be present and the role of these dislocations would be to maintain the spacing of the (hkl) planes at the boundary. These dislocations would correspond to the strain fields with periodicity d of the moiré fringes in the plane matching model.

Pumphrey tested his plane matching model on three different boundaries in which he had observed only one periodicity in electron microscope images. In one case (Pumphrey 1973) the boundary was a low-angle boundary, and in the other two cases (Pumphrey 1975) the boundaries were both close to low-Σ CSL orientations and in addition the line directions found for the dislocations were not contained in the determined boundary planes. Further, in no case was diffraction evidence sought to detect the presence of other periodicities which may not have been resolved in images. Thus, in the absence of convincing evidence for only one periodicity, these experiments do not provide strong support for the plane matching model. However, Schlinder *et al* (1979) have produced grain boundary dislocation structures in fabricated thin-film bicrystals which are compatible with plane matching. They prepared thin-film bicrystals of gold with misorientations that were specifically selected to test whether dislocation structures associated with the mismatching of (002), (220) and (420) planes could be identified. The fabricated boundaries spanned a range of tilt and twist deviations from exact plane matching conditions for a variety of boundary planes. They found dislocation structures that could be associated with the mismatching of (002) and (220) planes, but no dislocation structures were found for the mismatching of (420) planes. Their observed dislocation structures were compatible with a plane matching model for tilt and twist deviations up to 4° from (002) and (220) plane matching boundaries, and for the (002) case the tilt deviation could be as large as 14°.

The dislocation structures observed by Schindler *et al* show that boundaries can exist in which the dislocation structure is compatible with the maintenance of the spacing of a set of low-index planes at the boundary, or equivalently accommodates part of the departure of the misorientation from a high-Σ CSL orientation.

5.4 RIGID-BODY DISPLACEMENTS BETWEEN NEIGHBOURING GRAINS IN POLYCRYSTALS

It was pointed out in section 5.2.3 that rigid-body displacements can occur between neighbouring grains in polycrystalline aggregates which move atoms away from coincident sites, so as to generate lower energy atomic configurations at boundaries. As discussed in section 2.4.5 such a rigid-body displacement, $\boldsymbol{R}$, is revealed by the presence of fringe contrast which occurs in same-$\boldsymbol{g}_c$ images when $\boldsymbol{g}_c \cdot \boldsymbol{R}$ is non-integral. Pond and Smith (1974) were the first to show experimentally that rigid-body displacements occur between grains in a polycrystal, when they found fringe contrast in same-$\boldsymbol{g}_c$ images of an incoherent Σ3 boundary in aluminium. The fact that rigid-body displacements are a general feature of the structure of grain boundaries in polycrystalline metals and alloys has been demonstrated by the authors' observations on near-Σ3, Σ9, $\Sigma 27_a$, $\Sigma 27_b$, $\Sigma 81_c$, $\Sigma 81_d$ and $\Sigma 243_a$ boundaries, and some of these observations are shown in figures 5.44 and 5.45 (Forwood and Clarebrough 1983). The images in figure 5.44(*a*) are simultaneous double two-beam electron micrographs of portions of near-Σ9, $\Sigma 27_a$, $\Sigma 27_b$ and $\Sigma 81_c$ boundaries in a Cu − 6 at% Si alloy taken with different diffracting vectors operating in each grain. These boundaries are further examples of high-angle boundaries containing networks of secondary grain boundary dislocations with DSC Burgers vectors. Figure 5.44(*b*) shows, for each of the boundaries, a same-$\boldsymbol{g}_c$ image with strong fringe contrast indicating that a rigid-body displacement is present in each boundary. The generality of rigid-body displacements at grain boundaries in polycrystals is further illustrated by the montage of electron micrographs in figure 5.45(*a*). This figure shows an inner grain 1 surrounded by five outer grains, 2–6, separated from each other by Σ3 boundaries and from the inner grain by

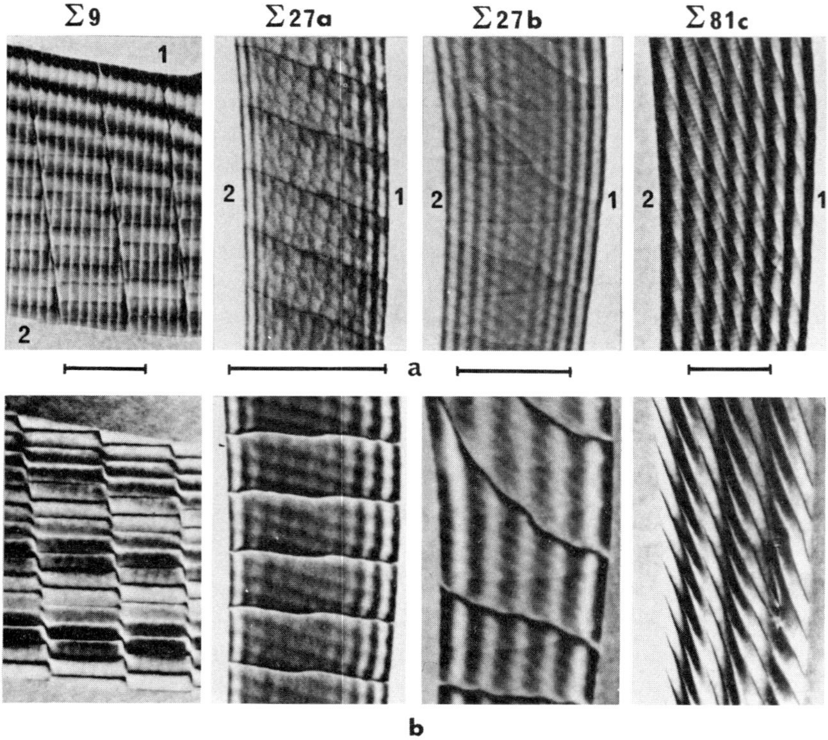

Figure 5.44 Double two-beam images showing secondary grain boundary dislocation structures and displacement fringe contrast for near-$\Sigma 9$, $\Sigma 27_a$, $\Sigma 27_b$, and $\Sigma 81_c$ boundaries in a Cu − 6 at% Si alloy. The images in (*a*) are for different diffracting vectors in each grain and those in (*b*) are for the same diffracting vector $\boldsymbol{g}_c$ in both grains. The lengths of the magnification marks correspond to 0.4 μm for all images and the diffracting vectors are: for $\Sigma 9$, $20\bar{2}_1$, $11\bar{1}_2$ in (*a*) and $\bar{1}3\bar{1}_{1/2}$ in (*b*); for $\Sigma 27_a$, $11\bar{1}_1$, 202_2 in (*a*) and $2\bar{2}0_{1/2}$ in (*b*); for $\Sigma 27_b$, $0\bar{2}0_1$, $1\bar{1}\bar{1}_2$ in (*a*) and $\bar{1}\bar{3}1_{1/2}$ in (*b*); for $\Sigma 81_c$, $\bar{1}\bar{1}1_1$, $\bar{1}\bar{1}\bar{1}_2$ in (*a*) and $13\bar{1}_{1/2}$ in (*b*).

$\Sigma 3$, $\Sigma 9$, $\Sigma 27_a$, and $\Sigma 27_b$ boundaries. The same 311 diffracting vector is operating in the inner grain and in grains 2–5. The image shows fringe contrast indicating rigid-body displacements between grains 1 and 3 ($\Sigma 9$), 1 and 4 ($\Sigma 27_b$), 1 and 5 ($\Sigma 9$). The same 311 diffracting vector is not operative in grain 6 so that rigid-body displacements between grains 2 and 6 ($\Sigma 9$), and 1 and 6 ($\Sigma 27_a$), are not revealed in this figure. The boundaries contain widely spaced secondary grain boundary dislocations which, with the exception of the $\Sigma 27_b$ boundary, have DSC Burgers vectors. The fringe contrast in the $\Sigma 27_b$ boundary shows the relatively rare occurrence of different

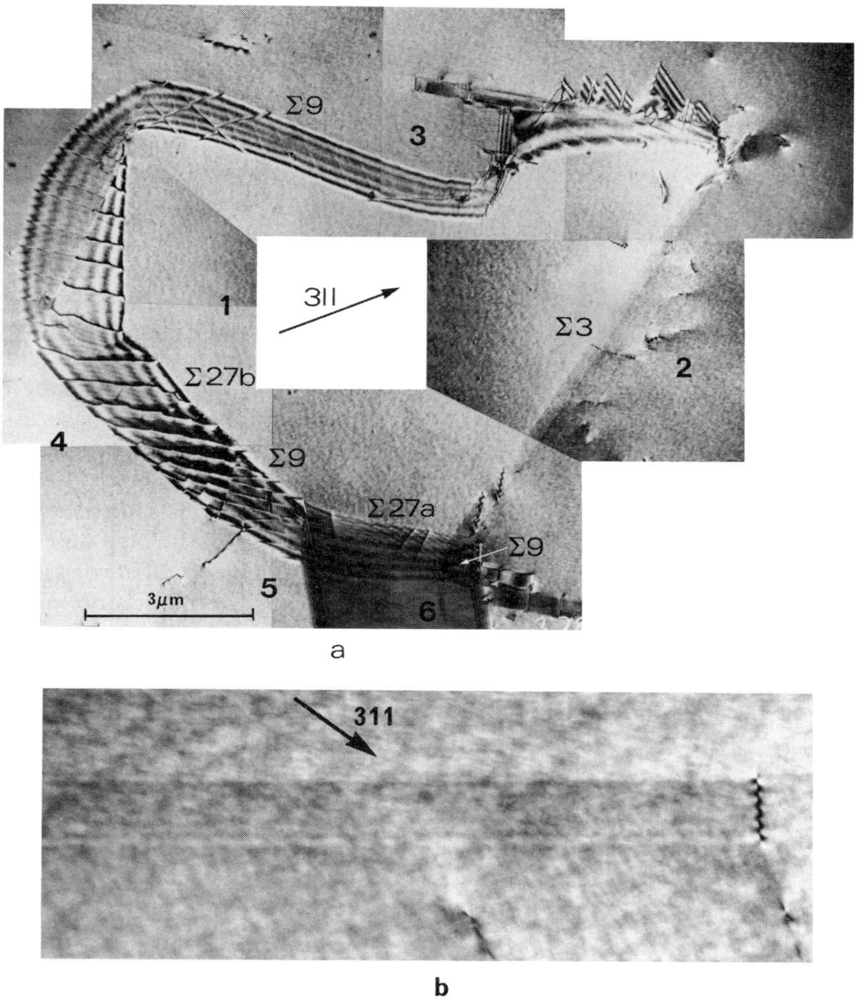

Figure 5.45 (*a*) Double two-beam image with the same 311 diffracting vector in an inner grain and four of the surrounding grains showing displacement fringe contrast at $\Sigma 3$, $\Sigma 9$ and $\Sigma 27_b$ boundaries. (*b*) Higher magnification image of a portion of the coherent-$\Sigma 3$ boundary between grains 1 and 2.

translation states separated by partial grain boundary dislocations with non-DSC Burgers vectors, and such arrays of partial dislocations will be discussed in section 5.4.2.

In polycrystalline specimens of FCC metals and alloys faint fringe contrast is sometimes observed in same-$\boldsymbol{g}_c$ images of $\{111\}$ coherent $\Sigma 3$ twin boundaries, and an example of such contrast is shown in the image of the boundary between grains 1 and 2 in figures 5.45(*a*) and (*b*). Fringing of this type has been reported by Pond and Smith (1976) in pure copper. In the experience of the authors, faint fringing in same-$\boldsymbol{g}_c$ images of {111} coherent Σ3 twin boundaries is most frequently observed when the boundaries form components of triple junctions involving combinations of boundaries such as Σ3–Σ27–Σ81 and Σ3–Σ9–Σ27, but is not observed when they are component boundaries of fully enclosed island twinned grains. Further, even when coherent Σ3 boundaries form triple junctions with other CSL boundaries, faint fringing is not always observed, e.g. for the boundaries in figure 5.45(*a*), faint fringing was observed between the grains 1 and 2 (which can be seen in more detail in figure 5.45(*b*)), but not at the boundaries between grains 5 and 6 and grains 4 and 5. In general the intensity of the faint fringing is close to the limit of visibility and is asymmetric about mid-specimen, so that it could well arise from a small difference in deviation from the Bragg condition w between the two grains. Such differences in w, due to small local strains, could well be present at triple junctions in specimens prepared by thinning bulk polycrystals. Faint fringing of the type observed can be simulated in computed images by including small differences in w between the two grains, differences which are within the uncertainty in the determination of w. Therefore in the absence of other evidence, it cannot be concluded that faint fringing is due to a small rigid-body displacement inherent in the structure of a {111} coherent Σ3 twin boundary.

5.4.1 Rigid-body Displacements and Partial Dislocations at Facet Intersections

Rigid-body displacements can be determined by comparing the intensity of the fringe contrast in same-$\boldsymbol{g}_c$ images of boundaries with fringe contrast predicted theoretically for different values of $\boldsymbol{R}$, and this can be done by comparing experimental images with computed intensity profiles or with computed images. If, for the boundary under examination, the lower grain with respect to the electron source is displaced by a vector $\boldsymbol{R}$ relative to the upper grain causing an offset of the $\boldsymbol{g}_c$ planes across the boundary, then, as discussed in section 2.4.5, fringe contrast will occur in a same-$\boldsymbol{g}_c$ image with an intensity determined by the phase change $2\pi\boldsymbol{g}_c \cdot \boldsymbol{R}$ between the transmitted and diffracted beams. As a consequence, by comparing theoretical and experimental fringe contrast, the quantity $(\boldsymbol{g}_c \cdot \boldsymbol{R} + n)$ can be determined in the range $-\frac{1}{2} \leqq (\boldsymbol{g}_c \cdot \boldsymbol{R} + n) \leqq \frac{1}{2}$ where n is an unknown integer, or equivalently $\boldsymbol{g}_c \cdot (\boldsymbol{R} + \boldsymbol{d})$ in the range $-\frac{1}{2} \leqq [\boldsymbol{g}_c \cdot (\boldsymbol{R} + \boldsymbol{d})] \leqq \frac{1}{2}$ where $\boldsymbol{d}$ is a vector of the DSC lattice (since the $\boldsymbol{g}_c$ lattice is reciprocal to the DSC lattice (Grimmer 1974)). Thus, by matching the fringe contrast in three images taken with three non-coplanar same-$\boldsymbol{g}_c$ diffracting vectors, $(\boldsymbol{R} + \boldsymbol{d})$ can be determined as a vector within the Wigner–Seitz cell of the DSC lattice, but

d remains an unknown DSC vector. It has been seen in section 5.2.1 that a DSC displacement between neighbouring grains merely causes a shift in the origin of the CSL. Therefore it follows that the unknown DSC vector ***d*** corresponds to an uncertainty in the level of the boundary plane in the array of lattice sites contained in the unit cell of the CSL. This uncertainty has to be taken into account because different levels for the boundary will correspond to different atomic configurations at the boundary, except in the special case when $1/\Sigma$ of the lattice sites in planes parallel to the boundary plane are in coincidence, because then the atomic arrangement for all possible levels of the boundary is the same in the unit cell of the CSL (Pond 1977). The way in which the atomic configuration at a boundary changes with the choice of level for the boundary plane can be seen from the simple example of a $\Sigma 3$ twin-boundary with a $\{111\}$ boundary plane common to both grains in an FCC bicrystal. For one level of the boundary, the FCC ABC ... stacking sequence of $\{111\}$ planes is

↓
A B C A B C B A C B A C

where the arrow indicates the level of the boundary, and the two alternative levels either side of this level give the sequences

↓
A B C A A C B A C B A C

↓
A B C A B C A A C B A C.

In this simple case it is clear that there is only one physically meaningful level for the boundary, namely the coherent boundary of the first sequence as the others involve forbidden AA stacking.

Incoherent Σ3 boundaries in aluminium

The first quantitative determinations of rigid-body displacements at grain boundaries, and the first observations of partial grain boundary dislocations, were those of Pond and Vitek (1977) and Pond (1977). They combined experimental observations with computer modelling to study $\Sigma 3$ boundaries in aluminium between a matrix M and a twinned grain T for the boundary planes $(\bar{1}55)_M/(171)_T$, $(\bar{2}11)_M/(\bar{1}2\bar{1})_T$ and $(1\bar{2}1)_M/(\bar{1}\bar{1}2)_T$. The intensity profiles of the fringe contrast in non-coplanar same-$\boldsymbol{g}_c$ images of these boundaries were determined by microdensitometer traces, and these profiles were compared with computed profiles for different values of $\boldsymbol{g}_c \cdot \boldsymbol{R}$ in the range $-\frac{1}{2} \leqq \boldsymbol{g}_c \cdot \boldsymbol{R} \leqq \frac{1}{2}$. In this way values of $\boldsymbol{g}_c \cdot \boldsymbol{R}$ were determined with an accuracy between ± 0.02 and ± 0.05. These experimental values of $\boldsymbol{g}_c \cdot \boldsymbol{R}$ were compared with calculated values of $\boldsymbol{g}_c \cdot \boldsymbol{R}_p$, where $\boldsymbol{R}_p$ is the component of $\boldsymbol{R}$ in the boundary plane predicted by the computer modelling method described in their paper. For each boundary plane, they found that the differences between the values of $\boldsymbol{g}_c \cdot \boldsymbol{R}$ and $\boldsymbol{g}_c \cdot \boldsymbol{R}_p$ were accounted for, within the limits of experimental error, by a constant expansion $\boldsymbol{R}_n$ normal to the boundary plane. For the $(\bar{1}55)_M/(171)_T$ plane $\boldsymbol{R}_n$ was $0.04a$ and for the $(\bar{2}11)_M/(\bar{1}2\bar{1})_T$ and $(1\bar{2}1)_M/(2\bar{1}\bar{1})_T$ planes $\boldsymbol{R}_n$ was $0.05a$, where a is the lattice parameter of aluminium. The fact that a constant value was obtained for the expansion normal to a given type of boundary plane was consistent with their computer model giving reliable calculated values for $\boldsymbol{R}_p$. However, calculated values of $\boldsymbol{R}_n$ were not reliable due to artificial constraints that had to be imposed in the computer model.

From their computer modelling Pond and Vitek found for the facet plane $(1\bar{2}1)_M/(\bar{1}\bar{1}2)_T$ that two different displacement states, α_1 and α_2, can occur which give the same low-energy arrangement of atoms at the boundary and these two displacement states are illustrated schematically in figure 5.46. Figure 5.46(a) shows the grains M and T in the reference state corresponding to exact coincidence where $\boldsymbol{R} = 0$, figure 5.46(b) shows grain T displaced relative to grain M by $(1/9)[111]_M + (1/4)[\bar{1}01]_M$ to give the α_2 state and figure 5.46(c) shows grain T displaced by $(2/3)[\bar{1}\bar{1}\bar{1}]_M + (1/9)[\bar{1}\bar{1}\bar{1}]_M + (1/4)[\bar{1}01]_M$ to give the α_1 state. These two displacement states with the same structure are simply related by a 180° rotation around $[111]_M/[111]_T$ which transforms one state into the other, and this rotation is a symmetry operation for the bicrystal in the coincident state with $\boldsymbol{R} = 0$, as represented in figure 5.46(a). If these two states with equal energy coexist in the same boundary plane, they will be separated by a partial dislocation with a Burgers vector given by the difference of the two rigid-body displacements which, for the example of figure 5.46, is $\boldsymbol{b} = \pm [(2/9)[111]_M + (2/3)[111]_M]$. However, if the α_1 and α_2 states occur on boundary planes which are at different levels, then the Burgers vector of the partial grain boundary dislocations will be $\boldsymbol{b} + \boldsymbol{d}$, where $\boldsymbol{d}$ is a DSC vector associated with a step height corresponding to the difference in the two levels. Thus, taking values of $\boldsymbol{d}$ as multiples of the basis DSC vector $(1/3)[111]_M$, the α_1 and α_2 states can be separated by partial grain boundary dislocations with the smaller Burgers vectors $\pm(2/9)[111]_M$ or $\pm(1/9)[111]_M$.

From their measurements of $\boldsymbol{g}_c \cdot \boldsymbol{R}$ on $(1\bar{2}1)_M/(\bar{1}\bar{1}2)_T$ facet planes, Pond and Vitek identified cases where α_1 and α_2 states of rigid-body displacement were present on different facets, and also cases where α_1 and α_2 states coexisted on the same facet. In the latter case, $\boldsymbol{g} \cdot \boldsymbol{b}$ criteria suggested that the partial grain boundary dislocation separating the two states had a Burgers vector of $(2/9)[111]_M$. They also found partial dislocations at facet intersections between $(\bar{1}55)_M/(171)_T$ planes and $(\bar{2}11)_M/(\bar{1}2\bar{1})_T$ planes and two cases were distinguished depending on whether the α_1 state or the α_2 state was on the $(\bar{2}11)_M/(\bar{1}2\bar{1})_T$ plane. The use of $\boldsymbol{g} \cdot \boldsymbol{b}$ criteria showed that the Burgers vectors

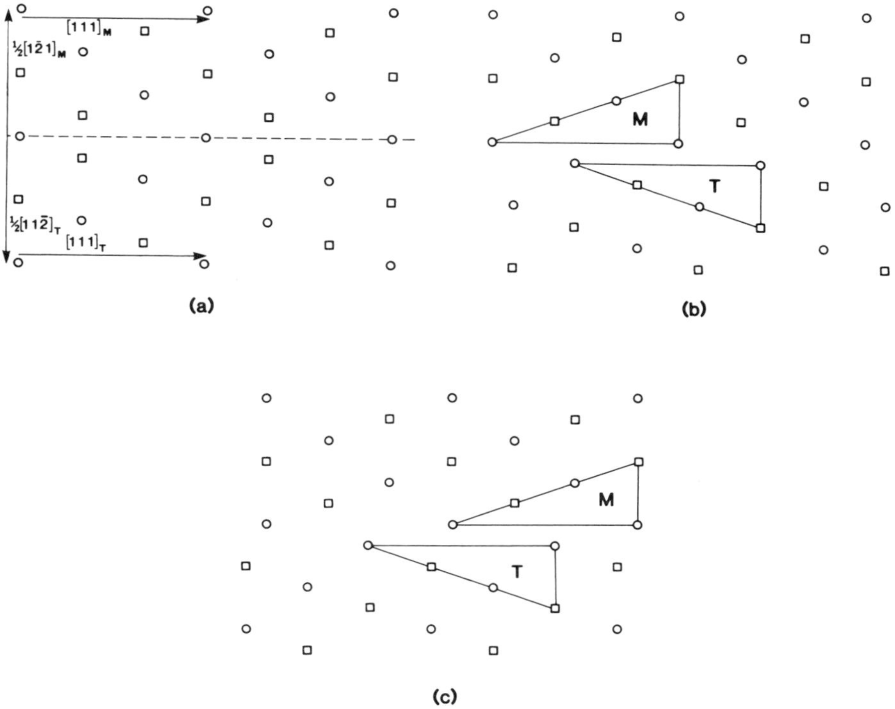

Figure 5.46 Schematic illustration of α_1 and α_2 displacement states at a $\Sigma 3$ twin-boundary on the $(1\bar{2}1)_M/(\bar{1}\bar{1}2)_T$ plane between grains M and T in an FCC structure. The diagrams are projections along $[\bar{1}01]_M/[\bar{1}10]_T$ and the lattice sites in the plane of the page are represented by □ and the sites $(1/4)[\bar{1}01]_M/(1/4)[\bar{1}10]_T$ below the plane of the page are represented by ○; (*a*) is the reference coincidence state in which the sites in the boundary plane are coincident sites and (*b*) and (*c*) represent the α_1 and α_2 states respectively.

of these partial dislocations at the facet intersections were consistent with the differences in the determined values of $\boldsymbol{R}$ for the different facets.

This work by Pond and Vitek led to a more general treatment by Pond and Bollmann (1979) and Pond and Vlachavas (1983) of the relationship between bicrystal symmetry and rigid-body displacements. They show, for a general interface, that the number of different rigid-body displacements which result in equivalent boundary structures is related to the symmetry elements in the coincident state destroyed by a particular displacement. For example, in the case of figure 5.46(*a*), in which a diad axis is destroyed, there are two different states of rigid-body displacement, namely α_1 and α_2, which lead to equivalent boundary structures.

Σ9 boundaries in a Cu–Si alloy

Although rigid-body displacements are a general feature of high-angle grain boundaries (Forwood and Clarebrough 1983), the only boundaries in FCC metals or alloys, other than Σ3 boundaries, in which such displacements have been determined from fringe contrast in same-$\boldsymbol{g}_c$ images are Σ9 boundaries in a Cu – 6 at% Si alloy (Forwood and Clarebrough 1985b)†. In the work of Forwood and Clarebrough five different Σ9 boundaries are involved in studies of ten boundary planes, some of which are boundary facets and others selected portions from curved boundaries. All the Σ9 boundaries are very close to the exact-Σ9 CSL orientation and the misorientation is always specified as a rotation of grain 1 with respect to grain 2 by 38.94° in a right-handed sense around the common axis [110], and basis DSC vectors for this Σ9 CSL orientation are $\boldsymbol{x} = (1/18)[\bar{1}14]_1$, $\boldsymbol{y} = (1/9)[2\bar{2}1]_1$ and $\boldsymbol{z} = (1/18)[\bar{7}\bar{2}1]_1$. Because the misorientation for all the Σ9 boundaries is specified in this consistent way, the rigid-body displacement $\boldsymbol{R}$ is also consistently defined here as a displacement of grain 2 with respect to grain 1. This means that grain 1 is the upper grain for some grain boundaries and grain 2 for others.

The Σ9 grain boundaries used in the investigation are shown in figures 5.47, 5.48(*a*) and (*b*) and figure 5.57. Figure 5.47 is a composite electron micrograph showing five of the boundary planes for which rigid-body displacements were determined. It shows four grains, 1, 2a, 2b and 3, where grains 2a and 2b are in the same orientation. The boundary along AB separating grains 1 and 2a is a faceted Σ9 boundary, and that along CD separating grains 1 and 2b is a curved Σ9 boundary. The boundary along BC separating grains 1 and 3 is a partly faceted $\Sigma 27_a$ boundary with incoherent Σ3 boundaries forming the balance at the triple junctions B and C. The micrograph in figure 5.47 was taken under simultaneous double two-beam diffraction conditions with the same-$\boldsymbol{g}_c = 220$ in all grains. It is clear from the presence of the fringe contrast for these diffracting conditions that rigid-body displacements are present between all grains. The fringe contrast at the curved Σ9 boundary CD shows no significant change with variation in boundary plane. However, the fringe contrast at the faceted Σ9 boundary AB shows marked changes with boundary plane, that is weak fringe contrast on facets X (close to $(\bar{1}14)_1$), strong fringe contrast on facets $Y((\bar{2}\bar{1}3)_1)$ and virtually no contrast on facets X' $((\bar{1}14)_1)$. Figure 5.48(*a*) is another example of a faceted Σ9 boundary from a different thin-foil specimen with the same types of facet X (close to $(1\bar{1}\bar{4})_1$) and $Y((21\bar{3})_1)$ as those in figure 5.47. The diffraction conditions for figure 5.48(*a*) are similar to those for figure 5.47 and the displacement fringes change in the same way for

† Determinations of rigid-body displacements have been made in germanium for a Σ5 boundary (Bacmann *et al* 1985) and for Σ9 and Σ11 boundaries (Papon *et al* 1984).

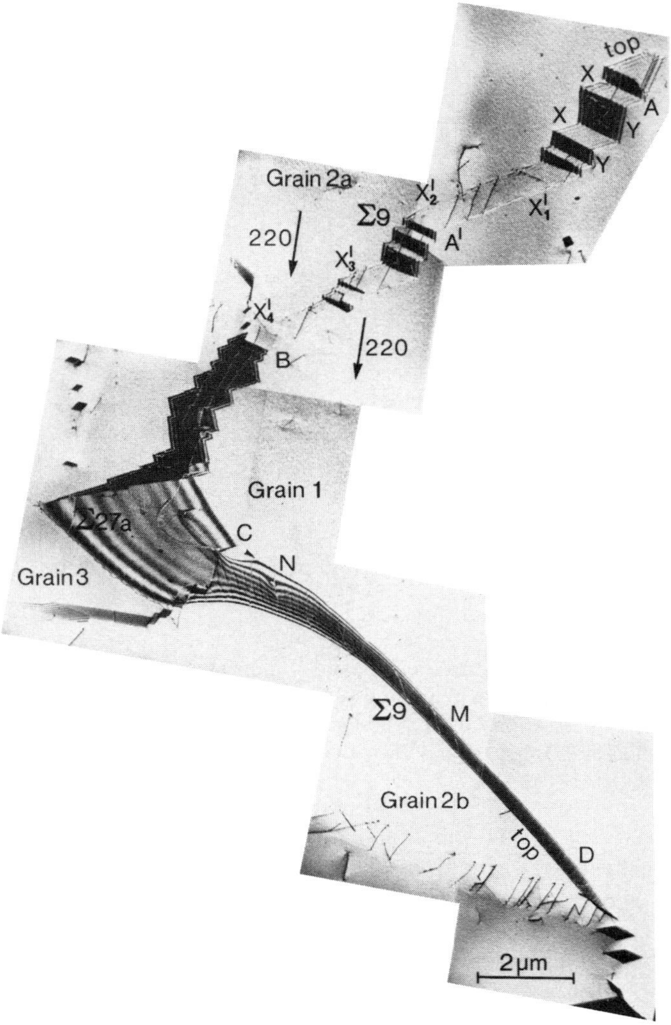

Figure 5.47 Double two-beam same-$\boldsymbol{g}_c$ image of faceted and curved Σ9 boundaries in a Cu − 6 at% Si alloy. The diffracting vectors and the intersections of the boundaries with the top surface of the specimen are indicated and $\boldsymbol{B}$ in the upper grain is $[1\ \bar{1}\ 14]_1$.

the two types of facet, indicating that the marked change in rigid-body displacement is a characteristic feature of the facet planes involved.

For all ten boundary planes investigated, which are listed in table 5.3, the values of $\boldsymbol{g}_c \cdot \boldsymbol{R}$ in the range $-\frac{1}{2} \leqq \boldsymbol{g}_c \cdot \boldsymbol{R} \leqq \frac{1}{2}$ were obtained from a comparison of experimental and theoretical images for both $+\boldsymbol{g}_c$ and $-\boldsymbol{g}_c$.

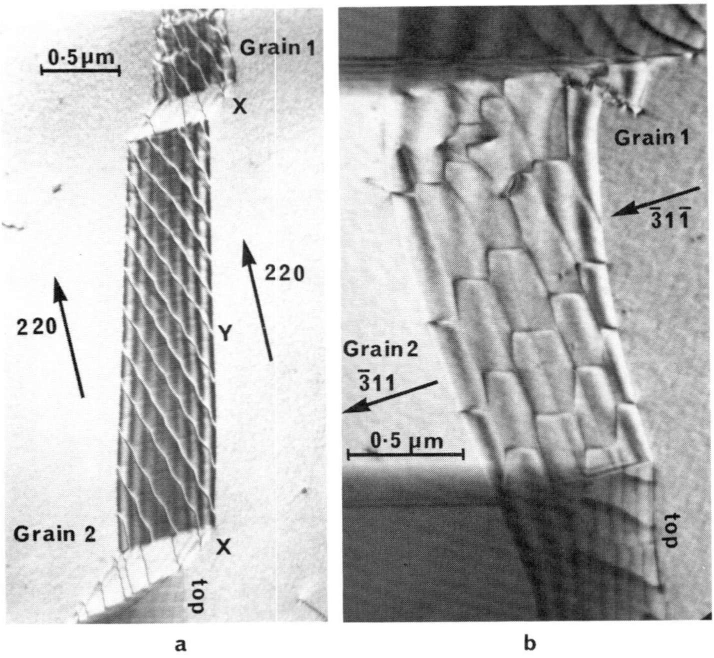

Figure 5.48 Double two-beam same-$\boldsymbol{g}_c$ images of two different $\Sigma 9$ boundaries (a) and (b) in a Cu – 6 at% Si alloy. The diffracting vectors and the intersections of the boundaries with the top surface of the specimen are indicated and $\boldsymbol{B}$ in the upper grain is $[3\bar{3}7]_2$ in (a) and close to $[\bar{1}\bar{4}1]_2$ in (b).

This enabled values of $\boldsymbol{g}_c \cdot \boldsymbol{R}$ to be specified to an accuracy which was typically ± 0.02; in general this is a narrower range than would have been possible had only $+\boldsymbol{g}_c$ images been used. Figure 5.49 is an example of the type of comparison made between experimental and computed images for the determination of $\boldsymbol{g}_c \cdot \boldsymbol{R}$. The experimental image in figure 5.49 is a portion of the Y facet of figure 5.48(a) and is taken with $\boldsymbol{g}_c = \overline{2}\overline{2}0_1/\overline{2}\overline{2}0_2$. A comparison of this image with theoretical images, computed using the experimentally determined diffraction parameters for different values of $\boldsymbol{g}_c \cdot \boldsymbol{R}$, shows that the double central dark fringe in the experimental image is not resolved in the theoretical images for values of $\boldsymbol{g}_c \cdot \boldsymbol{R}$ of 0.30 and 0.33, whereas it is resolved for values of $\boldsymbol{g}_c \cdot \boldsymbol{R}$ of 0.31 and 0.32. Thus for the diffraction parameters of figure 5.49, the value of $\boldsymbol{g}_c \cdot \boldsymbol{R}$ lies within the range 0.30–0.33. It should be noted that the contrast associated with the coarse network of secondary grain boundary dislocations does not interfere with the visual assessment of fringe contrast in the image matching process. The values of $\boldsymbol{g}_c \cdot \boldsymbol{R}$ determined in this way are listed in table 5.3 for the ten cases considered,

Table 5.3 Rigid-body displacements for Σ9 boundaries.

Boundary type	Case	Boundary normal $\boldsymbol{v}$ directed from grain 2 to grain 1	$-0.5 \leq \boldsymbol{g}_c \cdot \boldsymbol{R} \leq +0.5$			Components of $(\boldsymbol{R}+\boldsymbol{d})$† indexed in grain 1		
			$\boldsymbol{g}_c = 200_1$	$\boldsymbol{g}_c = 3\bar{1}1_1$	$\boldsymbol{g}_c = 024_1$	$(\boldsymbol{R}+\boldsymbol{d})_x \parallel [100]_1$	$(\boldsymbol{R}+\boldsymbol{d})_y \parallel [010]_1$	$(\boldsymbol{R}+\boldsymbol{d})_z \parallel [001]_1$
Faceted	(1) *X* facet	$(\bar{2}18)_1/(17\ \overline{26}\ 68)_2$	$+0.095 \pm 0.015$	$+0.015 \pm 0.015$	-0.20 ± 0.02	$+0.030 \pm 0.004$	$+0.017 \pm 0.006$	-0.059 ± 0.005
(see figures 5.47 and 5.51)	(2) *Y* facet	$(\bar{2}\bar{1}3)_1/(\bar{5}\ \overline{22}\ 25)_2$	-0.34 ± 0.02	-0.225 ± 0.015	$+0.34 \pm 0.02$	-0.070 ± 0.005	-0.100 ± 0.007	-0.115 ± 0.006
Faceted	(3) *X* facet	$(5\ \bar{2}\ \overline{23})_1/(\bar{2}3\bar{7})_2$	-0.10 ± 0.02	0.0 ± 0.02	$+0.20 \pm 0.02$	-0.028 ± 0.006	-0.022 ± 0.008	$+0.061 \pm 0.006$
(see figure 5.48(*a*))	(4) *Y* facet	$(21\bar{3})_1/(5\ 22\ \overline{25})_2$	$+0.315 \pm 0.015$	$+0.225 \pm 0.025$	-0.33 ± 0.01	$+0.065 \pm 0.006$	$+0.092 \pm 0.007$	$+0.121 \pm 0.004$
Faceted	(5) *X′* facet	$(\bar{1}14)_1/(1\bar{1}4)_2$	0.0	0.0	-0.21 ± 0.03	$+0.012 \pm 0.002$	-0.012 ± 0.002	-0.047 ± 0.007
(see figures 5.47 and 5.51)								
Curved	(6) region M	$(\bar{5}\bar{4}3)_1/(\overline{32}\ \overline{49}\ 25)_2$	-0.295 ± 0.015	-0.21 ± 0.02	-0.25 ± 0.01	-0.082 ± 0.005	-0.066 ± 0.007	-0.030 ± 0.004
(see figures 5.47 and 5.56)	(7) region N	$(\bar{3}\bar{6}5)_1/(10\ \overline{71}\ 23)_2$	-0.30 ± 0.01	-0.20 ± 0.02	-0.21 ± 0.01	-0.083 ± 0.005	-0.067 ± 0.006	-0.019 ± 0.004
Curved (see figure 5.48(*b*))	(8)	$(65\bar{6})_1/(29\ 70\ \overline{46})_2$	$+0.295 \pm 0.015$	$+0.145 \pm 0.015$	$+0.26 \pm 0.02$	$+0.067 \pm 0.004$	$+0.081 \pm 0.006$	$+0.025 \pm 0.005$
			$\boldsymbol{g}_c = 220_1$	$\boldsymbol{g}_c = 3\bar{1}1_1$	$\boldsymbol{g}_c = \bar{2}04_1$			
Curved	(9) region K	$(8\ \overline{89}\ \overline{14})_1/(\bar{1}\bar{8}6)_2$	-0.27 ± 0.01	-0.165 ± 0.015	-0.085 ± 0.015	-0.062 ± 0.004	-0.073 ± 0.005	-0.052 ± 0.004
(see figure 5.57)	(10) region L	$(\bar{1}\bar{4}1)_1/(8\ \overline{37}\ \bar{5})_2$	-0.28 ± 0.02	-0.16 ± 0.01	$+0.22 \pm 0.03$	-0.079 ± 0.004	$+0.061 \pm 0.008$	$+0.016 \pm 0.007$

† $(\boldsymbol{R}+\boldsymbol{d})$ is the displacement of grain 2 with respect to grain 1 and lies within the Wigner–Seitz cell of the DSC lattice. For the *Y* facets the values of $\boldsymbol{g}_c \cdot \boldsymbol{R}$ do not lead directly to values of $(\boldsymbol{R}+\boldsymbol{d})$ and $(\boldsymbol{R}+\boldsymbol{d})$ here was obtained by subtracting the DSC vector $(1/18)[\bar{1}14]_1$.

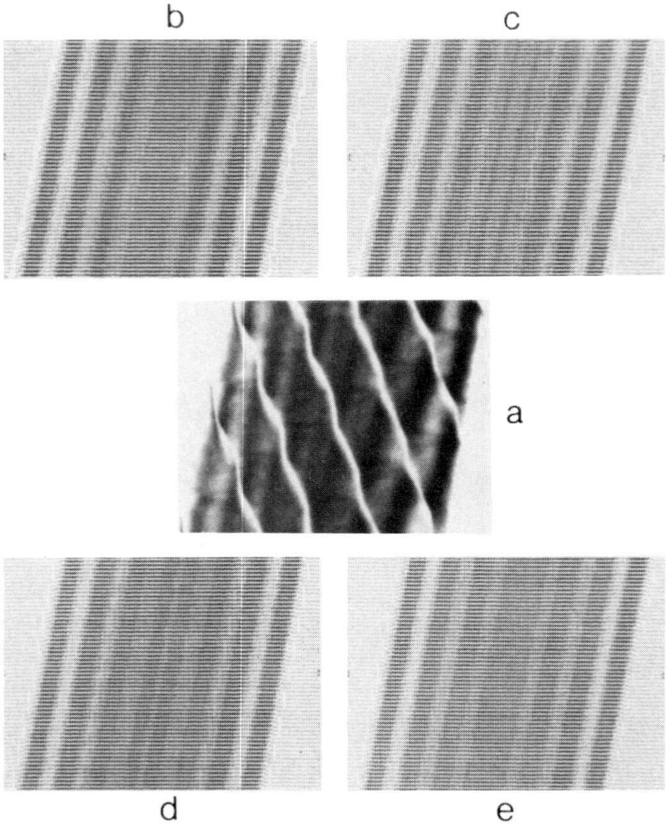

Figure 5.49 Comparison of an experimental same-$\boldsymbol{g}_c$ image (a) of a portion of facet Y of figure 5.48(a) with computed imates (b)–(e) for values of $\boldsymbol{g}_c \cdot \boldsymbol{R}$ of 0.30, 0.31, 0.32 and 0.33 respectively. The diffracting vector $\boldsymbol{g}_c = \bar{2}\bar{2}0_{1/2}$, the deviation parameters are $w_1 = 0.40$, $w_2 = 0.35$, the foil thickness is $5.3\xi_{220}$ and the beam direction in the upper grain is $[4\bar{4}9]_2$.

together with the derived components of $(\boldsymbol{R} + \boldsymbol{d})$, where in each case $\boldsymbol{d}$ is a DSC vector such that $(\boldsymbol{R} + \boldsymbol{d})$ lies in the Wigner–Seitz cell of the DSC lattice. The distribution of boundary planes listed in table 5.3 is shown in the stereographic projection of figure 5.50. It will now be shown how values of $\boldsymbol{R}$ can be determined from the values of $(\boldsymbol{R} + \boldsymbol{d})$ for the boundary planes corresponding to cases (1)–(5) in table 5.3 which are clearly defined facet planes. This will first involve establishing the values of $\boldsymbol{R}$ for the $(\bar{1}14)_1/(1\bar{1}4)_2$ plane of facets X' and for the closely related X facets, and then determining values of $\boldsymbol{R}$ for the interconnected Y facets by the identification of the partial grain boundary dislocations at the facet intersections.

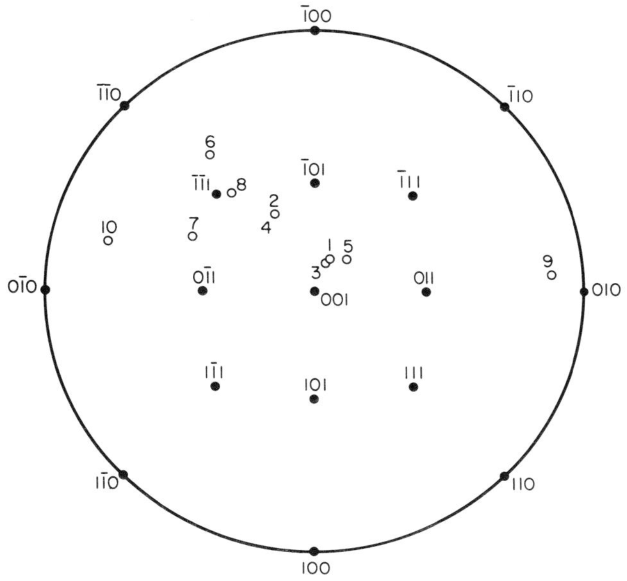

Figure 5.50 Stereographic projection of boundary plane normals (open circles) indexed with respect to grain 1 of cases (1)–(10) of table 5.3. In order to plot all boundary normals in the same half of the stereographic sphere the normals are negated for cases (3), (4), (8) and (9).

Figure 5.51 shows the portion AA′ of the faceted boundary AB in figure 5.47 at higher magnification for three different same-$\boldsymbol{g}_c$ diffracting vectors, illustrating the variation in fringe contrast on the facets X, Y and X'_1, where facets X lie on $(\bar{2}18)_1/(17\ \overline{26}\ 68)_2$ (case (1) of table 5.3), facets Y lie on $(\bar{2}\bar{1}3)_1/(\bar{5}\ \overline{22}\ 25)_2$ (case (2) of table 5.3) and facet X'_1 on $(\bar{1}14)_1/(1\bar{1}4)_2$ (case (5) of table 5.3). The $0\bar{2}\bar{4}_1/\bar{2}0\bar{4}_2$ image of figure 5.51(a) shows strong fringe contrast on all facets. However, for the $220_1/220_2$ and $3\bar{1}\bar{1}_1/3\bar{1}\bar{1}_2$ images, the fringe contrast is very weak on facets X (which are within 7° of $(\bar{1}14)_1/(1\bar{1}4)_2$) and strong on facets Y. For the facet X'_1 in figure 5.51 there is virtually no fringe contrast for the $220_1/220_2$ and $3\bar{1}1_1/3\bar{1}\bar{1}_2$ diffracting vectors which lie in the $(\bar{1}14)_1/(1\bar{1}4)_2$ plane of the facet, and this lack of contrast occurred for all the X' facets of figure 5.47.

The absence of fringe contrast on a $(\bar{1}14)_1/(1\bar{1}4)_2$ plane for in-plane same-$\boldsymbol{g}_c$ diffracting vectors is confirmed by results on other Σ9 boundaries. For example, figure 5.52 is a case where a Σ9 boundary on $(\bar{1}14)_1/(1\bar{1}4)_2$ joins two coherent Σ3 boundaries at a triple junction of grains labelled 1, 2 and 3. Apart from weak contrast effects at the lines of intersection of the Σ9 boundary with the surfaces of the specimen, no fringe contrast is present on this Σ9 boundary for the $\bar{2}\bar{2}0_1/\bar{2}\bar{2}0_2$ (figure 5.52(b)) and $\bar{3}1\bar{1}_1/\bar{3}11_2$ (figure 5.52(c)) in-plane diffracting vectors.

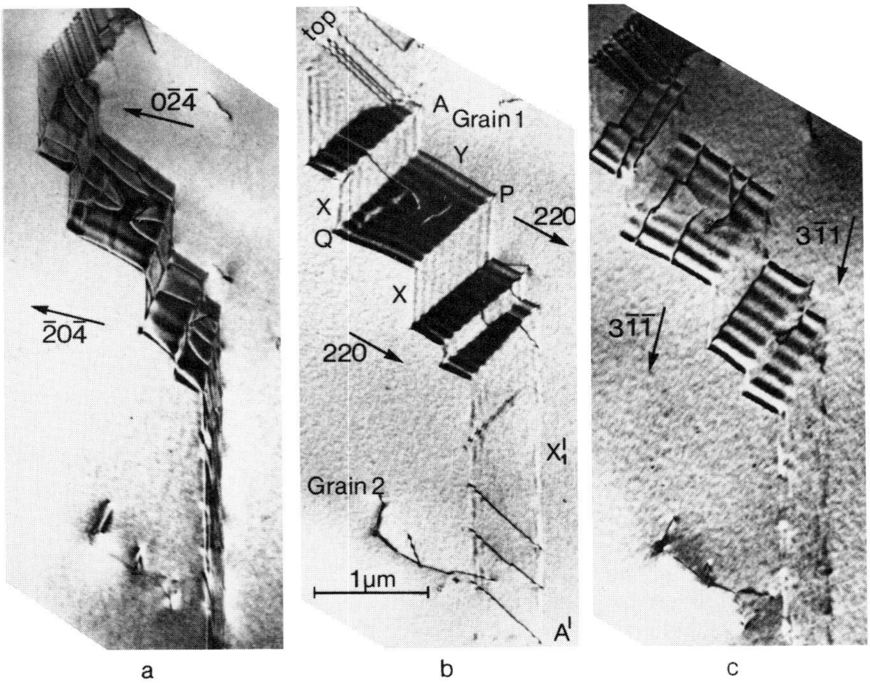

Figure 5.51 Three non-coplanar same-$\boldsymbol{g}_c$ images of facets X, X' and Y of figure 5.47. The diffracting vectors and the intersection of the boundary with the top surface of the specimen are indicated. $\boldsymbol{B}$ is $[\bar{2}\bar{6}3]_1$ in (a), $[1\ \bar{1}\ 14]_1$ in (b) and $[\bar{3}\bar{4}5]_1$ in (c).

The results in figures 5.51 and 5.52 show that for a $\Sigma 9$ boundary on $(\bar{1}14)_1/(1\bar{1}4)_2$, $(\boldsymbol{R}+\boldsymbol{d})$ has no in-plane component and is solely a displacement normal to the boundary. The magnitude of $(\boldsymbol{R}+\boldsymbol{d})$ normal to the boundary is determined from a comparison of experimental and computed same-$\boldsymbol{g}_c \pm (024_1/204_2)$ images as shown in figure 5.53 for one of the positions X' of figure 5.51(a) (see case (5) of table 5.3). This value of $(\boldsymbol{R}+\boldsymbol{d})$ represents a displacement of grain 2 away from grain 1 of approximately $(1/86)[1\bar{1}\bar{4}]_1$, that is a small expansion of 0.049a at the boundary, where a is the lattice parameter of the Cu–Si alloy equal to 3.6191 Å.

The value of $\boldsymbol{R}$ for the $(\bar{1}14)_1/(1\bar{1}4)_2$ $\Sigma 9$ boundary can be found by comparing the different atomic models that are obtained when different DSC vectors (values of $\pm\boldsymbol{d}$) are added to or subtracted from the experimentally determined value of $(\boldsymbol{R}+\boldsymbol{d}) = (1/86)[1\bar{1}\bar{4}]_1$. In order to make meaningful comparisons, it is necessary to start with an initial reference structure where the two lattices are in the coincident state from which the rigid-body displacement is to be measured, and to choose a level for the boundary plane in the coincident site lattice. The reference structure chosen is shown in figure

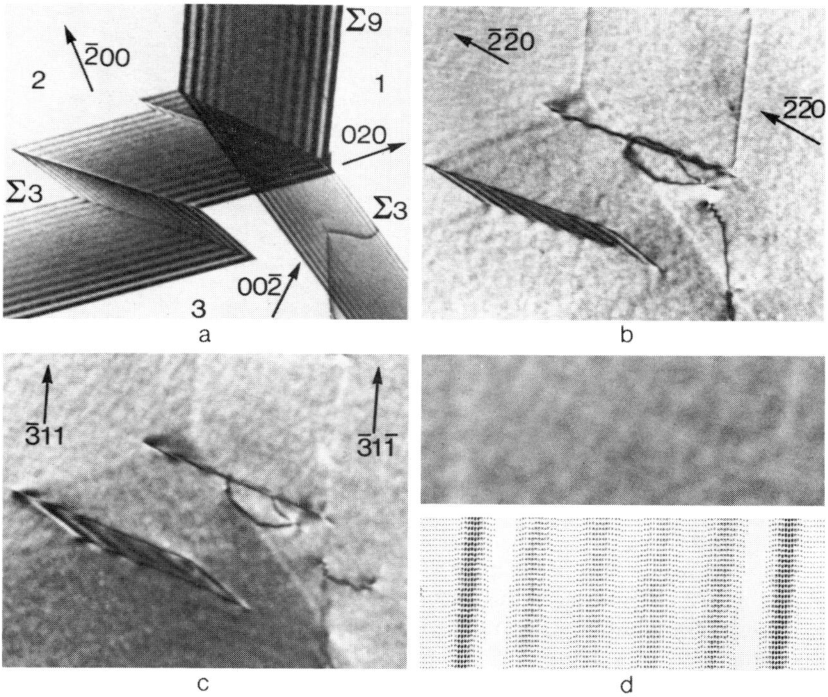

Figure 5.52 Images used to demonstrate that there is no in-plane translation for a Σ9 boundary on a $(\bar{1}14)_1/(1\bar{1}4)_2$ plane: (*a*) a triple junction between the Σ9 boundary and two coherent-Σ3 boundaries where each grain satisfies double two-beam diffraction conditions, with beam directions $[\bar{1}02]_1/[0\bar{1}2]_2/[\bar{2}10]_3$; (*b*) and (*c*) same-$\boldsymbol{g}_c$ images with $\boldsymbol{B}$ as $[\bar{1}\ 1\ 10]_1$ in (*b*) and $[\bar{3}\bar{1}8]_1$ in (*c*); (*d*) a portion of the image of the $(\bar{1}14)_1/(1\bar{1}4)_2$ plane of (*c*), at higher magnification, compared with a computed image for an in-plane displacement of 0.62*a* along $[\bar{2}2\bar{1}]_1$.

5.54(*a*), and is the simple juxtaposition of two grains at the Σ9 orientation meeting at the $(\bar{1}14)_1/(1\bar{1}4)_2$ plane, where the coincident sites in the boundary are marked *c*. The simplest changes needed to the structure of figure 5.54(*a*) to correct the misfit of atoms closer than the $(1/2)\langle 110 \rangle$ interatomic distance and to produce an expansion normal to the boundary are: (i) a localised expansion normal to the boundary of $(1/18)/[1\bar{1}\bar{4}]_1$ (0.236*a*) at atoms *a* (so that they are separated by the $(1/2)\langle 110 \rangle$ interatomic distance) and (ii) the movement of atoms *b* in grain 1 by approximately $(1/2)[1\bar{1}0]_1$, so that they lie on the boundary plane in the dilated regions near atoms *a*. If this modified structure relaxes so as to change the local expansion of $(1/18)[1\bar{1}\bar{4}]_1$ to an average expansion over the boundary as a whole equal to the measured expansion $(1/86)[1\bar{1}\bar{4}]_1$, then the structure shown in figure 5.54(*b*) is

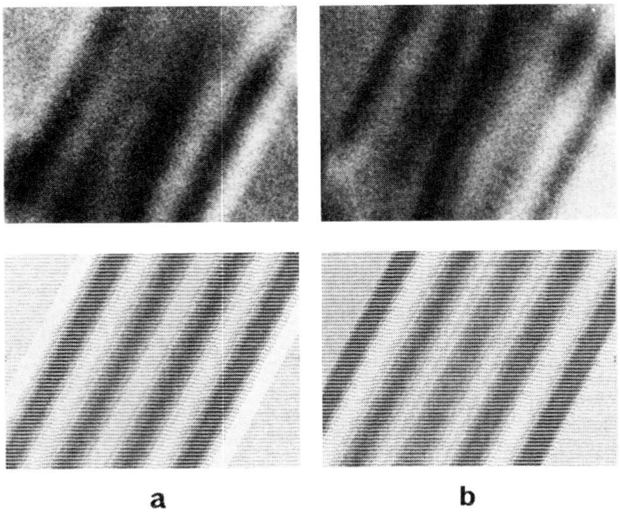

Figure 5.53 Comparison of same-$\boldsymbol{g}_c$ experimental and computed images of the region X'_3 of figure 5.47(a) for $\boldsymbol{g}_c = 0\bar{2}\bar{4}_1/\bar{2}0\bar{4}_2$ in (a) and $\boldsymbol{g}_c = 024_1/204_2$ in (b). In this figure, grain 1, the upper grain with respect to the electron beam, is on the left. In the matching computed images the values of $\boldsymbol{g}_c \cdot \boldsymbol{R}$ are 0.21 in (a) and -0.21 in (b), the deviation parameters are $w_1 = 0.30$ and $w_2 = 0.24$ in (a) and $w_1 = 0.0$ and $w_2 = 0.25$ in (b), $\boldsymbol{B}$ is $[\bar{2}\bar{6}3]_1$ in (a) and (b) and the specimen thickness is $3.74\xi_{420}$. The experimental images contain a segment of grain boundary dislocation which is approximately parallel to the fringes near mid-specimen.

obtained. If this structure is a valid model of the boundary, then $\boldsymbol{d} = 0$ so that $\boldsymbol{R} = (1/86)[1\bar{1}\bar{4}]_1$, i.e. the observed average rigid-body displacement between the grains. However, other models need to be considered before making a choice as to the most appropriate one. If the level of the boundary plane in the reference structure of figure 5.54(a) is kept fixed, eight other types of juxtaposition of the two grains are obtained from figure 5.54(a) by the displacement of grain 2 with respect to grain 1 by the in-plane DSC vectors $\boldsymbol{d} = (n/9)[\bar{2}2\bar{1}]_1$, where n takes the values 1–8. Alternatively these eight structures could be obtained with $\boldsymbol{d} = 0$ by changing the level of the boundary plane for the juxtaposed grains of figure 5.54(a). For each of these eight structures, corrections of the atomic misfit (where atoms are closer than $(1/2)\langle 110 \rangle$) cannot be made without either removing atoms, which results in a collapse rather than the observed expansion at the boundary, or resorting to displacements which have non-DSC in-plane components, contrary to observation. Therefore within the limits of this simple modelling, the structure

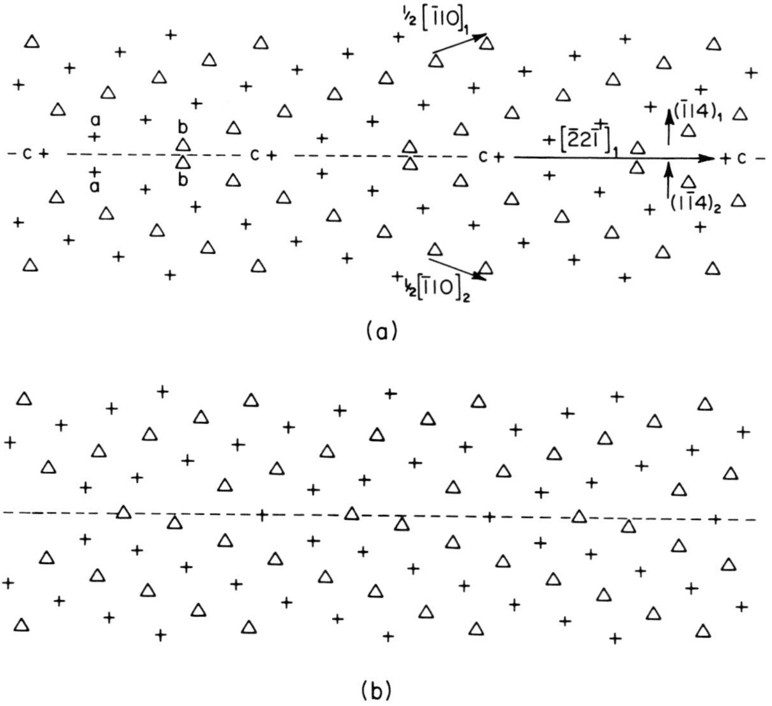

Figure 5.54 Hard-sphere models of a $\Sigma 9$ symmetric-tilt boundary on $(\bar{1}14)_1/(1\bar{1}4)_2$ in an FCC structure: (*a*) $[\bar{1}\bar{1}0]$ projection of two crystals juxtapored at the $(\bar{1}14)_1/(1\bar{1}4)_2$ plane (+represents atoms in the plane of the figure and $\triangle$ atoms at $\pm(1/4)[110]$ normal to the plane of the figure); (*b*) structure of (*a*) after atomic displacements to eliminate interatomic distances closer than $(1/2)\langle 110\rangle$.

of figure 5.54(*b*) is the best representation of a $\Sigma 9$ $(\bar{1}14)_1/(1\bar{1}4)_2$ boundary so that $\boldsymbol{R} = (1/86)[1\bar{1}\bar{4}]_1$.

Pond *et al* (1979), using their computer modelling technique, have calculated the structure of a $\Sigma 9$ $(\bar{1}14)_1/(1\bar{1}4)_2$ boundary in aluminium, and their calculated configuration of atoms at the boundary is similar to the model shown in figure 5.54(*b*). However, they found an in-plane translation of $0.62a$ parallel to $[\bar{2}2\bar{1}]_1$ in addition to a local expansion of $0.208a$. Their local expansion is close to the value of $0.236a$ in figure 5.54(*b*), but their in-plane shear is not observed in the Cu–Si alloy. For example, figure 5.52(*d*) shows a comparison of a portion of the experimental image of the $\Sigma 9$ boundary in figure 5.52(*c*) with a theoretical image computed for the same diffraction conditions, but with the inclusion of an in-plane shear of $0.62a$. Clearly, such a shear gives detectable fringes which are not observed in the experimental image. However, had the in-plane shear found by Pond *et al* been the DSC displacement 0.667a, corresponding to the vector $(2/9)[\bar{2}2\bar{1}]_1$, then no fringe contrast would have been obtained for the diffraction conditions of figure 5.52(c), and their structure would have been essentially the same as that in figure 5.54(b).

In figure 5.51 the fringe contrast on the X facets is not very different from that on the $(\bar{1}14)_1/(1\bar{1}4)_2$ X' facets for all three non-coplanar same-$\boldsymbol{g}_c$ images (see the values of $\boldsymbol{g}_c \cdot \boldsymbol{R}$ in table 5.3). In addition all the planes X are within 7° of X'. Therefore it will be assumed that the rigid-body displacements of the X and X' facets differ only marginally, and that for the X facets, as for the X' facets, $\boldsymbol{d} = 0$, so that the rigid-body displacements $\boldsymbol{R}$ for X facets are the experimental values $(\boldsymbol{R} + \boldsymbol{d})$ in table 5.3.

The determination of the rigid-body displacement $\boldsymbol{R}$ for the Y facets (case (2) of table 5.3) can now be considered by relating this displacement to that on the X facets. In order to determine the relation between the displacements on the X and Y facets, it is necessary to determine the Burgers vectors of the partial dislocations along the facet intersections. The partial dislocations at P and Q of figure 5.51(b) were selected because none of the secondary grain boundary dislocations which were approximately parallel to facet intersections were lying close to them. The Burgers vectors, $\boldsymbol{b}_\mathrm{P}$ and $\boldsymbol{b}_\mathrm{Q}$, of these dislocations (with the positive sense of the line directions taken from the bottom to the top of the specimen in figure 5.51(b)), specified in accordance with the FS/RH convention and with the procedure of Thompson (1953) for designating the Burgers vectors of partial dislocations, are related to the rigid-body displacements, $\boldsymbol{R}_X$ and $\boldsymbol{R}_Y$, on the X and Y facets by

$$\boldsymbol{b}_\mathrm{P} = -\boldsymbol{b}_\mathrm{Q} = \boldsymbol{R}_X - \boldsymbol{R}_Y.$$

In this equation $\boldsymbol{R}_X$ is the value of $(\boldsymbol{R} + \boldsymbol{d})$ for the X facet in case (1) of table 5.3, i.e.

$$\boldsymbol{R}_X = [0.030 \;\; 0.017 \;\; -0.059]_1$$

and $\boldsymbol{R}_Y$ can be expressed in terms of the value of $(\boldsymbol{R} + \boldsymbol{d})$ for the Y facet of case (2) of table 5.3 as

$$\boldsymbol{R}_Y = [-0.070 \;\; -0.100 \;\; -0.115]_1 - \boldsymbol{d}.$$

Therefore,

$$\boldsymbol{b}_\mathrm{P} = -\boldsymbol{b}_\mathrm{Q} = [0.030 \;\; 0.017 \;\; -0.059]_1 - [-0.070 \;\; -0.100 \;\; -0.115]_1 + \boldsymbol{d}$$

that is

$$\boldsymbol{b}_\mathrm{P} = -\boldsymbol{b}_\mathrm{Q} = [0.100 \;\; 0.117 \;\; 0.056]_1 + \boldsymbol{d}. \tag{5.8}$$

The value of $\boldsymbol{d}$ in equation (5.8) can be found by determining the Burgers vectors of the partial dislocations along P and Q using the technique of image matching for simultaneous double two-beam images. To compute images of partial grain boundary dislocations at facet intersections the program PCGBD described in Chapter 2 was modified to take into account the appropriate geometry and elastic displacement field for a dislocation along the line of intersection of boundary planes. The displacement field used was that derived for elastic anisotropy by Bonnet (1982). To identify the partial dislocations at P and Q, theoretical images were computed for 17 possible partial dislocations along P and along Q using equation (5.8) with values of $\boldsymbol{d}$ of 0, $\pm(1/18)[\bar{1}14]_1$, $\pm(1/9)[2\bar{2}1]_1$, $\pm(1/18)[\bar{7}\bar{2}1]_1$ and linear combinations of these basis DSC vectors up to the magnitude of $a/\sqrt{6}$. From a comparison of experimental and theoretical images, it was found that the theoretical images for all values of $\boldsymbol{d}$, except $\boldsymbol{d} = 0$, disagreed with the experimental images. The matching set of experimental and theoretical images for $\boldsymbol{b}_P = -\boldsymbol{b}_Q = [0.100\ 0.117\ 0.056]_1$ are shown in figure 5.55. From these

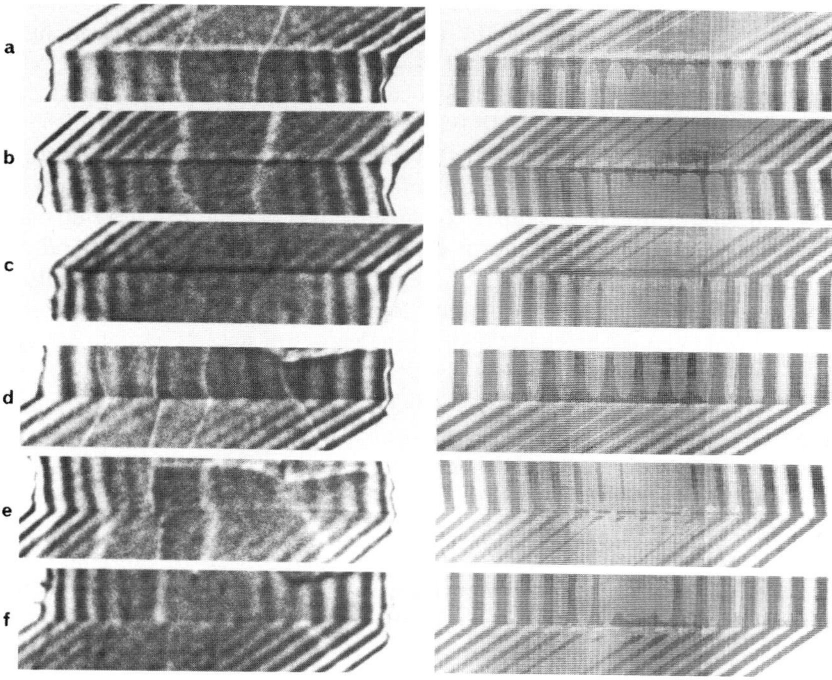

Figure 5.55 Comparison of matching double two-beam experimental and computed images of the partial grain boundary dislocations along the lines of intersection P, (a)–(c), and Q, (d)–(f), of the facet planes X and Y of figure 5.51(b), where $\boldsymbol{b}_P = -\boldsymbol{b}_Q = a[0.10, 0.117, 0.056]_1$. In this figure grain 1 is on the left. The diffraction parameters are as follows: for (a) and (d) $\boldsymbol{g}_1 = \bar{1}\bar{1}\bar{1}_1$, $\boldsymbol{g}_2 = \bar{2}00_2$, $w_1 = 0.48$, $w_2 = 0.41$ and $\boldsymbol{B}$ is $[\bar{3}\bar{4}7]_1$; for (b) and (e) $\boldsymbol{g}_1 = 020_1$, $\boldsymbol{g}_2 = \bar{1}11_2$, $w_1 = 0.28$, $w_2 = 0.23$ and $\boldsymbol{B}$ is $[\bar{1}09]_1$; for (c) and (f) $\boldsymbol{g}_1 = 111_1$, $\boldsymbol{g}_2 = \bar{1}\bar{1}\bar{1}_2$, $w_1 = 0.19$, $w_2 = 0.27$ and $\boldsymbol{B}$ is $[\bar{5}\bar{2}7]_1$.

results it is concluded that the rigid-body displacement $\boldsymbol{R}_Y$ at the Y facet is given by the value of $(\boldsymbol{R}+\boldsymbol{d})$ for the Y facet in table 5.3 (case (2)), i.e. $\boldsymbol{R}_Y = [-0.070\ \ -0.100\ \ -0.115]_1$.

This method of determining the rigid-body displacement on the facet plane $(\bar{2}\bar{1}3)_1$ is as valid as that for the $(\bar{1}14)_1/(1\bar{1}4)_2$ plane, provided the partial dislocation identified at each facet intersection arises solely from the difference in the rigid-body displacements on the facets concerned. However, it could always be argued that there is a possibility that an additional dislocation, with a DSC Burgers vector, running parallel to the partial dislocation throughout the thickness of the specimen, might be present close to such an intersection and be difficult to detect as a separate entity in electron micrographs. For example, such a dislocation could be an element of the secondary grain boundary dislocation network. If this were the case, the determination of $\boldsymbol{R}_Y$ would be invalidated. However, it is unlikely that such a dislocation would go undetected in the image matching process and, in addition, for the geometry of the facet intersection at P and Q in figure 5.51(b), image matching shows that $\boldsymbol{b}_P = -\boldsymbol{b}_Q$, so that for the present result to be invalidated two additional dislocations of opposite sign, one at each facet intersection, would have to be present and to go undetected.

For the example of figure 5.48(a) it can be seen from table 5.3 (cases (3) and (4)) that apart from a reversal of sign in the boundary normals, the planes X and Y are very similar to those treated in the example of figure 5.47 (cases (1) and (2)). Further it can be seen that the values of $\boldsymbol{g}_c\cdot\boldsymbol{R}$ and $(\boldsymbol{R}+\boldsymbol{d})$, apart from a change in sign, are equal within the limits of experimental error to those obtained for the case of figure 5.47. Therefore for the facets X and Y of figure 5.48(a) the values of $\boldsymbol{R}$ can be taken as equal to $(\boldsymbol{R}+\boldsymbol{d})$.

The results for the rigid-body displacements on the X, Y and X' facets (cases (1)–(5)) of table 5.3 are expressed in table 5.4 in terms of in-plane and out-of-plane components of $\boldsymbol{R}$. A comparison of the results for the X' and X facets shows that as soon as the boundary plane departs from the symmetric-tilt boundary on $(\bar{1}14)_1/(1\bar{1}4)_2$, an in-plane component of $\boldsymbol{R}$ is incurred, together with a slight increase in expansion at the boundary. For the Y facets the in-plane components are an order of magnitude greater than for the X facets and the expansion at the boundary is slightly less.

The remaining cases in table 5.3, namely cases (6)–(10), are for the curved boundary CD in figures 5.47 and 5.56, and the curved boundaries in figure 5.48(b) and in figure 5.57. It is not possible to determine $\boldsymbol{R}$ from the experimental determinations of $(\boldsymbol{R}+\boldsymbol{d})$ for these boundaries in the manner described for cases (1)–(5) of table 5.3, since they cannot be related to a reference state such as the $(\bar{1}14)_1/(1\bar{1}4)_2$ boundary. A common feature of the results found for curved boundaries is that $(\boldsymbol{R}+\boldsymbol{d})$ corresponds to a movement of grains 1 and 2 towards one another. Clearly, this is contrary to the normal expectation of an expansion at a grain boundary and implies

components, with the in-plane component along the [111] direction. Further, they argue that the rigid-body displacement could be similar on {213} and {110} planes.

This lack of experimental information led Forwood and Clarebrough (1988) to make a detailed investigation of rigid-body displacements at {112} and {213} boundary planes in a $\Sigma3$ boundary in iron (99.963% Fe, nominally containing 0.004 at% C, 0.003 at% N and a total of 0.03 at% substitutional impurities). The results show that {112} facets are symmetrical boundaries with only an expansive rigid-body displacement normal to the boundary plane, while the {213} facets have a rigid-body displacement parallel to the ⟨211⟩ direction nearest their boundary normals; in no case is there a component of displacement along the ⟨111⟩ rotation axis contained in the facet planes.

The faceted $\Sigma3$ boundary studied in detail is shown in the montage of electron micrographs in figure 5.58. The boundary which separates grains 1 and 2 is shown in two parts and extends from A in figure 5.58(*a*) to L in figure 5.58(*b*), and the two figures have a common region at ZZ′. The $\Sigma3$ CSL misorientation between the two grains is specified by a 60° rotation of the upper grain 1 with respect to the lower grain 2 in a right-handed sense around an axis with common indices [111] in both grains, and the rigid-body displacements are specified by displacements of grain 2 relative to grain 1. The region of the boundary A–B (labelled facet 1) lies on $(\bar{1}2\bar{1})_1/(\bar{2}11)_2$, the region C–D (labelled facet 2) lies on $(\bar{2}11)_1/(\bar{1}\bar{1}2)_2$, the region E–F (labelled facet 3) lies on $(\bar{2}\bar{1}3)_1/(1\bar{3}2)_2$, and the region F–G (labelled facet 4) lies on $(\bar{3}12)_1/(\bar{1}\bar{2}3)_2$. The region of the boundary G–I consists of facets very close to the planes of facets 3 and 4. Thus from A–I the various facet plane normals are all contained in the zone of the [111] rotation axis. However, in the region J–L the boundary is curved and the boundary normals depart markedly from the [111] zone. Facet 1 has already been discussed in connection with the determination of the Burgers vectors of secondary grain boundary dislocations in section 5.3.2.

Figure 5.59 shows a series of same-$\boldsymbol{g}_c$ images for facets 1–4, where facet 1 is shown in column (i), facet 2 in column (ii), and facets 3 and 4 in column (iii). For all facets no fringe contrast is present in images with $\boldsymbol{g}_c = 222_1/222_2$ (figure 5.59(*d*) (i), (ii), (iii)). Despite the large extinction distance for a 222 diffracting vector, displacement fringe contrast is readily detectable for non-integral values of $\boldsymbol{g}_c \cdot \boldsymbol{R}$ as illustrated in figure 5.60. In this figure the region of the boundary J–L which departs from the [111] zone shows strong displacement fringe contrast, whereas the region H–I, which is close to the plane of facet 4, shows no fringe contrast. In fact computed images show that, for the diffraction conditions used with $\boldsymbol{g}_c = 222_1/222_2$, fringe contrast would have been detected for facets 1–4 for $\boldsymbol{g}_c \cdot \boldsymbol{R}$ values as small as 0.02. Thus, within the limits of experimental error, it is concluded that there is no in-plane component of rigid-body displacement parallel to the [111] rotation

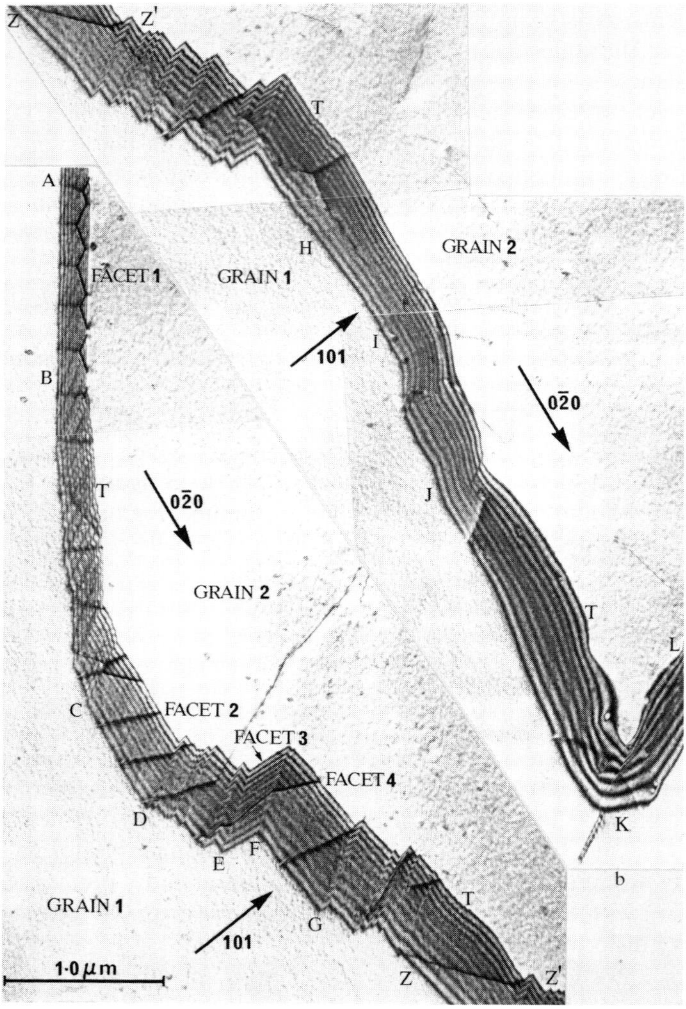

Figure 5.58 Montage of double two-beam images of a faceted Σ3 grain boundary in iron shown in two parts (*a*) and (*b*). In grain 1 ***B*** is close to $[\bar{2}32]_1$. The intersection of the boundary with the top surface of the specimen is marked T, and the diffracting vectors are indicated.

axis for each of the facets 1–4. Further it can be seen from figure 5.59 that fringe contrast is also absent for:
facet 1 with $\boldsymbol{g}_c = \bar{1}01_1/0\bar{1}1_2$ (figure 5.59(*f*)(i)),
facet 2 with $\boldsymbol{g}_c = 01\bar{1}_1/1\bar{1}0_2$ (figure 5.59(*a*)(ii)),
facet 3 with $\boldsymbol{g}_c = \bar{1}10_1/\bar{1}01_2$ (figure 5.59(*c*)(iii)) and
facet 4 with $\boldsymbol{g}_c = 01\bar{1}_1/1\bar{1}0_2$ (figure 5.59(*a*)(iii)).

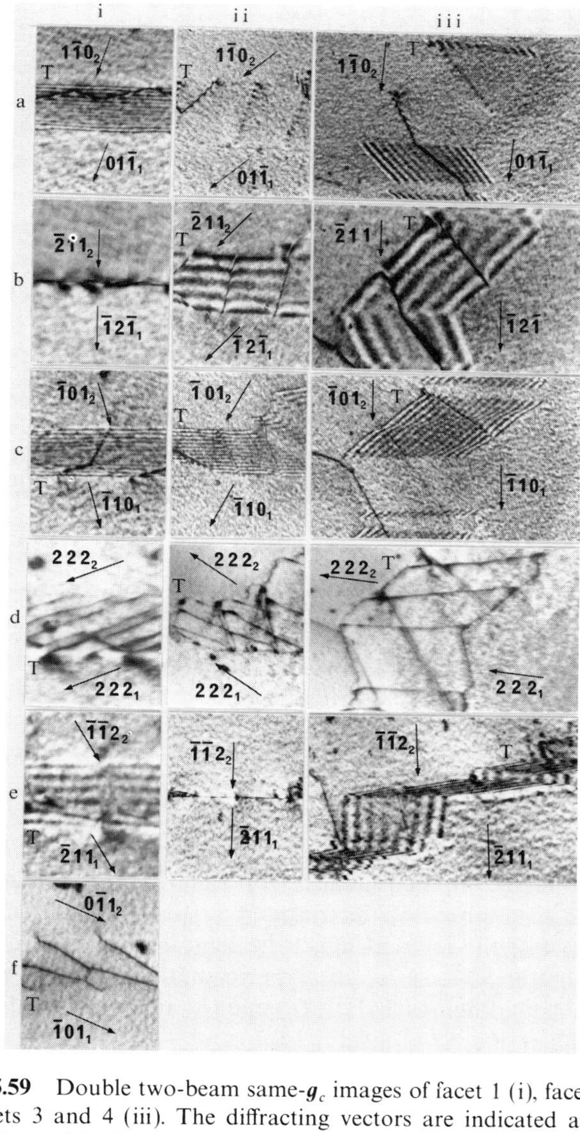

Figure 5.59 Double two-beam same-$\boldsymbol{g}_c$ images of facet 1 (i), facet 2 (ii) and facets 3 and 4 (iii). The diffracting vectors are indicated and the intersections of the facets with the top surface of the specimen are marked T. In grain 1 $\boldsymbol{B}$ is close to $[\bar{5}66]_1$ in (*a*), $[\bar{3}15]_1$ in (*b*), $[\bar{1}\bar{1}2]_1$ in (*c*), $[\bar{1}\bar{1}2]_1$ in (*d*), $[3\bar{2}8]_1$ in (*e*), and $[212]_1$ in (*f*) so that grain 1 is the upper grain in all cases except for facet 1 images (*c*), (*d*), (*e*) and (*f*) where grain 2 is the upper grain.

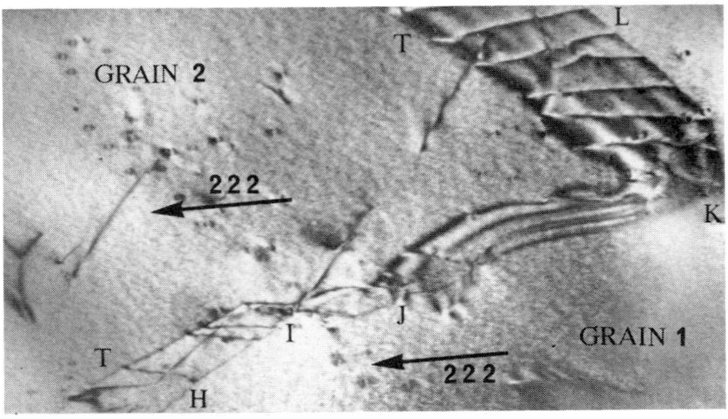

Figure 5.60 Double two-beam same-$\boldsymbol{g}_c$ image of the boundary region HKL with the diffraction conditions of figure 5.59(d).

Since, for each of the facets 1–4, fringe contrast is absent in images with two different same-$\boldsymbol{g}_c$, the rigid-body displacements, $(\boldsymbol{R}+\boldsymbol{d})$, must be parallel to $\pm[1\bar{2}1]_1/[2\bar{1}\bar{1}]_2$ for facet 1, $\pm[2\bar{1}\bar{1}]_1/[11\bar{2}]_2$ for facet 2, $\pm[11\bar{2}]_1/[\bar{1}2\bar{1}]_2$ for facet 3 and $\pm[2\bar{1}\bar{1}]_1/[11\bar{2}]_2$ for facet 4, where $\boldsymbol{d}$ has its usual meaning, i.e. it is the DSC vector such that $(\boldsymbol{R}+\boldsymbol{d})$ lies in the Wigner–Seitz cell of the DSC lattice.

Both the sense and magnitude of the displacement $(\boldsymbol{R}+\boldsymbol{d})$ for each facet were determined by image matching using experimental images taken in $\pm\boldsymbol{g}_c$ and corresponding computed images with values of $\boldsymbol{g}_c\cdot\boldsymbol{R}$ in the range $-\frac{1}{2}\leqq\boldsymbol{g}_c\cdot\boldsymbol{R}\leqq\frac{1}{2}$ and the results are given in table 5.5. Figure 5.61 shows two examples of this image matching procedure. Figure 5.61(a) is an experimental image of facet 1 with $\boldsymbol{g}_c=01\bar{1}_1/1\bar{1}0_2$ and figures 5.61(c), (d) and (e) are corresponding computed images for values $\boldsymbol{g}_c\cdot\boldsymbol{R}$ of -0.10, -0.14 and -0.18, respectively. The computed images (c) and (e) are mismatches to the experimental images because the fringes in (c) are not sufficiently developed near the centre of the foil and in (e) the central fringes are not sufficiently resolved. Figure 5.61(b) is an experimental image of facet 3 with $\boldsymbol{g}_c=\bar{1}21_1/\bar{2}11_2$ and figures 5.61(f), (g) and (h) are corresponding computed images for values of $\boldsymbol{g}_c\cdot\boldsymbol{R}$ of $+0.10$, $+0.14$ and $+0.18$. The computed images (f) and (h) are mismatches to the experimental image because the dark fringes are too weak in (f) and too strong in (h). In the results of table 5.5 all the determined values of $\boldsymbol{g}_c\cdot\boldsymbol{R}$ are the same (≈ 0.14), within the limits of experimental error, except for facet 4 case (e) where $\boldsymbol{g}_c\cdot\boldsymbol{R}=0.28$. The change in fringe contrast corresponding to this difference can be seen in figure 5.62, which shows a comparison of experimental and computed images for $\boldsymbol{g}_c=2\bar{1}\bar{1}_1/11\bar{2}_2$ which gives $\boldsymbol{g}_c\cdot\boldsymbol{R}=0.14$ for facet 3 and $\boldsymbol{g}_c\cdot\boldsymbol{R}=0.28$ for facet 4. From the results in table 5.5, the experimentally determined values of rigid-body displacement $(\boldsymbol{R}+\boldsymbol{d})$ of grain 2 relative to grain 1 at all facets are expansive. For facets 1 and 2 they correspond to simple

Table 5.5 Experimental values of $\boldsymbol{g}_c \cdot \boldsymbol{R}$ for the facet planes 1, 2, 3 and 4†.

$\boldsymbol{g}_c$	$\boldsymbol{g}_c \cdot \boldsymbol{R}$			
	Facet 1 $(\bar{1}2\bar{1})_1/(\bar{2}11)_2$	Facet 2 $(\bar{2}11)_1/(\bar{1}\bar{1}2)_2$	Facet 3 $(\bar{2}\bar{1}3)_1/(1\bar{3}2)_2$	Facet 4 $(\bar{3}12)_1/(\bar{1}\bar{2}3)_2$
(a) $0\bar{1}1_1/\bar{1}10_2$	$+0.14 \pm 0.02$	0	-0.15 ± 0.03	0
$01\bar{1}_1/1\bar{1}0_2$	-0.14 ± 0.02	0	$+0.15 \pm 0.03$	0
(b) $\bar{1}2\bar{1}_1/\bar{2}11_2$	edge on	-0.15 ± 0.03	$+0.14 \pm 0.02$	-0.14 ± 0.02
$1\bar{2}1_1/2\bar{1}\bar{1}_2$		$+0.15 \pm 0.03$	-0.16 ± 0.03	$+0.14 \pm 0.02$
(c) $1\bar{1}0_1/10\bar{1}_2$	-0.15 ± 0.03	$+0.14 \pm 0.02$	0	$+0.15 \pm 0.02$
$\bar{1}10_1/\bar{1}01_2$	$+0.14 \pm 0.02$	-0.14 ± 0.02	0	-0.16 ± 0.04
(d) $222_1/222_2$	0	0	0	0
$\bar{2}\bar{2}\bar{2}_1/\bar{2}\bar{2}\bar{2}_2$	0	0	0	0
(e) $\bar{2}11_1/\bar{1}\bar{1}2_2$	$+0.14 \pm 0.02$	edge on	-0.14 ± 0.02	-0.28 ± 0.02
$2\bar{1}\bar{1}_1/11\bar{2}_2$	—		$+0.14 \pm 0.02$	$+0.28 \pm 0.02$
(f) $\bar{1}01_1/0\bar{1}1_2$	0	—	—	—
	Resultant values of $(\boldsymbol{R} + \boldsymbol{d})$			
	$\frac{0.14 \pm 0.02}{3}[1\bar{2}1]_1/[2\bar{1}\bar{1}]_2$	$\frac{0.14 \pm 0.02}{3}[2\bar{1}\bar{1}]_1/[11\bar{2}]_2$	$\frac{0.14 \pm 0.02}{3}[11\bar{2}]_1/[\bar{1}2\bar{1}]_2$	$\frac{0.14 \pm 0.02}{3}[2\bar{1}\bar{1}]_1/[11\bar{2}]_2$

† All facet plane normals are specified as pointing from grain 2 into grain 1, but $\boldsymbol{R}$ is always defined as a displacement of the lower grain relative to the upper grain. The beam directions of the images used in determining the values of $\boldsymbol{g}_c \cdot \boldsymbol{R}$ are such that grain 1 is the upper grain in all cases except for (c), (d), (e), and (f) of facet 1, where grain 2 is the upper grain. However, the resultant rigid-body displacements $(\boldsymbol{R} + \boldsymbol{d})$ are all given as displacements of grain 2 with respect to grain 1.

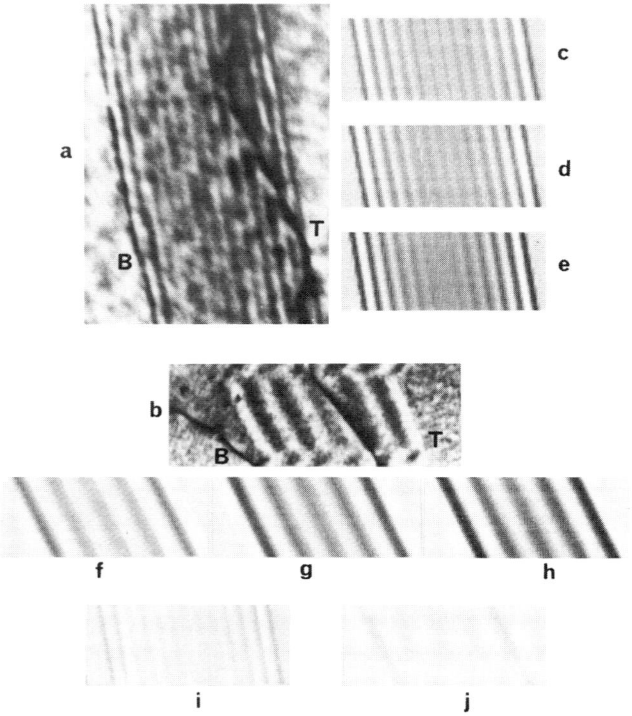

expansions normal to the facet planes of $(0.047 \pm 0.007)\,[1\bar{2}1]_1/[2\bar{1}\bar{1}]_2$ and $(0.47 \pm 0.007)[2\bar{1}\bar{1}]_1/[11\bar{2}]_2$ respectively, and for facets 3 and 4 they are inclined to the facet planes and are $(0.047 \pm 0.007)\,[11\bar{2}]_1/[\bar{1}2\bar{1}]_2$ and $(0.047 \pm 0.007)[2\bar{1}\bar{1}]_1/[11\bar{2}]_2$ respectively. All these displacements have a magnitude of (0.33 ± 0.05) Å and in no case is there a displacement parallel to the rotation axis.

The experimentally determined values of $(\boldsymbol{R} + \boldsymbol{d})$ will be discussed in terms of hard-sphere modelling of $\{112\}$ and $\{213\}$ boundary planes with the $(\bar{1}2\bar{1})_1/(\bar{2}11)_2$ boundary plane of facet 1 and the $(\bar{2}\bar{1}3)_1/(1\bar{3}2)_2$ boundary plane of facet 3 taken as specific examples of each type of plane. Figure 5.63 shows a projection along the $[111]$ rotation axis of the BCC lattices of grains 1 and 2, where grain 1 is rotated 60° in a right-handed sense with respect to grain 2, and the two lattices are juxtaposed at the $(1\bar{2}1)_1/(2\bar{1}\bar{1})_2$ boundary plane of facet 1. Circles indicate lattice sites in the plane of the projection, triangles those lattice sites in the plane spaced $(1/6)[\bar{1}\bar{1}\bar{1}]$ below the plane of the projection, squares those lattice sites spaced $(1/3)[\bar{1}\bar{1}\bar{1}]$ below the

can still be used to provide some information concerning the Burgers vectors of the secondary grain boundary dislocations.

It can be seen from figure 5.65(a) (with reference to the schematic diagram of figure 5.66(c)) that the secondary grain boundary dislocation structure consists of an hexagonal network with segments along $\overrightarrow{pu}$, $\overrightarrow{st}$, and $\overrightarrow{tu}$ in strong contrast, together with additional dislocations parallel to $\overrightarrow{vw}$ in weaker contrast. The sense of these dislocations is from the bottom to the top of the specimen and their Burgers vectors are designated $\boldsymbol{A}$, $\boldsymbol{B}$, $\boldsymbol{C}$, and $\boldsymbol{D}$ where $\boldsymbol{B} = \boldsymbol{A} + \boldsymbol{C}$. The dislocations with Burgers vector $\boldsymbol{D}$ divide each hexagonal cell of the type *pqrstu* into two pentagonal cells of the types *qrswv* and *pvwtu*. The micrograph of figure 5.65(a) is a simultaneous double two-beam image with different diffracting vectors operating in each grain, so that it does not show fringe contrast associated with states of different rigid-body displacement between the grains, but only shows diffraction contrast associated with the strain fields of the grain boundary dislocations. Figure 5.65(b), however, is a simultaneous double two-beam image with the same $0\bar{2}\bar{2}$ diffracting vector operating in both grains and therefore shows diffraction fringe contrast associated with the different states of rigid-body displacement between the grains in addition to the strain contrast associated with the grain boundary dislocations. It can be seen from this image that two different states of rigid-body displacement exist within the network and that these correspond to the neighbouring pentagonal cells in figure 5.66(c). For example, cells of the type *qrswv* (designated α cells) show stronger fringe contrast than cells of the type *pvwtu* (designated β cells). Partial grain boundary dislocations separate the α and β cells and correspond to the segments with Burgers vectors $\boldsymbol{A}$, $\boldsymbol{C}$ and $\boldsymbol{D}$ of figure 5.66(c), whereas the dislocation segments with Burgers vector $\boldsymbol{B}$ separate cells showing the same fringe contrast and, therefore, $\boldsymbol{B}$ will be a DSC vector. The periodic network of secondary grain boundary dislocations in figure 5.65 is specified in figure 5.66(c) by periodic repeat vectors $\overrightarrow{\mathrm{OX}}$ and $\overrightarrow{\mathrm{OY}}$ between neighbouring hexagonal cell centres. For example, if one hexagonal cell in figure 5.66(c) containing α and β pentagons, e.g. *pqrstu*, is taken as the motif of the periodic dislocation structure, the whole periodic network of dislocations can be generated by operating with the repeat vectors $\overrightarrow{\mathrm{OX}}$ and $\overrightarrow{\mathrm{OY}}$. Further, since the points O, X and Y lie in cells of the network showing the same fringe contrast, grain boundary dislocations intersected along each of the intervals $\overrightarrow{\mathrm{OX}}$ and $\overrightarrow{\mathrm{OY}}$ will have a Burgers vector sum equal to a vector of the DSC lattice, i.e. $\boldsymbol{A} + \boldsymbol{D}$ as well as $\boldsymbol{B}$ will be DSC vectors. Therefore, from purely a geometric point of view, the observed grain boundary dislocation structure can be considered as having developed from two periodic arrays of grain boundary dislocations parallel to $\overrightarrow{\mathrm{OX}}$ and $\overrightarrow{\mathrm{OY}}$ in figure 5.66(a). In figure 5.66(a) the set of grain boundary dislocations parallel to $\overrightarrow{\mathrm{OX}}$ with a periodic spacing $\overrightarrow{\mathrm{OY}}$ and with the DSC Burgers vector $\boldsymbol{A} + \boldsymbol{D}$, intersects the other set

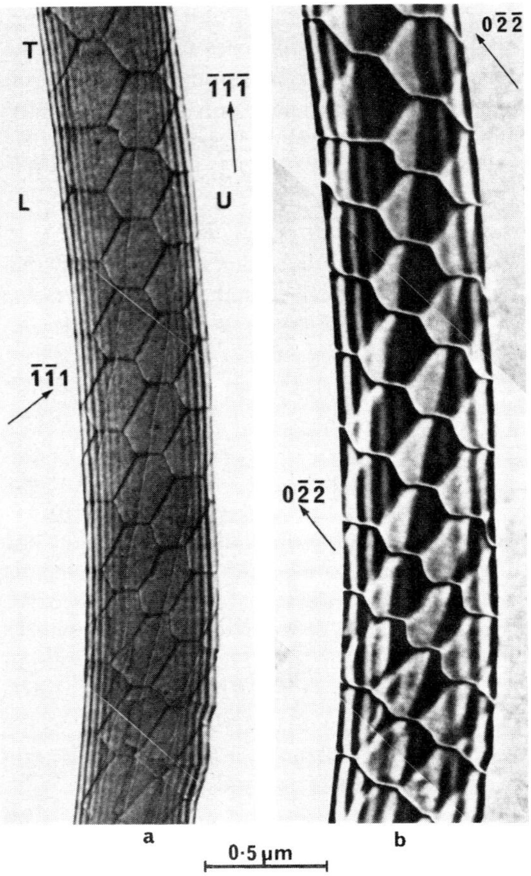

Figure 5.65 Double two-beam images of a near-Σ9 boundary in a Cu – 6 at% Si alloy with different diffracting vectors in each grain (*a*), and with the same diffracting vector in each grain (*b*); the intersection of the boundary with the top of the specimen is marked T.

parallel to $\overrightarrow{OY}$ with a periodic spacing $\overrightarrow{OX}$ and with the DSC Burgers vector ***B***. These sets of grain boundary dislocations generate a periodic array of parallelogram cell centres with the same geometry as those of the observed hexagonal cell centres. Further, the total DSC Burgers vector sum encountered between parallelogram cell centres is the same as that between the observed cell centres. Therefore, the periodic array of parallelograms will accommodate the same small angular departure of the misorientation of the grains from the exact-Σ9 orientation as the observed hexagonal array. Figure 5.66(*b*) shows the dislocation array after interaction has occurred at the fourfold

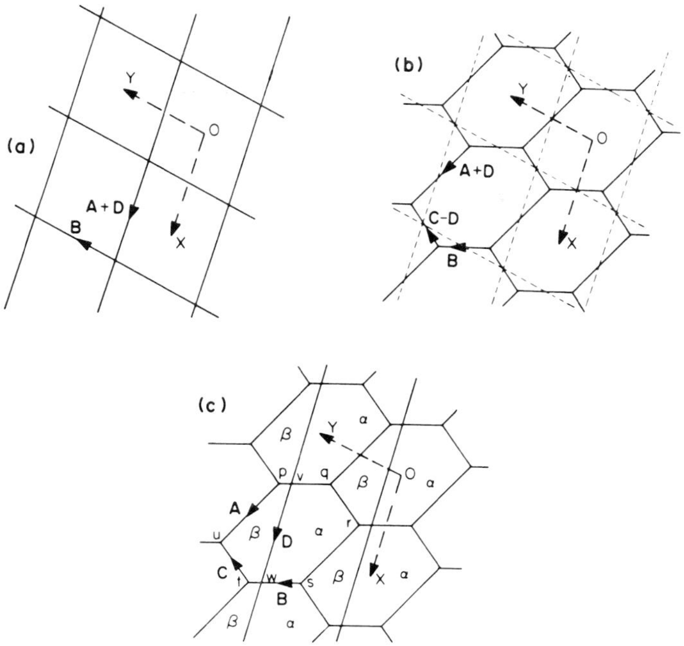

Figure 5.66 Schematic diagrams illustrating that the secondary grain boundary dislocation network (c) can be developed geometrically from two sets of grain boundary dislocations with DSC Burgers vectors $\boldsymbol{B}$ and $\boldsymbol{A} + \boldsymbol{D}$ (a). In (b) the dislocations of (a) (dashed lines) have interacted to form the hexagonal net.

nodes to give an hexagonal network, where the new segment of dislocation has a DSC Burgers vector of $\boldsymbol{B} - (\boldsymbol{A} + \boldsymbol{D}) = \boldsymbol{C} - \boldsymbol{D}$. Finally, in figure 5.66(c) the observed network is developed by dissociation of the segments with Burgers vectors $\boldsymbol{A} + \boldsymbol{D}$ and $\boldsymbol{C} - \boldsymbol{D}$ to give the lower energy configuration of grain boundary dislocations containing partial dislocations with Burgers vectors $\boldsymbol{A}$, $\boldsymbol{C}$ and $\boldsymbol{D}$, and two different states of rigid-body displacement corresponding to the α and β cells.

For this near-$\Sigma 9$ boundary only one suitable pair of low-index same-$\boldsymbol{g}_c$ images ($\boldsymbol{g}_c = \pm 022$) is available, so that the experimental information is limited to the components of $\boldsymbol{R}_\alpha$ and $\boldsymbol{R}_\beta$ parallel to [022], where $\boldsymbol{R}_\alpha$ and $\boldsymbol{R}_\beta$ are the rigid-body displacements of grain L with respect to grain U in the α and β cells respectively. From a comparison of the fringe contrast in these experimental same-$\boldsymbol{g}_c$ images with that in theoretical images computed for a range of values of $\boldsymbol{R}_\alpha$ and $\boldsymbol{R}_\beta$, the components $022 \cdot \boldsymbol{R}_\alpha$ and $022 \cdot \boldsymbol{R}_\beta$ are identified as

$$022 \cdot (\boldsymbol{R}_\alpha + \boldsymbol{d}_\alpha) = 0.30 \pm 0.05$$

and

$$022 \cdot (\boldsymbol{R}_\beta + \boldsymbol{d}_\beta) = 0.21 \pm 0.05$$

where $\boldsymbol{d}_\alpha$ and $\boldsymbol{d}_\beta$ are DSC vectors such that $(\boldsymbol{R}_\alpha + \boldsymbol{d}_\alpha)$ and $(\boldsymbol{R}_\beta + \boldsymbol{d}_\beta)$ lie in the Wigner–Seitz cell of the DSC lattice. This accuracy of ± 0.05 is less than that quoted for the determination of $\boldsymbol{g}_c \cdot \boldsymbol{R}$ in the previous examples and is due, in this case, to the restricted regions of fringe contrast available in the α and β cells. The difference $022 \cdot (\boldsymbol{R}_\alpha - \boldsymbol{R}_\beta + \boldsymbol{d}_\alpha - \boldsymbol{d}_\beta)$ is determined with greater accuracy than is implied by the errors in the determination of the individual components, and this is achieved by matching theoretical images to the characteristic change in fringe contrast that occurs in the experimental images on going from α to β cells. The result obtained is $022 \cdot (\boldsymbol{R}_\alpha - \boldsymbol{R}_\beta + \boldsymbol{d}_\alpha - \boldsymbol{d}_\beta) = 0.090 \pm 0.015 \approx 1/11$. Some examples showing a comparison of such fringe contrast in experimental and theoretical images can be seen later in figures 5.68(*c*), 5.69(*b*) and 5.70(*b*) where the theoretical images are computed for $022 \cdot \boldsymbol{R}_\beta = 2/9$ and $022 \cdot (\boldsymbol{R}_\alpha - \boldsymbol{R}_\beta) = 1/11$.

The Burgers vectors of the partial grain boundary dislocations separating the α and β cells are given by the difference between the rigid-body displacements $\boldsymbol{R}_\alpha$ and $\boldsymbol{R}_\beta$. Thus, following the procedure of Thompson (1953), for the line directions of the dislocations given in figure 5.66(*c*), the partial dislocation with Burgers vector $\boldsymbol{A}$ must satisfy the condition $022 \cdot \boldsymbol{A} = 022 \cdot (\boldsymbol{R}_\beta - \boldsymbol{R}_\alpha) \approx 022 \cdot (\boldsymbol{d}_\alpha - \boldsymbol{d}_\beta) - (1/11)$. Similarly the partial dislocations with Burgers vectors $\boldsymbol{C}$ and $\boldsymbol{D}$ must satisfy the conditions

$$022 \cdot \boldsymbol{C} \approx 022 \cdot (\boldsymbol{d}_\beta - \boldsymbol{d}_\alpha) + (1/11)$$

and

$$022 \cdot \boldsymbol{D} \approx 022 \cdot (\boldsymbol{d}_\beta - \boldsymbol{d}_\alpha) + (1/11).$$

The first step in the determination of the Burgers vectors $\boldsymbol{A}$, $\boldsymbol{B}$, $\boldsymbol{C}$ and $\boldsymbol{D}$ is to obtain information concerning the DSC Burgers vectors $\boldsymbol{B}$ and $\boldsymbol{A} + \boldsymbol{D}$. This is done by assuming that the geometric analysis for two arrays of grain boundary dislocations described in section 1.2.2 can be applied to the two arrays of secondary grain boundary dislocations, depicted schematically in figure 5.66(*a*), on the basis that these dislocations accommodate the small angular departure $(\boldsymbol{q}, \varphi)$ from the exact-$\Sigma 9$ CSL orientation. If $\boldsymbol{q}$ is taken as a unit vector parallel to $\boldsymbol{B} \wedge (\boldsymbol{A} + \boldsymbol{D})$, then the directions of the grain boundary dislocations are given by expressions (1.16) as

$$\overrightarrow{\mathrm{OX}} \parallel \boldsymbol{v} \wedge \boldsymbol{B} \tag{5.9}$$

and

$$\overrightarrow{\mathrm{OY}} \parallel (\boldsymbol{A} + \boldsymbol{D}) \wedge \boldsymbol{v} \tag{5.10}$$

and the spacings of the dislocations are given by equations (1.20) and (1.21) as

$$|\overrightarrow{OX}| = \left(2\sin(\varphi/2)\frac{|\boldsymbol{v}\cdot\boldsymbol{q}|}{|\boldsymbol{v}\wedge\boldsymbol{B}|}\right)^{-1} \tag{5.11}$$

$$|\overrightarrow{OY}| = \left(2\sin(\varphi/2)\frac{|\boldsymbol{v}\cdot\boldsymbol{q}|}{|\boldsymbol{v}\wedge(\boldsymbol{A}+\boldsymbol{D})|}\right)^{-1} \tag{5.12}$$

where $\boldsymbol{v}$ is the boundary normal.

The boundary normal $\boldsymbol{v}$ is $(\cos 71.5^\circ \pm 2^\circ \cos 101.1^\circ \pm 2^\circ \cos 21.8^\circ \pm 2^\circ)_L \approx (8\ \bar{5}\ 23)_L$, and the measured directions and magnitudes of $\overrightarrow{OX}$ and $\overrightarrow{OY}$ are listed in table 5.6, where the quoted errors reflect the lack of regularity of the network. On the assumption that the observed grain boundary dislocation network can be described by the two-dislocation model of figure 5.66 then, from expression (5.9), the DSC Burgers vector $\boldsymbol{B}$ must lie in the plane normal to the experimentally observed direction $\overrightarrow{OX}$ in table 5.6. This defines, within the limits of experimental error, the set of DSC vectors $\pm(1/18)[\bar{5}18]_L$, $\pm(1/6)[2\bar{1}1]_L$, $\pm(1/6)[\bar{2}15]_L$, and higher-order combinations as possible solutions for $\boldsymbol{B}$. From this analysis no sense can be attributed to the possible solutions for $\boldsymbol{B}$ because, as already pointed out, the small angular departure $(\boldsymbol{q}, \varphi)$ from the $\Sigma 9$ CSL orientation lies within the limits of experimental error and cannot be specified. The actual Burgers vector $\boldsymbol{B}$ was determined from the above set of possibilities by a comparison of simultaneous double two-beam experimental images with theoretical images. Only theoretical images with $\boldsymbol{B} = (1/18)[5\bar{1}\bar{8}]_L$ gave satisfactory matches to the experimental images and figure 5.67 shows a comparison of three of the experimental images with the corresponding theoretical images. In this comparison two segments of dislocation with $\boldsymbol{B} = (1/18)[5\bar{1}\bar{8}]_L$ are involved: one on the right starting at a node very close to the surface of the specimen, and the other on the left near the middle of the specimen. These two segments are separated in the experimental images by other components of the network which are not taken into account in the theoretical images.

Expression (5.10) could be used in a similar manner to expression (5.9) to give information on possible Burgers vectors $\boldsymbol{A} + \boldsymbol{D}$. However, in this case,

Table 5.6 Observed directions and spacings of the network of figure 5.65.

	Direction (ρ°)†	Magnitude (Å)
$\overrightarrow{OX}$	-9 ± 5	3180 ± 300
$\overrightarrow{OY}$	$+105 \pm 3$	3340 ± 400

† The angles ρ° are measured from $[001]_L \wedge \boldsymbol{v}$ in a right-handed sense around an axis parallel to $\boldsymbol{v}$.

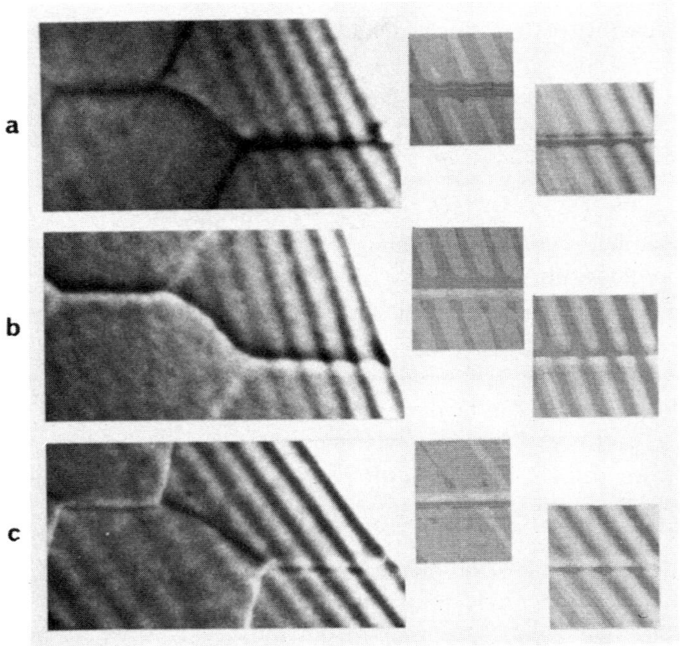

Figure 5.67 Three experimental double two-beam images of a portion of the grain boundary dislocation network arranged so that the segments with Burgers vector $\boldsymbol{B}$ are horizontal and the lower grain is on the right. The matching computed images are for the Burgers vector $\boldsymbol{B} = (1/18)[5\bar{1}\bar{8}]_L$. The beam directions, $\boldsymbol{B}$, and diffracting vectors, $\boldsymbol{g}$, are: for (*a*) $\boldsymbol{B} = [0\bar{1}9]_U$ with $\boldsymbol{g}_U = \bar{2}00_U$ and $\boldsymbol{g}_L = 020_L$; for (*b*) $\boldsymbol{B} = [\bar{2}\bar{3}5]_U$ with $\boldsymbol{g}_U = \bar{1}1\bar{1}_U$ and $\boldsymbol{g}_L = 1\bar{1}\bar{1}_L$ and for (*c*) $\boldsymbol{B} = [4\bar{3}7]_U$ with $\boldsymbol{g}_U = \bar{1}11_U$ and $\boldsymbol{g}_L = 0\bar{2}0_L$.

image matching could not be used to determine $\boldsymbol{A} + \boldsymbol{D}$, since $\boldsymbol{A}$ and $\boldsymbol{D}$ occur separately in the observed network. Therefore the approach adopted was to obtain a value for the partial Burgers vector $\boldsymbol{D}$ by image matching, then use this value, in combination with different values for $\boldsymbol{A} + \boldsymbol{D}$ suggested from expression (5.10), to provide a set of partial DSC Burgers vectors from which $\boldsymbol{A}$ could be distinguished by image matching.

Possible values of the partial Burgers vector $\boldsymbol{D}$, which need to be distinguished by image matching, do not form a discrete set of values, but take the form of a continuous variable. The general form of $\boldsymbol{D}$ can be written as

$$\boldsymbol{D} = \boldsymbol{b}_n + \boldsymbol{b}_p + \boldsymbol{d}$$

where $\boldsymbol{b}_n$ and $\boldsymbol{b}_p$ are components normal and parallel to the boundary plane respectively and $\boldsymbol{d}$ is a vector of the DSC lattice, i.e. $\boldsymbol{d}$ represents the DSC component of $\boldsymbol{D}$, and $\boldsymbol{b}_n + \boldsymbol{b}_p$ represents the fractional DSC component. Then, using the result that the Burgers vector $\boldsymbol{D}$ must satisfy the condition $022 \cdot \boldsymbol{D} \approx 022 \cdot (\boldsymbol{d}_\beta - \boldsymbol{d}_\alpha) + (1/11)$, the fractional DSC component of $\boldsymbol{D}$ must satisfy the condition $022 \cdot (\boldsymbol{b}_\mathrm{n} + \boldsymbol{b}_\mathrm{p}) \approx 1/11$. Images were computed with values of $\boldsymbol{D}$ satisfying the above condition, and they showed that the DSC component $\boldsymbol{d}$ had to be zero. In fact, the inclusion of a DSC vector always resulted in theoretical images which were very much stronger than the consistently weak contrast observed in experimental images with different diffraction vectors in each grain. This weak contrast, together with the fact that the possible values of $\boldsymbol{D}$ take the form of a continuous variable, meant that the identification of $\boldsymbol{D}$ by image matching could not be as definitive as that for $\boldsymbol{B}$. In fact theoretical images satisfying the condition $022 \cdot \boldsymbol{D} = 1/11$ were found to give acceptable matches to the experimental images for a range of values of $\boldsymbol{D}$. This range of acceptability corresponded to Burgers vectors lying within a solid angle with semicone angle of 15° centred on the value $\boldsymbol{D} = \boldsymbol{b}_\mathrm{n} = (1/396)[8\ \bar{5}\ 23]_\mathrm{L}$ for which $\boldsymbol{b}_\mathrm{p} = 0$. Figures 5.68(*a*), (*b*) and (*c*) show the agreement between theoretical and experimental images for this value of $\boldsymbol{D}$, where images (*a*) and (*b*) have different diffracting vectors in each grain and image (*c*) has the same $0\bar{2}\bar{2}$ diffracting vector in both grains. The image in (*a*) shows characteristically weak contrast, that in (*b*) is out of contrast, while the image in (*c*) is dominated by the fringe contrast arising from the difference in rigid-body displacement between the α and β cells. An example of the type of disagreement, between experimental and theoretical images, which enabled the 15° region of acceptability for the Burgers vector $\boldsymbol{D}$ to be defined, can be seen from a comparison of the experimental image in figure 5.68(*b*) with the theoretical images in figures 5.68(*d*) and (*e*). The theoretical image in figure 5.68(*d*) corresponds to a Burgers vector which is 15° from $[8\ \bar{5}\ 23]_\mathrm{L}$ (i.e. on the limit of acceptability), and already some disagreement with the experimental image can be seen. This takes the form of a break along the line of the dislocation in one of the dark fringes. The disagreement between experimental and theoretical images becomes more pronounced for Burgers vectors outside the 15° limit as illustrated in figure 5.68(*e*), where the Burgers vector is 23° from $[8\ \bar{5}\ 23]_\mathrm{L}$.

The difference in rigid-body displacement between the α and β cells, $\boldsymbol{R}_\alpha - \boldsymbol{R}_\beta$, is defined by the sense and magnitude of the separating partial dislocation with Burgers vector $\boldsymbol{D}$. Thus, for the range of Burgers vectors found for $\boldsymbol{D}$, the difference in rigid-body displacement has a major component $\boldsymbol{b}_\mathrm{n}$ normal to the boundary such that the displacement between the grains in the β cells relative to that in the α cells is a displacement of the lower grain away from the upper grain by approximately 0.23 Å. Further, if any shear component $\boldsymbol{b}_\mathrm{p}$ contributes to the difference in rigid-body displacement between the α and β cells it will be less than 0.06 Å as given by the uncertainty of 15° in the semicone angle.

By using $\boldsymbol{D} = (1/396)[8\ \bar{5}\ 23]_\mathrm{L}$, possible values for $\boldsymbol{A}$ were obtained from the set of discrete DSC Burgers vectors for $\boldsymbol{A} + \boldsymbol{D}$, suggested by expression

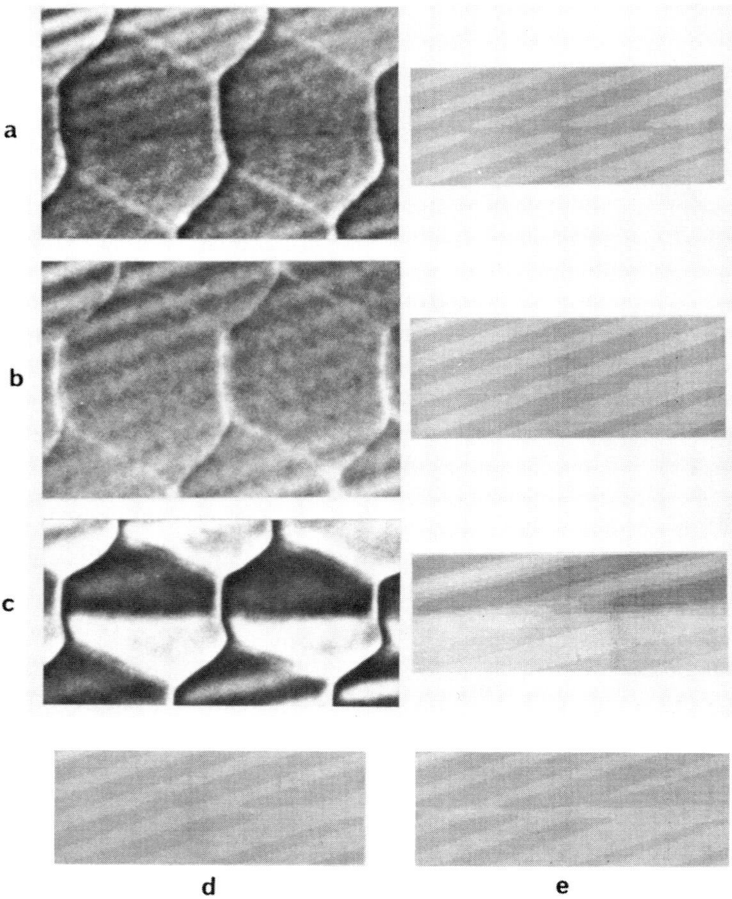

Figure 5.68 Three double two-beam images of a portion of the grain boundary dislocation network arranged so that the segments with Burgers vector $\boldsymbol{D}$ are horizontal and the lower grain is on the right. The matching computed images are for the Burgers vector $\boldsymbol{D} = (1/396)[8\ \bar{5}\ 23]_L$. The beam directions, $\boldsymbol{B}$, and the diffracting vectors, $\boldsymbol{g}$, are: for (*a*) $\boldsymbol{B} = [\bar{2}\bar{3}5]_U$ with $\boldsymbol{g}_U = \bar{1}\bar{1}1_U$ and $\boldsymbol{g}_L = 1\bar{1}\bar{1}_L$; for (*b*) $\boldsymbol{B} = [0\bar{1}7]_U$ with $\boldsymbol{g}_U = 200_U$ and $\boldsymbol{g}_L = 0\bar{2}0_L$ and for (*c*) $\boldsymbol{B} = [2\bar{7}7]_U$ with $\boldsymbol{g}_U = 0\bar{2}\bar{2}_U$ and $\boldsymbol{g}_L = 0\bar{2}\bar{2}_L$. The computed images (*d*) and (*e*) are for the diffracting conditions of (*b*) with Burgers vector $(1/462)[16\ \overline{25}\ 46]_L$ in (*d*) and $(1/132)[8\ \overline{17}\ 23]_L$ in (*c*).

(5.10) from the experimentally determined direction $\overrightarrow{OY}$. However, when the set of DSC vectors for $\boldsymbol{A} + \boldsymbol{D}$ was restricted to those permissible within the limits of experimental error for $\overrightarrow{OY}$, i.e. $\pm(1/18)[\bar{2}\bar{5}5]_L$, $\pm(1/18)[745]_L$, $\pm(1/2)[\bar{1}\bar{1}0]_L$ and higher-order combinations, the resulting values of $\boldsymbol{A}$ gave theoretical images which were not in agreement with the experimental images. In fact, acceptable agreement between experimental and theoretical images could only be obtained for the value of $\boldsymbol{A} = (1/396)[124\ 137\ \overline{155}]_L$, which results from the DSC vector $\boldsymbol{A} + \boldsymbol{D} = (1/3)[11\bar{1}]_L$. Moreover, this value of $\boldsymbol{A} + \boldsymbol{D}$ was also required to obtain matching images for $\boldsymbol{A}$ for other values of $\boldsymbol{D}$ within the uncertainty limit of $\pm 15°$. The DSC vector $\boldsymbol{A} + \boldsymbol{D} = (1/3)[11\bar{1}]_L$ corresponds to a direction for $\overrightarrow{OY}$ which is 6° outside the experimental error of table 5.6 and this discrepancy will be discussed in more detail below.

Figure 5.69 shows a comparison of experimental and theoretical images for $\boldsymbol{A} = (1/396)[124\ 137\ \overline{155}]_L$ involving four segments of the dislocation network extending from the bottom of the specimen to beyond mid-specimen. The agreement between the theoretical and experimental images is good, and although the theoretical images in figure 5.69(b) appear to exaggerate some of the features in the experimental images, all the other possibilities for $\boldsymbol{A}$, obtained from DSC vectors $\boldsymbol{A} + \boldsymbol{D}$ other than $(1/3)[11\bar{1}]_L$, gave theoretical images which failed to reproduce the characteristic features of contrast in the experimental image of figure 5.69(b).

A further check on the identification of the Burgers vectors $\boldsymbol{A}$ and $\boldsymbol{B}$ and therefore $\boldsymbol{D}$ (since $\boldsymbol{D}$ is used in the determination of $\boldsymbol{A}$) can be made by a comparison of theoretical and experimental images for the segments of the grain boundary dislocation network with Burgers vector $\boldsymbol{C}$, where $\boldsymbol{C} = \boldsymbol{B} - \boldsymbol{A} = (1/396)[\overline{14}\ \overline{159}\ \overline{21}]_L$. This comparison is shown in figure 5.70 for three segments of this dislocation extending from close to the bottom surface of the specimen on the left, to beyond mid-specimen on the right. It can be seen that there is good agreement between all three experimental images and their corresponding theoretical images.

In summary, image matching has shown that it is possible to describe the different segments of the secondary grain boundary dislocation network by the Burgers vectors $\boldsymbol{A} = (1/396)[124\ 137\ \overline{155}]_L$, $\boldsymbol{B} = (1/18)[5\bar{1}\bar{8}]_L$, $\boldsymbol{C} = (1/396)[\overline{14}\ \overline{159}\ \overline{21}]_L$ and $\boldsymbol{D} = (1/396)[8\ \bar{5}\ 23]_L$. However, with the exception of $\boldsymbol{B}$, the above indices should not be considered as exact because of the uncertainty in the determination of $\boldsymbol{D}$ which is also reflected in the values of $\boldsymbol{A}$ and $\boldsymbol{C}$.

As already pointed out, expression (5.10) did not predict the DSC vector $\boldsymbol{A} + \boldsymbol{D} = (1/3)[11\bar{1}]_L$ within the experimental error for $\overrightarrow{OY}$. Therefore it is of interest to examine the extent to which other measured parameters of the network can be predicted by the identified Burgers vectors using the two-dislocation model of figure 5.66 defined by $(\boldsymbol{q}, \varphi)$ and expressions a

b

c

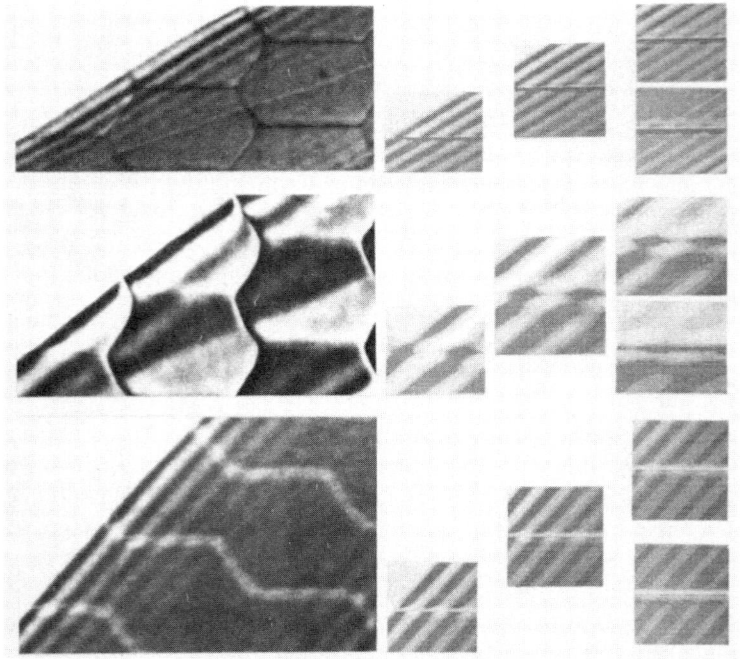

Figure 5.69 Three double two-beam images of portion of the grain boundary dislocation network arranged so that the segments with Burgers vector $\boldsymbol{A}$ are horizontal and the upper grain is on the left. The matching computed images are for the Burgers vector $\boldsymbol{A} = (1/396)[124\ 137\ \overline{155}]_{\mathrm{L}}$. The beam directions, $\boldsymbol{B}$, and diffracting vectors, $\boldsymbol{g}$, are: for (a) $\boldsymbol{B} = [1\bar{3}2]_{\mathrm{U}}$ with $\boldsymbol{g}_{\mathrm{U}} = \bar{1}1\bar{1}_{\mathrm{U}}$ and $\boldsymbol{g}_{\mathrm{L}} = \bar{1}\bar{1}1_{\mathrm{L}}$; for ($b$) $\boldsymbol{B} = [2\bar{7}7]_{\mathrm{U}}$ with $\boldsymbol{g}_{\mathrm{U}} = 0\bar{2}\bar{2}_{\mathrm{U}}$ and $\boldsymbol{g}_{\mathrm{L}} = 0\bar{2}\bar{2}_{\mathrm{L}}$ and for (c) $\boldsymbol{B} = [\bar{3}\ 0\ 11]_{\mathrm{U}}$ with $\boldsymbol{g}_{\mathrm{U}} = 020_{\mathrm{U}}$ and $\boldsymbol{g}_{\mathrm{L}} = 0\bar{2}0_{\mathrm{L}}$.

(5.9)–(5.12). The axis $\boldsymbol{q}$ is given by the unit vector parallel to $\boldsymbol{B} \wedge (\boldsymbol{A} + \boldsymbol{D})$ i.e. parallel to $[3\bar{1}2]_{\mathrm{L}}$. When this value of $\boldsymbol{q}$ is taken with the known $\Sigma 9$ orientation and the measured misorientation $(\boldsymbol{u}, \theta)$, φ is found to lie in the range $0.037^\circ \leqq \varphi \leqq 0.044^\circ$. For this range of φ equations (5.11) and (5.12) give the range of spacings of $\overrightarrow{\mathrm{OX}}$ and $\overrightarrow{\mathrm{OY}}$ listed in table 5.7, which also includes the directions of $\overrightarrow{\mathrm{OX}}$ and $\overrightarrow{\mathrm{OY}}$ obtained from expressions (5.9) and (5.10). It can be seen from a comparison of tables 5.6 and 5.7 that, apart from the direction of $\overrightarrow{\mathrm{OY}}$, the Burgers vectors arrived at by image matching predict a geometry for the network in the grain boundary which is in agreement with the observed network. Thus, despite the fact that only two or three repeat distances can be observed across the thickness of the specimen, the geometry of the secondary grain boundary dislocation network is close to that predicted by a two-dislocation theory.

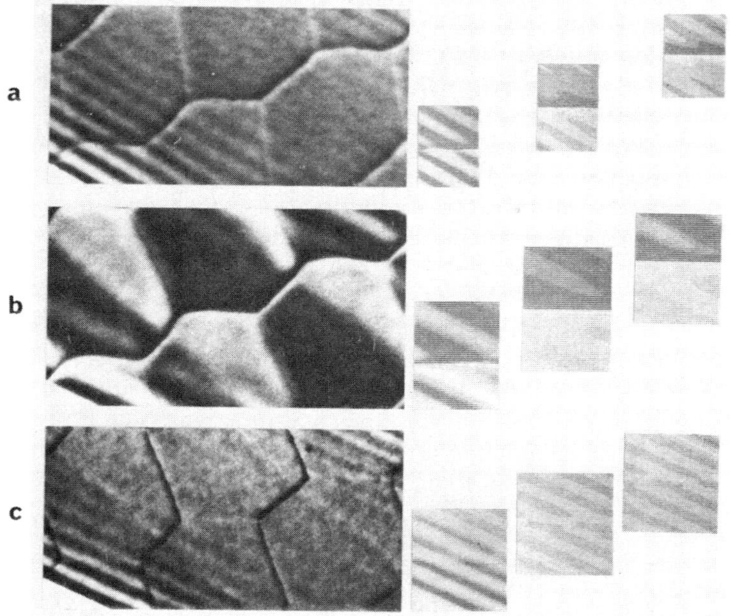

Figure 5.70 Three double two-beam images of a portion of the grain boundary dislocation network arranged so that the segments with Burgers vector $\boldsymbol{C}$ are horizontal and the upper grain is on the left. The matching computed images are for the Burgers vector $\boldsymbol{C} = (1/396)[\overline{14}\ \overline{159}\ \overline{21}]_L$. The beam directions, $\boldsymbol{B}$, and diffracting vectors, $\boldsymbol{g}$, are: for (a) $\boldsymbol{B} = [235]_U$ with $\boldsymbol{g}_U = \bar{1}1\bar{1}_U$ and $\boldsymbol{g}_L = 1\bar{1}\bar{1}_L$; for ($b$) $\boldsymbol{B} = [2\bar{7}7]_U$ with $\boldsymbol{g}_U = 0\bar{2}\bar{2}_U$ and $\boldsymbol{g}_L = 0\bar{2}\bar{2}_L$ and for (c) $\boldsymbol{B} = [1\bar{2}3]_U$ with $\boldsymbol{g}_U = 1\bar{1}\bar{1}_U$ and $\boldsymbol{g}_L = 11\bar{1}_L$.

The analysis of the network has shown that the partial grain boundary dislocations separate two distinct states of rigid-body displacement, and the difference between the states is predominantly a difference in expansion between the grains of approximately 0.23 Å normal to the boundary. Therefore it is likely that these different states of rigid-body displacement

Table 5.7 Predicted directions and spacings of the network of figure 5.65.

	Direction ($\rho°$)†	Spacing (Å)
$\overrightarrow{OX}$	-14 ± 2	2970–2490
$\overrightarrow{OY}$	$+116 \pm 2$	3580–3010

† The angles $\rho°$ are defined as for table 5.6 and the errors arise from the experimental errors in the determination of $\boldsymbol{v}$.

are associated with different arrangements of atoms at the boundary and that the grain boundary energy will be different for the α and β cells. Thus, these states of rigid-body displacement are different from symmetry-related states with the same structure of the type identified by Pond and Vitek (1977) in aluminium. The results suggest that the observed dislocation network containing partial grain boundary dislocations, which is depicted in figure 5.66(c), is a low-energy network resulting from the dissociation of the grain boundary dislocations with DSC Burgers vectors as depicted in figure 5.66(b).

Σ3 boundaries

Networks of secondary grain boundary dislocations containing partial dislocations, such as that analysed for the Σ9 boundary, also occur in incoherent-Σ3 boundaries. It will be demonstrated here how the analysis of the Burgers vectors of the dislocations and the rigid-body displacements in such incoherent-Σ3 boundaries enables the difference in grain boundary energy between two different states of rigid-body displacement to be determined.

Electron micrographs of two incoherent-Σ3 boundaries in a Cu – 6 at% Si alloy, both of which contain secondary grain boundary dislocation networks with partial dislocations, are shown in figures 5.71(a) (case 1) and 5.72(a) (case 2). The incoherent boundary of figure 5.71(a) consists of three main facets (marked 1, 2 and 3) which are terminated by steeply inclined coherent-Σ3 boundaries along IA and FE. The boundary facets intersect the top of the specimen along ABCD and the bottom of the specimen along FGHI, so that the upper grain is grain U and the lower grain is grain L. The misorientation between the two grains is very close to an exact-Σ3 orientation corresponding to a 60° rotation of grain U in a right-handed sense with respect to grain L around an axis with common indices, [111], in both grains. Basis DSC vectors of the DSC lattice corresponding to this Σ3 CSL are $\boldsymbol{x} = (1/6)[11\bar{2}]_U/(1/6)[\bar{1}2\bar{1}]_L$, $\boldsymbol{y} = (1/6)[1\bar{2}1]_U/(1/6)[2\bar{1}\bar{1}]_L$ and $\boldsymbol{z} = (1/3)[111]_U/(1/3)[111]_L$. The incoherent-Σ3 boundary of figure 5.72(a) consists of four families of incoherent facets, numbered 4, 5, 6 and 7, where facets 4 and 5 terminate at (111) coherent facets along CB and ED respectively. This boundary intersects the top of the specimen along ABCDEFGI and the bottom of the specimen along JKM, so that it separates an upper included grain U from a lower surrounding grain L, where the misorientation between the grains U and L is specified in the same way as that for the Σ3 boundary of figure 5.71(a).

The planes of facets 1 and 2 of case 1 are $(\bar{4}21)_U/(\bar{8}\ \bar{5}\ 10)_L$ and $(\overline{10}\ 8\ 5)_U/(\bar{2}\bar{1}4)_L$ respectively, with their line of intersection HB along $[\bar{1}\bar{5}6]_U/[5\bar{6}1]_L$, which is also the line of intersection of facet 1 with the (111) coherent interface along IA. Facet 3 is curved and for most of its length is within 4° of the plane of facet 1. The planes of facets 4, 5, 6 and 7 of case 2 are $(\overline{10}\ 5\ 8)_U/(\bar{1}\bar{2}4)_L$, $(\bar{4}12)_U/(\bar{5}\ \bar{8}\ 10)_L$, $(\bar{2}\bar{1}4)_U/(5\ \overline{10}\ 8)_L$ and

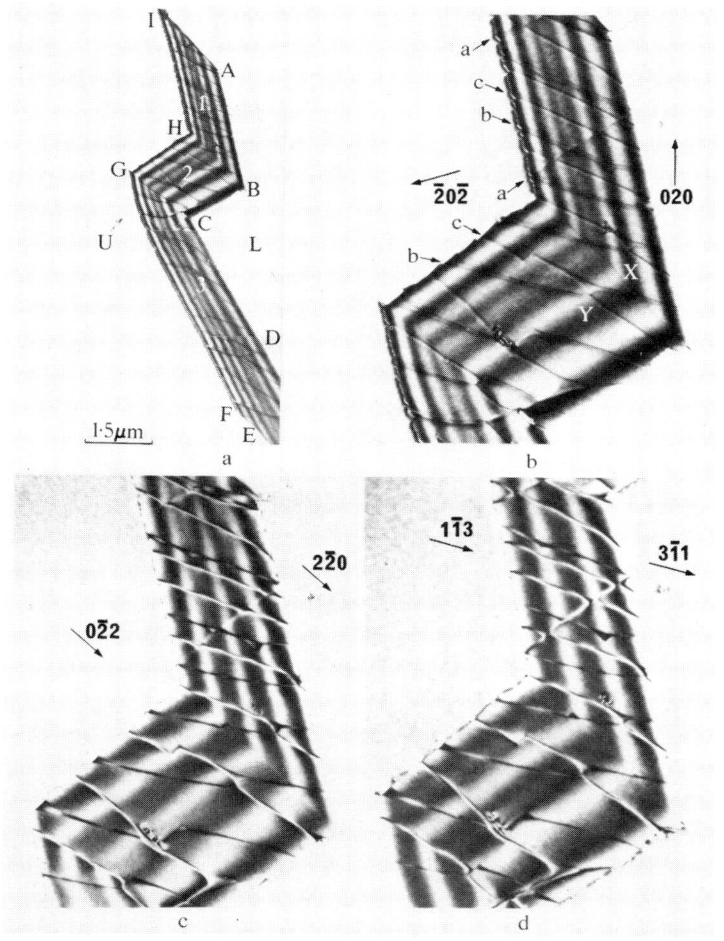

Figure 5.71 Double two-beam images of a faceted incoherent-Σ3 boundary in a Cu − 6 at% Si alloy; (*b*) is a higher magnification image of a portion of (*a*) and the beam direction ***B*** is close to $[\bar{2}32]_U$; (*c*) and (*d*) are same-$\boldsymbol{g}_c$ images with ***B*** close to $[\bar{3}22]_U$ and $[\bar{7}54]_U$ respectively; the diffracting vectors are indicated.

$(\bar{8}\ \bar{5}\ 10)_U/(1\bar{4}2)_L$ respectively. Thus, with the exception of facet 3 of case 1, the incoherent facets of both cases 1 and 2 are all Σ3 boundaries of the type $\{421\}/\{10\ 8\ 5\}$, where the specific crystallographic indices show that the facets 1 and 2 of case 1 are both of one handedness (say right-handed), whereas facets 4, 5, 6 and 7 of case 2 are all of the opposite handedness (left-handed). Thus, disregarding the change in handedness, it would be expected that the local atomic structure of each facet, in regions between secondary grain boundary dislocations, would be the same.

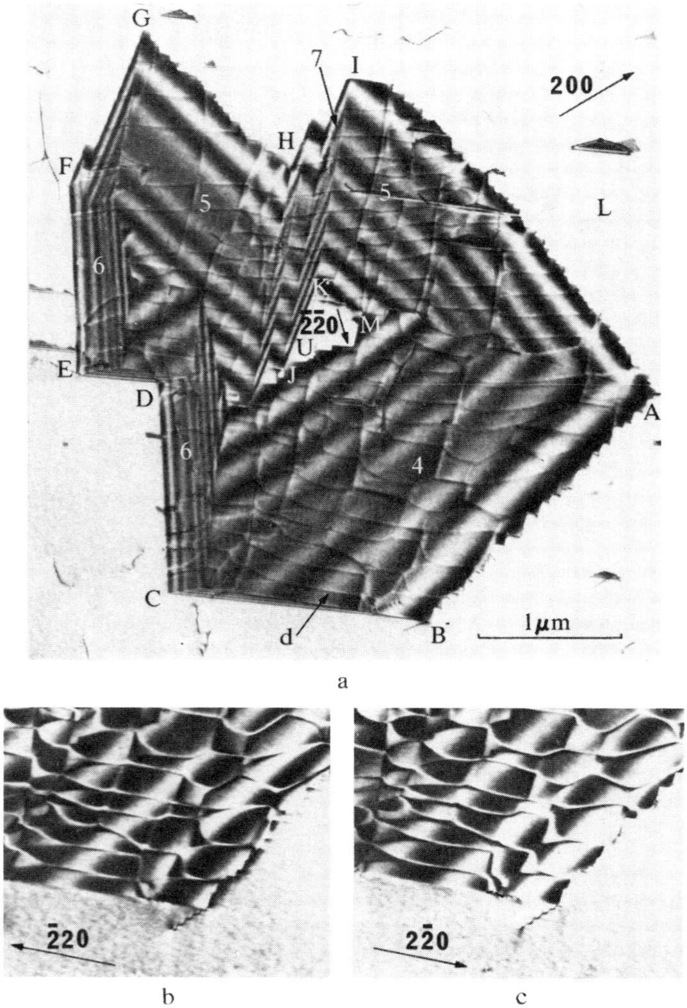

Figure 5.72 Double two-beam images of a faceted incoherent-$\Sigma 3$ boundary in a Cu − 6 at% Si alloy. In (*a*) ***B*** is close to $[\bar{2}23]_U$; (*b*) and (*c*) are same $\pm \boldsymbol{g}_c$ images of a portion of facet 4 and ***B*** is close to $[\bar{2}11]_U$; the diffracting vectors are indicated.

The secondary grain boundary dislocations on facets 1, 2 and 3 show contrast, in images taken over a wide range of diffracting conditions, which indicates that they form a network consisting of three arrays of dislocations with different Burgers vectors and that the Burgers vectors of the elements in each array are the same on each facet. The network of secondary grain boundary dislocations on facets 1 and 2 of figure 5.71(*a*) can be seen more

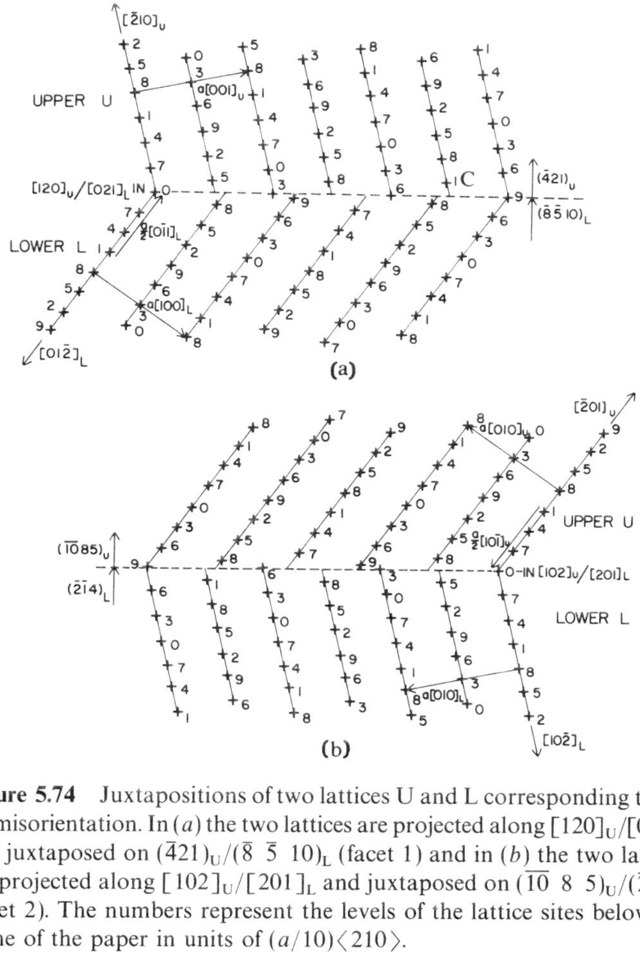

Figure 5.74 Juxtapositions of two lattices U and L corresponding to the $\Sigma3$ misorientation. In (*a*) the two lattices are projected along $[120]_U/[021]_L$ and juxtaposed on $(\bar{4}21)_U/(\bar{8}\ \bar{5}\ 10)_L$ (facet 1) and in (*b*) the two lattices are projected along $[102]_U/[201]_L$ and juxtaposed on $(\overline{10}\ 8\ 5)_U/(\bar{2}\bar{1}4)_L$ (facet 2). The numbers represent the levels of the lattice sites below the plane of the paper in units of $(a/10)\langle 210\rangle$.

between the structures of facets 1 and 2, an alternative but equivalent rigid-body displacement needs to be used. This is to displace the upper grain relative to the lower grain of figure 5.75(*b*) by $(1/6)[10\bar{1}]_U/(1/6)[01\bar{1}]_L$ (i.e. parallel to the $(1/2)[10\bar{1}]_U$ vector marked on figure 5.74(*b*)). This alternative displacement, together with the removal of sites closer than $(1/2)\langle 110\rangle$ and a contraction normal to the plane of the boundary of $0.046a$, results in a model from figure 5.74(*b*) for facet 2 which is crystallographically equivalent to that obtained from figure 5.74(*a*) for facet 1. This equivalence is apparent from figures 5.74(*a*) and (*b*) because figure 5.74(*b*), and the rigid-body displacements associated with it, will become identical with figure 5.74(*a*) and its associated rigid-body displacements, if figure 5.74(*b*) is rotated through 180° and re-indexed to have the same indices as figure 5.74(*a*). Thus it can be concluded from the determined values of rigid-body displacement $\boldsymbol{R}_\alpha$ that the main parts of the boundary at facets 1 and 2 are crystallographically equivalent and can be described by the same atomic model; therefore they can be considered as having the same local atomic structure.

In order to determine the rigid-body displacement $\boldsymbol{R}_\beta$, associated with regions of the boundary between the partial dislocations, it is necessary to determine the Burgers vectors of these partial dislocations. In discussing the determination of these Burgers vectors attention will be confined initially to elements *a*, *b* and *c* of the network on facet 1 for which the total Burgers vector associated with each element is a DSC vector. Figure 5.75 is a same-$\boldsymbol{g}_c$ electron micrograph ($\boldsymbol{g}_c = \bar{1}\bar{1}\bar{1}_U/\bar{1}\bar{1}\bar{1}_L$) where the displacement fringe contrast is weak, so that the dislocation contrast associated with elements *a*, *b* and *c* of the secondary dislocation network can be taken as a guide to the values of $\boldsymbol{g}_c \cdot \boldsymbol{b}$ associated with their total DSC Burgers vectors. The strong contrast of element *a* in figure 5.75 suggests that it has a DSC Burgers vector with a component of $\pm(1/3)[111]_U/(1/3)[111]_L$ for which $\boldsymbol{g}_c \cdot \boldsymbol{b} = \pm 1$. On the other hand, the very weak contrast associated with elements *b* and *c* suggests that the total DSC Burgers vector(s) associated with each of them is/are either a basis vector of the DSC lattice of the type $(1/6)\langle 112 \rangle$ in the $(111)_U/(111)_L$ plane, or a linear combination of basis vectors of this type, for which $\boldsymbol{g}_c \cdot \boldsymbol{b} = 0$. This information is used as a guide in the initial selection of possible Burgers vectors to be tested by image matching.

For each element *a*, image matching showed that there was no dissociation into partial dislocations, and the Burgers vector is identified as the basis DSC

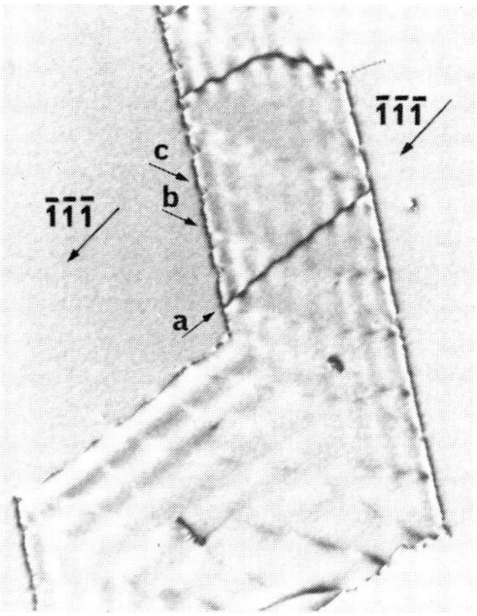

Figure 5.75 Same-$\boldsymbol{g}_c = \bar{1}\bar{1}\bar{1}_{U/L}$ image of facets 1 and 2 of figure 5.71(*a*) for $\boldsymbol{B} = [\bar{5}23]_U$.

vector $z = (1/3)[111]_U/(1/3)[111]_L$, confirming the suggestion made in connection with figure 5.75. An example showing the type of agreement and disagreement between an experimental image and theoretical images used in the identification is given in Chapter 2 in figure 2.12.

Unlike elements *a*, each element *b* is dissociated into two partial dislocations, and this is illustrated schematically in figure 5.76. Figure 5.76 represents the boundary plane of facet 1 with the upper grain above on the left and the lower grain below on the right, with the positive sense of the line direction of the partial dislocations with Burgers vectors $\boldsymbol{b}_1$ and $\boldsymbol{b}_2$ indicated by the arrows, and the rigid-body displacements of the lower grain relative to the upper grain indicated as $\boldsymbol{R}_\alpha$ and $\boldsymbol{R}_\beta$. The sum of the two partial Burgers vectors $\boldsymbol{b}_1 + \boldsymbol{b}_2 = \boldsymbol{B}$ must be a DSC vector so that, in accordance with the FS/RH convention and the procedure of Thompson (1953) for specifying the Burgers vectors of partial dislocations, there are two possible dissociation reactions giving rise to the Burgers vectors $\boldsymbol{b}_1$ and $\boldsymbol{b}_2$, namely

$$\boldsymbol{b}_1 = \boldsymbol{R}_\beta - \boldsymbol{R}_\alpha \text{ with } \boldsymbol{b}_2 = \boldsymbol{B} - (\boldsymbol{R}_\beta - \boldsymbol{R}_\alpha) \tag{5.13}$$

or

$$\boldsymbol{b}_1 = \boldsymbol{B} - (\boldsymbol{R}_\alpha - \boldsymbol{R}_\beta) \text{ with } \boldsymbol{b}_2 = \boldsymbol{R}_\alpha - \boldsymbol{R}_\beta. \tag{5.14}$$

Since the Burgers vectors of such partial grain boundary dislocations are not quantised to a set of lattice vectors, as are DSC Burgers vectors, it is necessary to assume values of $(\boldsymbol{R}_\alpha - \boldsymbol{R}_\beta)$ in reactions (5.13) and (5.14) in order to obtain a set of discrete possibilities for $\boldsymbol{b}_1$ and $\boldsymbol{b}_2$ to be tested by image matching. As pointed out earlier, fringe contrast suggests that $(\boldsymbol{R}_\alpha - \boldsymbol{R}_\beta)$ is approximately equal to $2\boldsymbol{R}_\alpha$, with an uncertainty of a DSC vector $\boldsymbol{d}$, and to satisfy this condition it is assumed that $\boldsymbol{R}_\beta$ has a major component which, with an uncertainty of a DSC vector, is the negative of the major component of $\boldsymbol{R}_\alpha$, and a minor component which is the same as the minor component of $\boldsymbol{R}_\alpha$. On this basis

$$\boldsymbol{R}_\alpha - \boldsymbol{R}_\beta = (1/3)[\bar{1}01]_U/(1/3)[0\bar{1}1]_L + \boldsymbol{d} \approx 2\boldsymbol{R}_\alpha + \boldsymbol{d}. \tag{5.15}$$

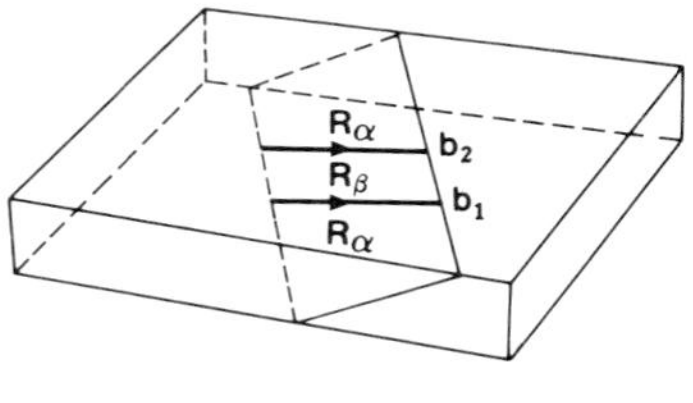

Figure 5.76 Schematic representation of partial grain boundary dislocations with Burgers vectors $\boldsymbol{b}_1$ and $\boldsymbol{b}_2$ and associated rigid-body displacements $\boldsymbol{R}_\alpha$ and $\boldsymbol{R}_\beta$.

Using this value of $(\boldsymbol{R}_\alpha - \boldsymbol{R}_\beta)$ and the knowledge that $\boldsymbol{B}$ is a DSC vector which probably lies in the $(111)_U/(111)_L$ plane, theoretical images of pairs of partial dislocations for comparison with each experimental image were computed in batches of 12. The 12 possibilities for each batch arise from dissociation reactions (5.13) and (5.14) with values of $\boldsymbol{B}$ of $\pm(1/6)[11\bar{2}]_U/(1/6)[\bar{1}2\bar{1}]_L$, $\pm(1/6)[1\bar{2}1]_U/(1/6)[2\bar{1}\bar{1}]_L$ and $\pm(1/6)[\bar{2}11]_U/(1/6)[\bar{1}\bar{1}2]_L$ and each batch of 12 involves a different value of $\boldsymbol{d}$ in equation (5.15) for $(\boldsymbol{R}_\alpha - \boldsymbol{R}_\beta)$. Only one of these possibilities gave theoretical images which matched the experimental images, namely that computed for dissociation reaction (5.14) with $\boldsymbol{b}_1 = (1/6)[1\bar{1}0]_U/(1/6)[10\bar{1}]_L$, $\boldsymbol{b}_2 = (1/6)[10\bar{1}]_U/(1/6)[01\bar{1}]_L$ (i.e. $\boldsymbol{B} = (1/6)[2\bar{1}\bar{1}]_U/(1/6)[11\bar{2}]_L$) and a value of $\boldsymbol{d} = (1/2)[10\bar{1}]_U/(1/2)[01\bar{1}]_L$. Thus $(\boldsymbol{R}_\alpha - \boldsymbol{R}_\beta) = (1/6)[10\bar{1}]_U/(1/6)[01\bar{1}]_L$ and this gives $\boldsymbol{R}_\beta = (1/3)[\bar{1}01]_U/(1/3)[01\bar{1}]_L + (1/100)[\bar{4}21]_U/(1/300)[\bar{8}\ \bar{5}\ 10]_L + \boldsymbol{d}$.

Figure 5.77 shows a set of experimental images involving three non-coplanar diffracting vectors of an element *b*†, together with a matching set of theoretical images computed for the above values of $\boldsymbol{b}_1$ and $\boldsymbol{b}_2$ with a separation of the partial grain boundary dislocations of 140 Å. Equally good agreement was obtained for separations in the range 120–160 Å. An example of matching and mismatching between an experimental image and theoretical images computed for separations of 100 Å, 140 Å, and 180 Å is shown in figure 5.78. The main feature of the contrast in the experimental image of the element *b* in figure 5.78 is that it consists of three lines of below-background intensity, i.e. a central weak line bordered by two strong lines. Good agreement is obtained with this experimental image for a separation in the theoretical image of 140 Å, whereas for a separation of 100 Å the three lines are not clearly resolved and for a separation of 180 Å the central line is too broad and the overall image too wide.

To test the validity of the assumptions concerning $\boldsymbol{R}_\beta$, which are used in the determinations of the Burgers vectors $\boldsymbol{b}_1$ and $\boldsymbol{b}_2$ of elements *b*, images were computed in which the magnitude of these vectors was varied. It was found that $\boldsymbol{b}_1$ and $\boldsymbol{b}_2$ could not be altered in theoretical images by more than $(1/60)\langle 100 \rangle$ without causing mismatches with the experimental images. The fact that matching theoretical and experimental images could only be obtained over a narrow range of values of $\boldsymbol{b}_1$ and $\boldsymbol{b}_2$ indicates that there is only a small uncertainty arising from the assumptions made concerning $\boldsymbol{R}_\beta$. This uncertainty in $\boldsymbol{R}_\beta$, like that in $\boldsymbol{R}_\alpha$, will be considered as an uncertainty in the minor component of $\boldsymbol{R}_\beta$.

As in the case of elements *b*, each element *c* is dissociated into two partial dislocations and exactly the same assumptions and procedure adopted for element *b* were used in arriving at possible Burgers vectors for the

† In this element there is a short length over which the separation of the partial grain boundary dislocations is perturbed and no account is taken of this in theoretical images.

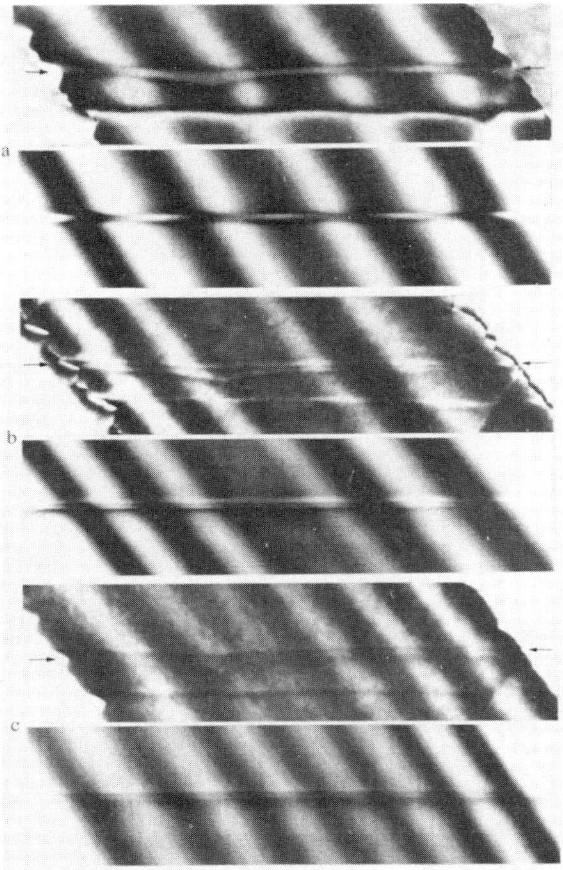

Figure 5.77 Comparison of experimental and computed double two-beam images of an element *b* on facet 1 (indicated by the arrows in the experimental images). The diffraction parameters for the experimental and computed images, computed for $\boldsymbol{b}_1 = (1/6)[1\bar{1}0]_U$ and $\boldsymbol{b}_2 = (1/6)[\bar{1}0\bar{1}]_U$ with a separation, *s*, of 140 Å, are: (*a*) $\boldsymbol{g}_U = 02\bar{2}$, $\boldsymbol{g}_L = \bar{2}20$, $\boldsymbol{B} = [\bar{3}22]_U$, $w_U = w_L = 0.33$, $\boldsymbol{g}_c \cdot \boldsymbol{R}_\alpha = -0.30$, $\boldsymbol{g}_c \cdot \boldsymbol{R}_\beta = -0.37$; (*b*) $\boldsymbol{g}_U = \bar{2}0\bar{2}$, $\boldsymbol{g}_L = 0\bar{2}0$, $\boldsymbol{B} = [\bar{2}32]_U$, $w_U = w_L = 0.10$; (*c*) $\boldsymbol{g}_U = 002$, $\boldsymbol{g}_L = 022$, $\boldsymbol{B} = [\bar{4}10]_U$, $w_U = 0.25$, $w_L = 0.65$.

partial dislocations $\boldsymbol{b}_1$ and $\boldsymbol{b}_2$ of element *c*. Matching theoretical images were obtained in this case for $\boldsymbol{b}_1 = (1/6)[01\bar{1}]_U/(1/6)[\bar{1}10]_L$ and $\boldsymbol{b}_2 = (1/6)[10\bar{1}]_U/(1/6)[01\bar{1}]_L$, with the same uncertainty as discussed for element *b*. These partial dislocations correspond to dissociation reaction (5.14) with a total DSC Burgers vector $\boldsymbol{B}$ of $(1/6)[11\bar{2}]_U/(1/6)[\bar{1}2\bar{1}]_L$ and a

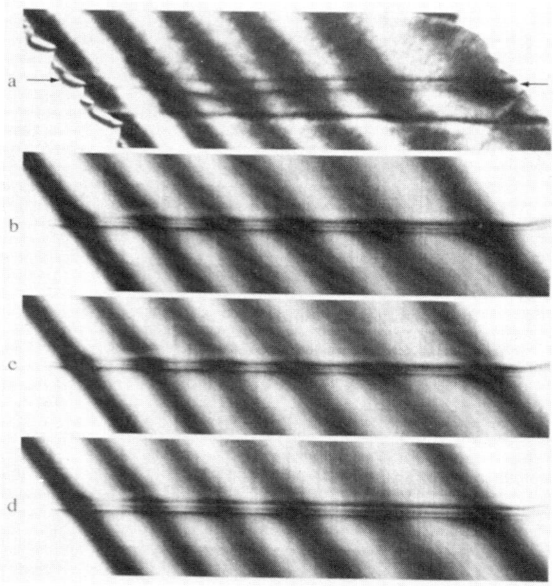

Figure 5.78 Comparison of experimental (a) with computed (b)–(d) double two-beam images of the same element b as in figure 5.77 (indicated by the arrows in the experimental image) showing the influence of a variation in the separation of $\boldsymbol{b}_1$ and $\boldsymbol{b}_2$ on the contrast in the computed images; $s = 140$ Å in (b), 100 Å in (c) and 180 Å in (d). The diffraction parameters for the experimental and computed images are: $\boldsymbol{g}_{\mathrm{U}} = \bar{2}0\bar{2}$, $\boldsymbol{g}_{\mathrm{L}} = 020$, $\boldsymbol{B} = [\bar{2}32]_{\mathrm{U}}$, $w_{\mathrm{U}} = 0.60$ and $w_{\mathrm{L}} = 0.30$.

value of $\boldsymbol{d} = (1/2)[10\bar{1}]_{\mathrm{U}}/(1/2)[01\bar{1}]_{\mathrm{L}}$ giving the same values of $(\boldsymbol{R}_\alpha - \boldsymbol{R}_\beta)$ and $\boldsymbol{R}_\beta$ as found for element b with the same uncertainties. A matching set of experimental and theoretical images for an element c, for a separation of the partial grain boundary dislocations in the computed images of 100 Å, is given in figure 5.79, and this good agreement between the experimental and theoretical images was obtained for separations in the range 80–120 Å.

In the case of facet 2, images of the partial grain boundary dislocations comprising elements b and c show that they have the same Burgers vectors as elements b and c on facet 1. Further, as already shown, the main parts of facets 1 and 2 are crystallographically equivalent. Therefore the state of rigid-body displacement in the regions between the partial dislocations on facet 2 is also crystallographically equivalent to that between the partial dislocations on facet 1 and, under the same assumptions made for facet 1, can be expressed as

$$\boldsymbol{R}_\beta = (1/3)[\bar{1}01]_{\mathrm{U}}/(1/3)[0\bar{1}1]_{\mathrm{L}} + (1/300)[\overline{10}\ 8\ 5]_{\mathrm{U}}/(1/100)[\bar{2}\bar{1}4]_{\mathrm{L}} + \boldsymbol{d}.$$

Figure 5.79 Experimental double two-beam images of two elements *c* on facet 1 and computed images for one of the elements. The diffraction parameters for the experimental and computed images, computed for $\boldsymbol{b}_1 = (1/6)[01\bar{1}]_U$ and $\boldsymbol{b}_2 = (1/6)[10\bar{1}]_U$ with $s = 100$ Å, are: (*a*) $\boldsymbol{g}_U = \bar{1}\bar{3}1$, $\boldsymbol{g}_L = 1\bar{3}\bar{1}$, $\boldsymbol{B} = [\overline{16}\ 7\ 5]_U$, $w_U = w_L = -0.23$, $\boldsymbol{g}_c \cdot \boldsymbol{R}_\alpha = 0.31$, $\boldsymbol{g}_c \cdot \boldsymbol{R}_\beta = 0.64$; (*b*) $\boldsymbol{g}_U = \bar{2}0\bar{2}$, $\boldsymbol{g}_L = 020$, $\boldsymbol{B} = [\bar{2}32]_U$, $w_U = 0.60$, $w_L = 0.30$ and (*c*) $\boldsymbol{g}_U = 002$, $\boldsymbol{g}_L = 022$, $\boldsymbol{B} = [\bar{4}10]_U$, $w_U = 0.25$, $w_L = 0.65$.

This differs from the value for $\boldsymbol{R}_\beta$ for facet 1 in exactly the same way as the value of $\boldsymbol{R}_\alpha$ differs between facets 1 and 2, i.e. the minor components of $\boldsymbol{R}_\beta$ correspond to contractions along the different facet normals.

These results indicate that only two different states of rigid-body displacement are associated with the boundary structure at each of the facets 1 and 2, one for the main parts of the boundary and the other for the regions between the partial dislocations. That these two states are generally applicable to $\Sigma 3$ boundary facets of the type $\{421\}/\{10\ 8\ 5\}$ was shown by analysing case 2 of figure 5.72, which involves boundary facets that are crystallographically equivalent (apart from a change in handedness) to those of case 1. In this analysis, it was demonstrated by using image matching that the Burgers vectors of the partial dislocations in array d on facet 4 of case 2 were simply those obtained from values of $\boldsymbol{R}_\alpha$ and $\boldsymbol{R}_\beta$ which were the appropriate crystallographically equivalent values to those found for case 1. The Burgers vectors obtained in this way were $\boldsymbol{b}_1 = (1/6)[\bar{1}10]_U/(1/6)[\bar{1}01]_L$ and $\boldsymbol{b}_2 = (1/6)[01\bar{1}]_U/(1/6)[\bar{1}10]_L$. These values of $\boldsymbol{b}_1$ and $\boldsymbol{b}_2$ correspond to the dissociation of an element d with the DSC Burgers vector $\boldsymbol{B} = (1/6)[\bar{1}2\bar{1}]_U/(1/6)[\bar{2}11]_L$ according to dissociation reaction (5.13) with

$$\boldsymbol{R}_\alpha = (1/6)[\bar{1}10]_U/(1/6)[\bar{1}01]_L + (1/300)[\overline{10}\ 5\ 8]_U/(1/100)[\bar{1}\bar{2}4]_L + \boldsymbol{d}$$

and

$$\boldsymbol{R}_\beta = (1/3)[\bar{1}10]_U/(1/3)[\bar{1}01]_L + (1/300)[\overline{10}\ 5\ 8]_U/(1/100)[\bar{1}\bar{2}4]_L + \boldsymbol{d}$$

which are the crystallographically equivalent values to those of $\boldsymbol{R}_\alpha$ and $\boldsymbol{R}_\beta$ of facet 1.

Figure 5.80 shows examples of matching experimental and theoretical images, involving three non-coplanar diffracting vectors, of the type used to identify the partial dislocations of an element d on the $(\overline{10}\ 5\ 8)_U/(\bar{1}\bar{2}4)_L$ plane of facet 4. The experimental images in parts (i) of figures 5.80(a) and (b) each show two elements d with a variation in the separation of the partial dislocations. The theoretical images in parts (ii) of figures 5.80(a) and (b) are for an element d with a separation of the partial dislocations of 60 Å and match the right-hand-side of each of the elements d in the experimental images. In the theoretical images of parts (iii) of figures 5.80(a) and (b) the partial dislocations are separated by 1000 Å and, although no account is taken of the experimental variation in line direction, the detail of the contrast matches that of the individual partial dislocations on the left-hand-side of the experimental images. For example, in (i) of figure 5.80(a) the contrast of the upper partial grain boundary dislocation is weak and double, and this contrast is matched in the theoretical image of (iii) of figure 5.80(a).

A feature of the secondary grain boundary dislocation structure of all the incoherent facets of both cases 1 and 2 is the wide separation of the partial dislocations which occurs where elements of different arrays intersect. Such configurations have been analysed for case 1 in terms of the Burgers vectors of the partial dislocations that have been identified. Figure 5.81(a) is an image of region Y of facet 2 of case 1, where it can be seen that a common partial dislocation forms the upper border of the widely dissociated configuration. The Burgers vector of this partial dislocation is $(1/6)[10\bar{1}]_U/(1/6)[01\bar{1}]_L$, i.e. the common Burgers vector $\boldsymbol{b}_2$ of elements b and c. From the schematic diagrams of figures 5.81(b) and (c), it can be seen that the widely dissociated configuration of figure 5.81(a) can be envisaged as arising

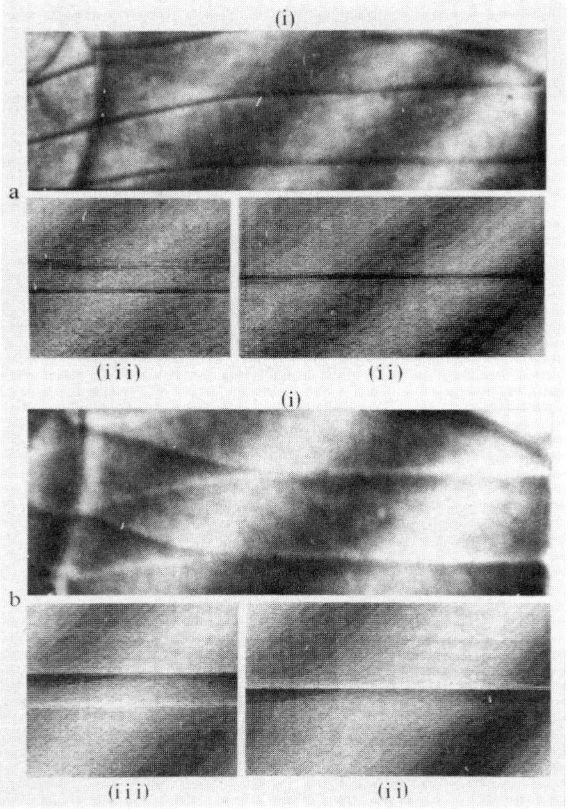

Figure 5.80 Comparison of experimental double two-beam images (i) containing two elements *d* on facet 4 with computed images for one of these elements computed for $\boldsymbol{b}_1 = (1/6)[\bar{1}10]_U$ and $\boldsymbol{b}_2 = (1/6)[01\bar{1}]_U$ with the separation, *s*, of the partial grain boundary dislocations being 60 Å in (ii) and 1000 Å in (iii). The diffraction parameters are (*a*) $\boldsymbol{g}_U = 020$, $\boldsymbol{g}_L = 202$, $\boldsymbol{B} = [\bar{4}01]_U$, $w_U = 0.30$, $w_L = 0.60$; (*b*) $\boldsymbol{g}_U = \bar{2}\bar{2}0$, $\boldsymbol{g}_L = 200$, $\boldsymbol{B} = [\bar{2}23]_U$, $w_U = 0.20$, $w_L = 0.30$.

from the simple crossover of the dissociated elements *b* and *c* of figure 5.81(*b*) which then react to form the configuration of figure 5.81(*c*). In figure 5.81(*c*) the partial dislocations of elements *b* and *c* with the common Burgers vector $\boldsymbol{b}_2$ have linked together, and the upper partial dislocation of this configuration has minimised its line length to form the widely dissociated region.

The Burgers vectors that have been identified for each pair of partial grain boundary dislocations, that form the elements of the secondary grain boundary dislocation networks on the different facets, are such that the elastic interaction between them is always mutually repulsive. For the temperature

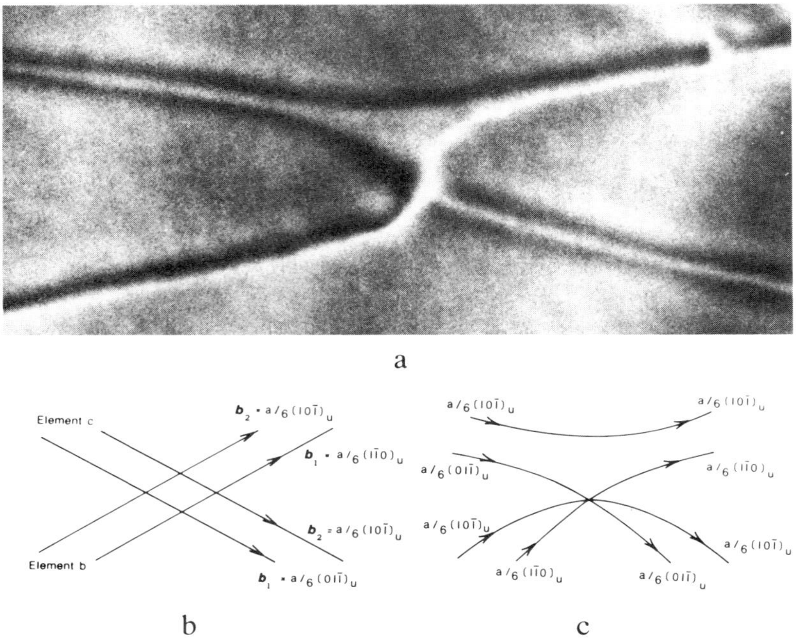

Figure 5.81 Nodal interaction between the partial grain boundary dislocation elements *b* and *c* in region *Y* (see figure 5.71(*b*)) of facet 2; (*a*) same-$\boldsymbol{g}_c$ image with $\boldsymbol{g}_c = 3\bar{1}1_U/31\bar{1}_L$ and $\boldsymbol{B}$ close to $[\bar{1}36]_U$; (*b*) and (*c*) are schematic diagrams of the array before and after interaction respectively.

at which the dislocation networks formed, it will be assumed that the only forces acting were the elastic repulsion between the partial dislocations and the attraction resulting from the difference in grain boundary energy between the two different states of rigid-body displacement, so that the separation of the partial dislocations represents a balance between these forces. This balance should not have been significantly influenced by any steps in the boundary associated with the partial dislocations, since such steps will contribute to the self-energy, but will only make a minor contribution to the long range forces of interaction. On this basis, the force of elastic interaction between parallel partial dislocations at a separation s will give the difference in grain boundary energy between that of the main parts of the boundary, γ_α, and that of the regions between the partial dislocations, γ_β.

The magnitude of the force of elastic interaction, F, acting in the boundary plane in a direction normal to the line direction of the partial dislocations can be computed in elastic anisotropy using the expression

$$F = b_i^{(2)}\,\sigma_{i2}^{(1)}$$

where the line direction of the partial dislocations is parallel to the Ox_3 axis, the boundary normal is parallel to the Ox_2 axis, $b_i^{(2)}$ are the three components of the Burgers vector $\boldsymbol{b}_2$ and $\sigma_{i2}^{(1)}$ are the components of the stress which acts across the boundary at the partial dislocation with Burgers vector $\boldsymbol{b}_2$ because of the presence of the partial dislocation with Burgers vector $\boldsymbol{b}_1$. The expression for $\sigma_{i2}^{(1)}$ is given in equation (1.50) and was evaluated using the subroutine ANCALC in the grain boundary image program PCGBD. The values obtained in this way for $(\gamma_\beta - \gamma_\alpha)$ are given in table 5.9 for two elements in the array *b*, two elements in the array *c* and an element in the array *d*. These values of $(\gamma_\beta - \gamma_\alpha)$ which have been calculated for different line directions, facet planes and Burgers vectors of the partial grain boundary dislocations, are constant within the limits of experimental error. Therefore they support the conclusion that only two different boundary states exist on $\{421\}/\{10\ 8\ 5\}$ facets, and indicate that these have a small difference in grain boundary energy of approximately 2.5 mJ m^{-2}.

So far the minor components of $\boldsymbol{R}_\alpha$ and $\boldsymbol{R}_\beta$ have been assumed to be displacements normal to the boundary, i.e. any in-plane components have been ignored. However, in hard-sphere modelling of the atomic structure associated with the two states of rigid-body displacement the relative contributions to $\boldsymbol{R}_\alpha$ and $\boldsymbol{R}_\beta$ of both normal and in-plane minor components need to be considered. Since the same two states are found on all $\{421\}/\{10\ 8\ 5\}$ facets, atomic modelling need only be considered for one particular facet geometry. Hard-sphere models will be developed from the starting state corresponding to the simple juxtaposition of the two lattices in figure 5.74(*a*) which is for the geometry of facet 1. The structure for the state of rigid-body displacement corresponding to the main part of the boundary will be described using figure 5.82(*a*) and for the regions of the boundary between the partial dislocations using figure 5.82(*b*).

Figure 5.82(*a*) is obtained from figure 5.74(*a*) by displacing the lower lattice relative to the upper lattice by $(1/6)[\bar{1}01]_U/(1/6)[0\bar{1}1]_L$, which is the major component of the determined rigid-body displacement $\boldsymbol{R}_\alpha$ for the main part of the boundary (this displacement corresponds to one third of the $(1/2)[0\bar{1}1]_L$ vector marked on figure 5.74(*a*)) and by removing those

Table 5.9 Difference in grain boundary energy.

	Facet 1		Facet 2		Facet 4	
	s (Å)	$(\gamma_\beta - \gamma_\alpha)$ (mJ m^{-2})	*s* (Å)	$(\gamma_\beta - \gamma_\alpha)$ (mJ m^{-2})	*s* (Å)	$(\gamma_\beta - \gamma_\alpha)$ (mJ m^{-2})
Element *b*	140 ± 20	2.9 ± 0.4	140 ± 20	2.9 ± 0.4	—	—
Element *c*	100 ± 20	2.4 ± 0.5	90 ± 20	2.2 ± 0.5	—	—
Element *d*	—	—	—	—	60 ± 20	3.0 ± 1.0

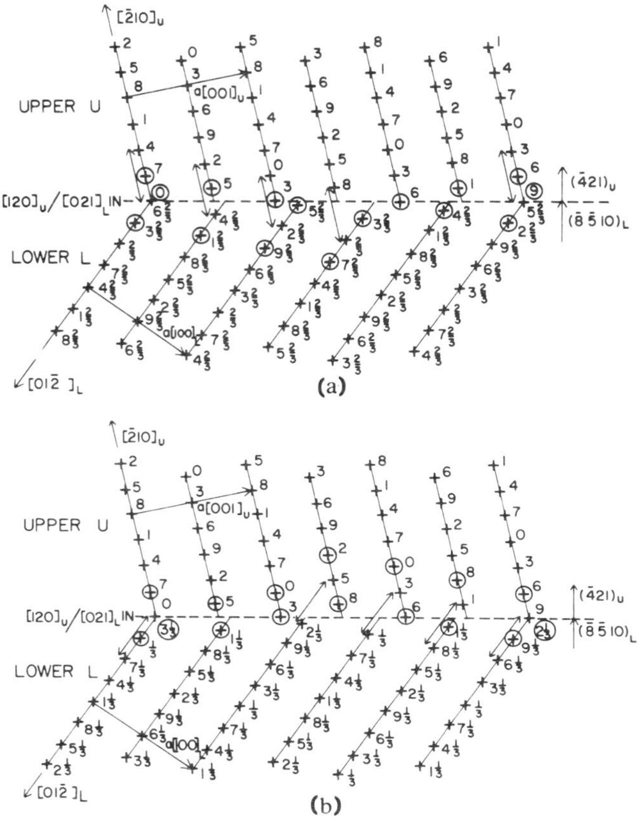

Figure 5.82 Hard-sphere models of two states of rigid-body displacement on facet 1, based on the reference structure of figure 5.74(*a*), after a rigid-body displacement of grain L relative to grain U of $(1/6)[\bar{1}01]_U/(1/6)[0\bar{1}1]_L$ in (*a*) and of $(1/3)[\bar{1}01]_U/(1/3)[0\bar{1}1]_L$ in (*b*).

lattice sites of the lower lattice which would otherwise interpenetrate the upper lattice. This displacement results in all the lattice sites in the lower lattice being moved up out of the page by $(1/30)[\bar{1}\bar{2}0]_U/(1/30)[0\bar{2}\bar{1}]_L$, as indicated by the new labelling (figure 5.82(*a*)) of the levels of the lattice sites in the lower lattice. In figure 5.82(*a*) some of the lattice sites in both lattices near the boundary are closer than the $(1/2)\langle 110 \rangle$ interatomic spacing. To correct this situation lattice sites need to be removed which will result, for this geometry, in interatomic spacings across the boundary being greater than $(1/2)\langle 110 \rangle$, and this will lead, in the model, to a rigid-body collapse of the lattices towards one another, corresponding to the minor component of $\boldsymbol{R}_\alpha$. The lattice sites removed in figure 5.82(*a*) are indicated by the symbols enclosed in circles and these sites have been selected so as to minimise the number of sites removed and the spacings across the boundary which are greater than $(1/2)\langle 110 \rangle$. In figure 5.82(a) the closest interatomic spacings across the boundary are indicated by the double arrows and these correspond to a spacing of $a\sqrt{(5/3)}$ along $[1\bar{1}0]_U/[10\bar{1}]_L$.

Figure 5.82(b) is obtained from figure 5.74(a) in a similar way to figure 5.82(a), but in this case the lower grain is displaced relative to the upper grain by $(1/3)[\bar{1}01]_U/(1/3)[0\bar{1}1]_L$ which is the main component of the rigid-body displacement $\boldsymbol{R}_\beta$ for the regions between the partial dislocations. Following the procedure outlined for figure 5.82(a) the lattice sites removed are again indicated by the symbols enclosed in circles and the double arrows again indicate the closest interatomic spacings across the boundary of $a\sqrt{(5/3)}$, but in this case these lie along $[021]_U/[012]_L$.

In the models of figures 5.82(a) and (b) the number of lattice sites removed is the same, but the pattern of sites removed differs and this results in different directions for the closest interatomic spacings as indicated by the double arrows. For both states there are two simple ways in which the minor components of $\boldsymbol{R}_\alpha$ and $\boldsymbol{R}_\beta$, i.e. the collapse of the lattices towards one another, can be considered; namely a collapse which is solely normal to the boundary plane, or a collapse which is along the directions of closest distance of approach between the atoms across the boundary. A collapse normal to the boundary would result in contractions of $0.042a$ for the main part of the boundary (figure 5.82(a)) and $0.087a$ for the regions between the partial grain boundary dislocations (figure 5.82(b)), so that the minor components of $\boldsymbol{R}_\alpha$ and $\boldsymbol{R}_\beta$ would not be equal. This difference in the minor components would lead to a situation where the excess volume associated with the boundary structure would be greater in the main part of the boundary, which would imply that the atomic fit there would be worse than for those regions between the partial dislocations. However, the experimental results show that the lower energy state, and therefore the better atomic fit, is associated with the main part of the boundary. Thus collapse of the grains which is solely normal to the boundary cannot give rise to appropriate atomic structures at the boundary. However, a collapse along the directions of closest distance of approach would result, for the main part of the boundary, in a contraction normal to the boundary plane of $0.035a$ together with an in-plane displacement of the lower grain with respect to the upper grain of $0.014a$ along $[13\bar{2}]_U/[\bar{5}\ 10\ 1]_L$ and, for the regions between the partial dislocations, in a contraction normal to the boundary plane of $0.019a$ together with an in-plane displacement of $0.033a$ along $[584]_U/[10\ 22\ 19]_L$. In this case the excess volume associated with the boundary structure would be greater in the regions of the boundary between the partial dislocations than in the main part of the boundary, and this is in agreement with the observed difference in energy between the two states. Thus a collapse of the grains in this way leads to a better description for the atomic structure associated with the two states of rigid-body displacement. It is clear from these hard-sphere atomic models that the minor components of $\boldsymbol{R}_\alpha$ and $\boldsymbol{R}_\beta$ probably contain small in-plane displacements, but these cannot be determined from the experimental results as they are contained within the range of the experimental errors.

The components of the displacements $\boldsymbol{R}_\alpha$ and $\boldsymbol{R}_\beta$ which were ignored in developing the models of figure 5.82 are the unknown DSC vectors $\boldsymbol{d}$. In relation to the hard-sphere models of the different states of displacement on the $\Sigma 3$ boundaries, all the possible values for $\boldsymbol{d}$ in the expressions for $\boldsymbol{R}_\alpha$ and $\boldsymbol{R}_\beta$ can be taken into account simply by considering three different levels for the choice of the position of the boundary plane within a unit cell of the CSL. These three different levels correspond to three different juxtapositions of the two lattices of the type illustrated for one level of the boundary in figure 5.74(a). The two alternative juxtapositions to that in figure 5.74(a) would correspond to two levels for the boundary between that shown in figure 5.74(a) and a level through the site marked C in the upper grain in figure 5.74(a). In the hard-sphere models of figure 5.82 for the two states of displacement, the atoms which are removed at the interface encompass all three closely spaced levels. Therefore, after removing these atoms, no distinction can be drawn between the three different levels, and the local atomic arrays at the boundary in figures 5.82(a) and (b) would be identical for starting conditions corresponding to the three different juxtapositions. Thus, for the type of modelling used here no significance can be associated with the components $\boldsymbol{d}$ in the expressions for $\boldsymbol{R}_\alpha$ and $\boldsymbol{R}_\beta$.

In contrast to the $\{211\}/\{211\}$ incoherent-$\Sigma 3$ boundaries discussed by Pond and Vitek (1977), the partial grain boundary dislocations on $\{420\}/\{10\ 8\ 5\}$ incoherent-$\Sigma 3$ facets discussed here separate states of different boundary structure, so that their mutual elastic repulsion is balanced by a difference in grain boundary energy and it is this which enables the difference in boundary energy to be estimated.

5.5 *n*-BEAM LATTICE IMAGES OF HIGH-ANGLE BOUNDARIES

Studies of high-angle grain boundaries using n-beam lattice imaging suffer from the restrictions discussed in sections 2.5 and 4.5, in that they are confined to pure tilt boundaries where the tilt axis is a low-index crystallographic direction with the electron beam aligned along this axis. Because of these restrictions n-beam lattice imaging usually involves specially prepared bicrystals. However, an obvious advantage of atomic column resolution in n-beam lattice images of grain boundaries is that the image of the boundary gives direct information on both grain boundary dislocations (primary and secondary) and the rigid-body displacement between neighbouring grains. Because atomic column resolution is more readily achieved in germanium and silicon than in metals and alloys, these materials have been favoured for detailed investigation of high-angle grain boundaries (see, for example, D'Anterroches and Bourret 1984, Skrotzki *et al* 1988), but only work on metals and alloys will be reviewed here.

In an early investigation using lattice imaging, Ichinose and Ishida (1981) studied a near-$\Sigma 11$ tilt boundary that occurred in combination with $\Sigma 3$ boundaries and a low-angle boundary in an evaporated thin-film polycrystal of gold. The tilt axis was $[110]$ and the boundary plane $(1\bar{1}3)$ and from measurements on the lattice image they found an expansion of the $(1\bar{1}3)$ planes close to the boundary of 0.3 Å. This expansion was the rigid-body displacement required on a simple hard-sphere model of the exact-$\Sigma 11$ boundary to allow nearest neighbour atoms across the boundary to have a separation of $(1/2)\langle 110 \rangle$. They also found a component of rigid-body displacement of 0.04 Å parallel to the direction $[3\bar{3}\bar{2}]$ in the boundary plane. Of course, no information could be obtained on the component of rigid-body displacement parallel to the $[110]$ tilt axis, as this axis is parallel to the beam direction, and this restriction on the determination of rigid-body displacements applies to all lattice imaging of tilt boundaries. Secondary grain boundary dislocations were observed in the boundary and, on the assumption that the secondary dislocations were edge dislocations with DSC Burgers vectors of $(1/11)[1\bar{1}3]$, the spacing of these dislocations was found to be compatible with the measured departure of 1.5° around the tilt axis of the misorientation from the exact-$\Sigma 11$ CSL orientation.

In later experiments Ichinose *et al* (1985) used n-beam lattice images to study rigid-body displacements in coherent $\{111\}$ and incoherent $\{112\}$ $\Sigma 3$ tilt boundaries with $\langle 110 \rangle$ tilt axes in polycrystalline thin films of gold. Excluding any possible displacement parallel to the tilt axis, they found no rigid-body displacement at the coherent-$\Sigma 3$ boundaries, but for the incoherent boundaries they found a displacement parallel to the in-plane $\langle 111 \rangle$ direction of $(1/6)\langle 111 \rangle$ at positions on the boundary well removed from junctions with coherent boundaries. Their result for the coherent-$\Sigma 3$ boundaries supports the conclusion in section 5.4 concerning the origin of faint fringe contrast which sometimes occurs in same-$\boldsymbol{g}_c$ images of coherent-$\Sigma 3$ boundaries.

Cosandey and Bauer (1981) used two-beam lattice images of (200) planes to study a wide range of near-CSL high-angle $[001]$ tilt boundaries in thin-film bicrystals of gold†. The near-CSL boundaries were $\Sigma 5$, $\Sigma 13$, $\Sigma 17$, $\Sigma 25$, $\Sigma 37$ and $\Sigma 41$ and the small angular departures from the exact-CSL orientations, determined to an accuracy of $\pm 0.5°$, were 1.6°, 0.1°, 0.6°, 1.3°, 0.6° and 1.2° respectively around the $[001]$ tilt axis. For all these boundaries they determined the spacings of the primary grain boundary dislocations from

† The misorientations used in their work included low-angle boundaries and these have already been discussed in section 4.4.

lattice images and, like the work of Tan *et al* (1975) and Sass *et al* (1975) on twist boundaries (see section 5.3), they found that these spacings agreed with the simple geometric description $d = |\boldsymbol{b}|/\theta$ for misorientations θ up to 36.9° on the basis that the Burgers vectors of the primary dislocations were $(1/2)\langle 110 \rangle$. They also observed periodically spaced centres of strain contrast in the boundaries which they were able to associate with secondary grain boundary dislocations. For example, for the near-$\Sigma 5$ boundary, they found agreement between the observed spacing of the strain centres and a spacing calculated on the assumption that the strain centres were edge dislocations with $(1/10)\langle 310 \rangle$ DSC Burgers vectors that were accommodating the 1.6° departure from the exact-$\Sigma 5$ CSL orientation.

n-Beam lattice images of $\{111\}$ planes have been used by Krakow *et al* (1986) to study a $\Sigma 19_a$ $[110]$ symmetric-tilt boundary on a $(3\bar{3}1)$ plane in gold. Two low-energy relaxed structures, which differed by only 4% in energy, were predicted by computer modelling for this $\Sigma 19$ boundary, and the only difference between these structures was a $(1/4)[110]$ rigid-body displacement along the tilt axis. Both structures gave computed lattice images which agreed with the experimental image, but no decision could be made as to which of these structures best represented the experimental observations, because the only difference between them could not be detected in the experimental lattice image, as it corresponded to a rigid-body displacement parallel to the beam direction. The results showed that the structure of the boundary consisted of capped trigonal prisms (see e.g. the structure depicted in figure 5.1(*a*)) separated by tetrahedra, and that the trigonal prisms correlated with the cores of primary $(1/2)\langle 110 \rangle$ grain boundary dislocations. The rigid-body displacement for this boundary was found to be a large expansion of 0.8 Å normal to the boundary plane, and this was localised to a region three $(3\bar{3}1)$ planes thick centred on the boundary.

In an investigation of a $\Sigma 5$ $[001]$ symmetric-tilt boundary on a (310) plane of gold, Cosandey *et al* (1988) were able to identify their *n*-beam lattice images of the boundary (obtained using (200) planes) with one of two structures which had been proposed previously by Vitek *et al* (1983) from computer modelling of this type of boundary. These models are shown in figures 5.83(*a*) and (*b*) where the bicrystal is illustrated schematically by columns of atoms parallel to the $[001]$ tilt axis which is normal to the page. The model in figure 5.83(*b*) is obtained from that in figure 5.83(*a*) by removing a (620) plane of atoms adjacent to the boundary and collapsing the structure by a (620) interplanar spacing. The difference in the arrangement of the columns of atoms in the vicinity of the boundary can be seen from a comparison of the repeat units of the two structures which are shown in the outlined blocks in figure 5.83. Figure 5.84 shows theoretical lattice images for these two structures (computed for the diffraction conditions of figure 5.85) where the image in figure 5.84(*a*) corresponds to the structure of figure 5.83(*a*), and that in figure 5.84(*b*) to the structure of figure 5.83(*b*). The fine

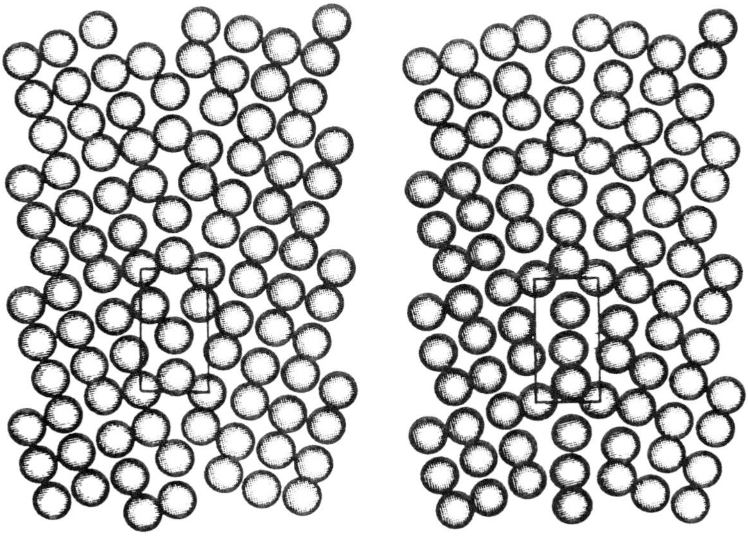

a b

Figure 5.83 Two models, (a) and (b), for the atomic structure of a Σ5 (310) symmetric-tilt boundary in gold viewed in projection along the [001] tilt axis (Cosandey *et al* 1988).

black dots in these computed images mark the exact positions of the columns of atoms. The experimental lattice image of Cosandey *et al* in figure 5.85 can be compared directly with their computed images in figures 5.84(a) and (b). The repeat unit of structure outlined in the experimental lattice image agrees with that outlined in figure 5.84(a) rather than that in figure 5.84(b),

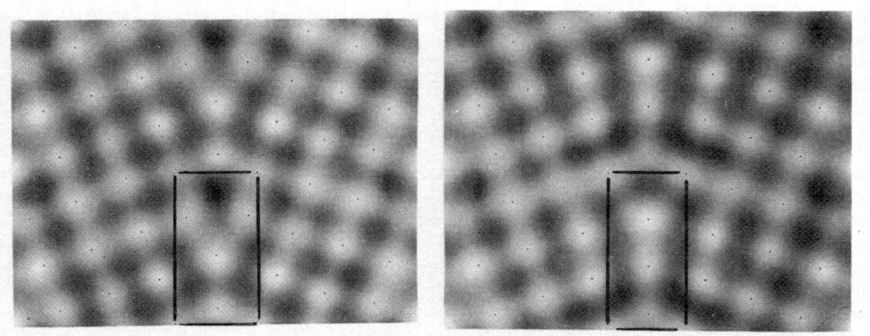

a b

Figure 5.84 Computed lattice images (a) and (b) corresponding to the models (a) and (b) of figure 5.83 where the black dots denote the positions of the atomic columns (Cosandey *et al* 1988).

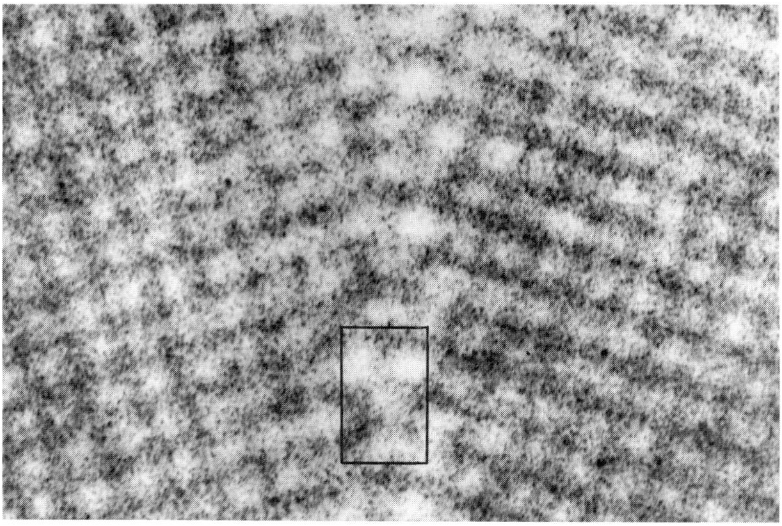

Figure 5.85 n-beam lattice image of a $\Sigma 5$ (310) symmetric-tilt boundary in gold where the atomic columns correspond to the white contrast (Cosandey *et al* 1988).

therefore the model for the boundary structure of figure 5.83(a) is preferred. Each unit of structure in figure 5.85 corresponds to the core of a primary grain boundary dislocation with a Burgers vector of [001], and the measured spacing of (6.4 ± 0.5) Å for these dislocations agrees with the calculated spacing of 6.5 Å obtained for a $\Sigma 5$ boundary from the simple geometric relationship $d = |\boldsymbol{b}|/\theta$. The rigid-body displacement found from measurements on lattice images for this boundary was an expansion of 20% normal to the (310) boundary plane and was localised to within four (620) planes centred on the boundary plane.

The results on n-beam lattice imaging of high-angle grain boundaries reviewed so far have demonstrated that the structures of high-angle symmetric-tilt boundaries, close to exact-CSL orientations, consist of repeating units of structure which, for each particular case, hard-sphere or computer modelling has shown to be characteristic of the CSL orientation involved. However, Penisson *et al* (1988), using n-beam lattice imaging, have demonstrated the presence of more than one type of structural unit in a tilt boundary in molybdenum, and their results have been discussed already in section 4.5. The misorientation of the boundary of Penisson *et al* was $14^\circ \pm 0.5^\circ$ around the [001] tilt axis, i.e. an orientation which is between that of $\Sigma 25$ and $\Sigma 41$, and departs by approximately 2° from each of these orientations. They found that the structure of the boundary consisted of a mixture of structural units computed for symmetric-tilt boundaries at the exact-$\Sigma 41$ and $\Sigma 25$ CSL orientations and these results support the structural unit model for tilt boundaries with misorientations intermediate between the two-CSL orientations (see section 5.1). No rigid-body displacements were detected for this symmetric-tilt boundary in molybdenum, within the limits of the accuracy of the measurements of ± 0.2 Å, either normal or parallel to the boundary plane.

5.6 HIGH-ANGLE GRAIN BOUNDARIES IN NON-CUBIC METALS AND ALLOYS

So far the discussion of the structure of high-angle grain boundaries has been concerned solely with boundaries in metals and alloys with cubic crystal structure. It was for cubic crystal structures that the CSL model of high-angle boundaries was originally developed and for which it has found wide application. However, for high-angle boundaries in other crystal structures, although the concept of a coincident site lattice can be applied, extra constraints are involved. In fact, in non-cubic crystals, exact three-dimensional CSLs can only be found for rotations about a very restricted number of crystal axes, except at special values of axial ratios. For example, in hexagonal close-packed (HCP) metals and alloys exact CSLs occur for rotations about [0001], and for other axes only when the axial ratio c/a is such that $(c/a)^2$ is a rational number (Bruggeman *et al* 1972).

The CSL model for the structure of high-angle boundaries in cubic crystals can be extended to boundaries in non-cubic crystals, in a general way, if the secondary grain boundary dislocations are taken as accommodating not only a small angular departure, but also the strain required to bring near coincident lattice sites in the interpenetrating lattices into a constrained coincidence, i.e. to form a constrained coincident site lattice (CCSL) (Chen and King 1988)†. This model has been developed by a number of investigators (see, for example, Bonnet *et al* 1981, Grimmer 1989), and tables that give the data for exact CSLs and CCSLs in hexagonal metals are given for different values of $(c/a)^2$ by Bonnet *et al* (1981), Chen and King (1987) and Grimmer (1989).

Very little experimental work has been done on the structure of high-angle grain boundaries in non-cubic metals and alloys and the only detailed experimental investigation of secondary grain boundary dislocations in such boundaries is that of Chen and King (1988), who studied high-angle boundaries in polycrystalline zinc. A difficulty they encountered in interpreting their experimental results was that for each case there were several CCSLs, within a small angular range of orientation, which were all equally good

† See also Chapter 7 on Interphase Interfaces.

possibilities for the interpretation of the experimental secondary grain boundary dislocation structure.

Chen and King used $\boldsymbol{g} \cdot \boldsymbol{b}$ criteria and image matching, with the program PCGBD modified for hexagonal crystals (Head *et al* 1973), to determine the Burgers vectors of the secondary grain boundary dislocations. The only boundary studied which could be analysed directly as a near CSL boundary involved a rotation of 9.75° around a $[1\ \bar{1}\ 0\ \overline{16}]$ axis, i.e. an orientation close to an exact-$\Sigma37$ orientation which is a rotation of 9.43° around $[000\bar{1}]$. For this near-$\Sigma37$ boundary only one system of secondary grain boundary dislocations was observed, and it was shown that the Burgers vector of these dislocations was either $[000\bar{1}]$ or $(1/3)[\bar{1}2\bar{1}\bar{3}]$, corresponding to a basis DSC vector or a combination of basis DSC vectors respectively. The other boundaries studied were treated as near-CCSL boundaries but only one case was amenable to detailed analysis and this boundary involved a rotation of 85.38° around $[98\ \bar{3}\ \overline{95}\ 0]$. Although this misorientation was close to a range of possible CCSLs (e.g. those corresponding to $\Sigma13$, $\Sigma15$, $\Sigma17$, $\Sigma24$ and $\Sigma28$)† they found, using $\boldsymbol{g} \cdot \boldsymbol{b}$ criteria, image matching and geometric analysis, that the two observed systems of secondary grain boundary dislocations agreed best with the CCSL corresponding to $\Sigma13$, even though this CCSL did not involve the lowest strain.

5.7 COMMENT

In commenting on the electron microscopy of high-angle boundaries in metals and alloys, both the progress achieved and the difficulties encountered will be considered, for although considerable advances have been made in the understanding of the structure of high-angle grain boundaries in cubic metals and alloys, there are still several outstanding experimental problems.

The work on fabricated thin-film bicrystals and the application of image matching to general boundaries in polycrystals with misorientations close to CSL orientations has demonstrated the applicability of the CSL model in these cases. Moreover, the demonstration, from the identification of the Burgers vectors of secondary grain boundary dislocations by image matching, that the geometric method can be applied to secondary grain boundary dislocation structures, has made the structure of boundaries in polycrystals, with misorientations that are far from CSL orientations, amenable to solution. Further it is now possible, from the application of high-resolution n-beam lattice imaging, to obtain solutions to problems of boundary structure for simple tilt boundaries at the atomic level.

† For non-cubic crystal lattices CSL orientations with even values of Σ are possible.

The outstanding problems are mainly concerned with the determination of the structure of general boundaries in polycrystals. Many high-angle grain boundaries in polycrystals do not show any periodic structure in either diffraction patterns or in images taken under two-beam conditions. This type of observation suggests that in such cases, either

(i) the boundary structure has no long-range periodicity, as for example would be the case if the structure was a non-ordered mix of structural units associated with favoured CSL orientations (Sutton 1988), or
(ii) the boundary structure consists of secondary grain boundary dislocations, with Burgers vectors too small to be detected, accommodating a departure from a high-Σ CSL orientation. The details of the structure of such boundaries will probably have to await an extension of the methods used for n-beam lattice images and their interpretation to high-angle boundaries in polycrystals which are not simple tilt boundaries.

Rigid-body displacements between neighbouring grains are a general feature of the structure of high-angle boundaries and the assumption that the displacement is localised at the boundary in the determination of such displacements from the fringe contrast in same-$\boldsymbol{g}_c$ images has been confirmed by n-beam lattice-imaging for a number of near-CSL orientations. Enough near-CSL boundaries have now been studied using fringe contrast in same-$\boldsymbol{g}_c$ images to demonstrate that the state of rigid-body displacement in a given metal or alloy is a characteristic of a particular boundary plane and CSL. However, even when the rigid-body displacement has been determined for a particular boundary plane and a given CSL orientation, there is still the problem of deciding on the level of the boundary in the unit cell of the CSL. This problem has been addressed by Papon *et al* (1984) and Bacmann *et al* (1985) in germanium, by using diffraction evidence to distinguish between different possible locations for the boundary and it may be possible to use such diffraction effects in a similar way in metals and alloys. The technique for determining rigid-body displacements from fringe contrast in same-$\boldsymbol{g}_c$ images for CSL orientations with $\Sigma > 15$ is virtually impossible to apply because the extinction distance for at least two of the three necessary non-coplanar same-$\boldsymbol{g}_c$ diffracting vectors becomes too large to have an appreciable effect on diffraction contrast. Other techniques, which have been little-used so far, such as the moiré fringe, convergent beam and Fresnel fringe methods suffer from the same difficulty, as they all rely on contrast effects arising from the degree of offset of $\boldsymbol{g}_c$ planes across the boundary. For high-Σ CSL orientations, n-beam lattice imaging, despite its limitations, is at present the only method potentially available for determining rigid-body displacements.

A locally expanded structure at a grain boundary creates favourable sites for segregation and precipitation, so that a method is required to distinguish between an expansion inherent in the boundary structure and one which is due to, or accentuated by, segregation, as was the case described for the boundaries in iron. It is possible that the Fresnel fringe method for determining rigid-body displacements (Boothroyd *et al* 1986) or the technique of Anstis and Thompson (1989), which uses residual fringe contrast arising from a segregated species in same-$\boldsymbol{g}_c$ images with $\boldsymbol{g}_c \cdot \boldsymbol{R}$ integral, may help in making such a decision.

6

Some Properties of High-angle Grain Boundaries

6.1 INTRODUCTION

Many processes that occur in polycrystalline metals and alloys, both stress-activated and thermally activated, depend on the properties of grain boundaries which in turn depend on boundary structure. This chapter will describe a small selection of these processes which have been investigated quantitatively using transmission electron microscopy processes such as—the interaction of glide dislocations with secondary grain boundary dislocation structures, the transfer of slip across grain boundaries, the generation of crystal defects by moving boundaries and the dissociation of grain boundaries. The chapter will start, however, with a demonstration of a basic property of grain boundary dislocations, namely that their Burgers vectors balance at nodes where the dislocations meet at triple junctions between grains in a polycrystal.

6.2 SECONDARY GRAIN BOUNDARY DISLOCATIONS AT THE JUNCTION OF THREE GRAINS

In polycrystalline aggregates the junction of three grains usually occurs along a line known as a triple line. Although it has been shown in section 5.3 that periodic arrays of secondary grain boundary dislocations accommodate the departure ($\boldsymbol{q}$, φ) of the misorientation ($\boldsymbol{u}$, θ) between two neighbouring grains in a polycrystal from a near-CSL orientation ($\boldsymbol{p}$, ω), there has been no discussion so far on how such secondary grain boundary dislocation structures are modified by reactions between the dislocations in neighbouring arrays at a triple line. This problem was first treated theoretically by Bilby (1955) and later in more detail by Bollmann (1981, 1984) in terms of the total Burgers vector content of the boundaries involved. Their analyses indicate that grain boundary dislocations accommodating the misorientations between the grains must form nodes at the triple line at which the different Burgers vectors involved balance to zero. While this result cannot readily be demonstrated experimentally for arrays of primary dislocations in high-angle grain boundaries, it has been demonstrated for arrays of secondary grain boundary dislocations in near-$\Sigma 27_b$, $\Sigma 9$ and $\Sigma 3$ grain boundaries meeting at a triple line in a polycrystalline Cu – 6 at% Si alloy (Clarebrough and Forwood 1987a).

Figure 6.1(*a*) is an electron micrograph where the junction of the near-$\Sigma 27_b$, $\Sigma 9$ and $\Sigma 3$ grain boundaries at the triple line can be clearly seen, since the diffraction conditions have been set so that all three grains are diffracting under two-beam conditions simultaneously. In the schematic diagram in figure 6.1(*b*) these boundaries are labelled together with the triple line *tl* at the junction of grains 1, 2 and 3, and the intersections T and B of the boundaries with the top and bottom surfaces of the specimen are also labelled. In the vicinity of the triple line, the boundary plane normals of the $\Sigma 9$ and $\Sigma 27_b$ boundaries are both close to $(121)_3$, and the $\Sigma 3$ boundary is a coherent boundary with a plane normal $(\bar{1}11)_3$.

The misorientations between the grains are very close ($\lesssim 0.05°$) to the exact-CSL orientations, which will be described by rotations around a [113] axis common to all three grains, with grain 2 being rotated anticlockwise by 67.11° with respect to grain 1 for the $\Sigma 9$ boundary, grain 3 being rotated clockwise by 146.44° with respect to grain 2 for the $\Sigma 3$ boundary, and grain 1 being rotated anticlockwise by 79.33° with respect to grain 3 for the $\Sigma 27_b$ boundary. These rotations are described by the matrix equation

$$\mathbf{R}_{13}\mathbf{R}_{32}\mathbf{R}_{21} = \mathbf{I} \tag{6.1}$$

where $\mathbf{I}$ is the identity and $\mathbf{R}_{ij}$ re-indexes a vector indexed with respect to crystal j into indices with respect to crystal i, i.e.

$$(1/27)\begin{pmatrix} 7 & -22 & 14 \\ 26 & 7 & -2 \\ -2 & 14 & 23 \end{pmatrix}(1/3)\begin{pmatrix} -2 & 2 & 1 \\ -1 & -2 & 2 \\ 2 & 1 & 2 \end{pmatrix}(1/9)\begin{pmatrix} 4 & -7 & 4 \\ 8 & 4 & -1 \\ -1 & 4 & 8 \end{pmatrix}$$

$$= \begin{pmatrix} 1 & 0 & 0 \\ 0 & 1 & 0 \\ 0 & 0 & 1 \end{pmatrix}.$$

For the theory of Bilby and Bollmann to apply, the fact that equation (6.1) holds means that the Burgers vectors of the primary dislocations for these exact-CSL orientations balance to zero at the triple line. Moreover, it follows

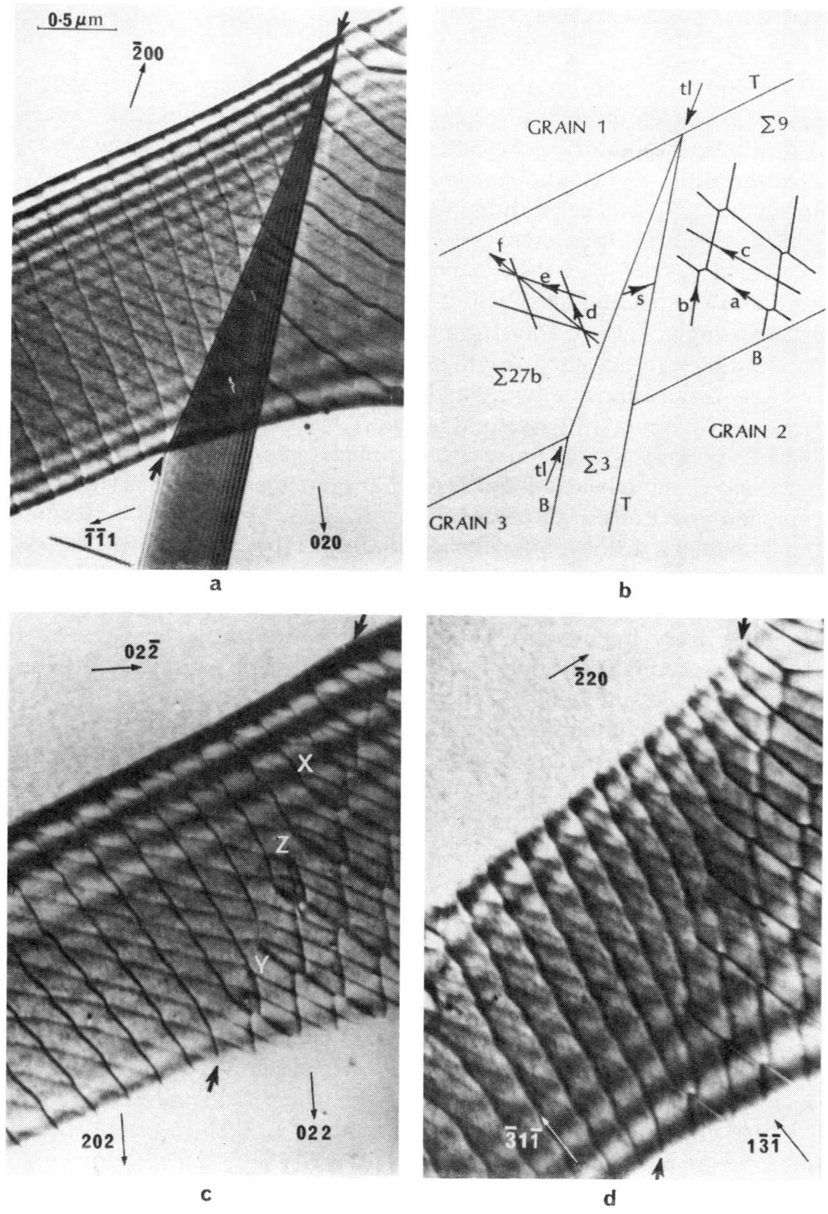

Figure 6.1 Electron micrographs (a), (c) and (d) showing a triple junction of three grains meeting along a triple line (indicated by arrows) which is marked tl in the schematic diagram (b). The nature of the boundaries, the labelling of the different components of the grain boundary dislocation arrays and intersections of the boundaries with the top T and the bottom B of the specimen are given in (b). The diffracting vectors, indexed in their respective grains, are indicated on the micrographs and the beam directions are close to $[123]_3$ in (a), $[17\bar{1}]_3$ in (c) and $[\bar{1}69]_3$ in (d).

that the Burgers vectors of the secondary grain boundary dislocations in the present example, in which the orientations depart from the exact CSL orientations, should also balance to zero at the triple line for the theory of Bilby and Bollmann to apply.

In figure 6.1(*a*), different diffracting vectors are operating in grains 2 and 3 so that strong fringe contrast is present in the coherent-$\Sigma3$ boundary whereas, in figure 6.1(*c*), the same diffracting vector is operating in grains 2 and 3 and the $\Sigma3$ boundary is out of contrast; a similar condition applies when looking at figure 6.1(*d*). From a comparison of (*a*), (*c*) and (*d*) in figure 6.1, it can be seen that there are three independent arrays of secondary grain boundary dislocations in the $\Sigma9$ and $\Sigma27_b$ boundaries and that in the region marked Z (figure 6.1(*c*)) a secondary grain boundary dislocation is present in the $\Sigma3$ boundary. These arrays of secondary grain boundary dislocations are indicated and labelled in figure 6.1(*b*), where the arrows denote the sense of the line directions of the dislocations. In the $\Sigma9$ boundary the secondary grain boundary dislocations labelled *a* and *b* interact where they cross to form an hexagonal network, while the secondary grain boundary dislocations labelled *c* cross this network without interacting. In the $\Sigma27_b$ boundary the secondary grain boundary dislocations labelled *d* and *e* cross each other without interacting, while the secondary grain boundary dislocations labelled *f* (figure 6.1(*d*)) interact and step where they cross dislocations *d*. The secondary grain boundary dislocations *f* always showed weak contrast over a wide range of diffracting conditions and it will be seen below that this arises from the small magnitude $(a/(3\sqrt{6}))$ of their Burgers vector. In the coherent-$\Sigma3$ boundary only one segment, labelled *s*, of the secondary grain boundary dislocation network is present in that region of the triple line contained in the specimen.

For a wide range of diffraction conditions no dislocation contrast is observed along the triple line, but four different types of nodal reactions between the secondary grain boundary dislocations can be distinguished at the triple line. Two of these involve dislocations *b* of $\Sigma9$, one of these occurring at the region Y and the other at the region Z in figure 6.1(*c*), which are the only places where dislocations *b* meet the triple line. The other two types occur at several places on the triple line, one where dislocations *a* of $\Sigma9$ meet the triple line, and the other where dislocations *c* of $\Sigma9$ meet the triple line (see, for example, region X in figure 6.1(*c*)). All the nodal reactions involve dislocations *f* of $\Sigma27_b$, with the exception of the reaction at Y, where dislocations *b* and *c* of $\Sigma9$ form a node at the triple line with dislocations *d* and *e* of $\Sigma27_b$ such that,

$$\boldsymbol{b} + \boldsymbol{c} = \boldsymbol{d} + \boldsymbol{e} \tag{6.2}$$

where $\boldsymbol{b}$, $\boldsymbol{c}$, $\boldsymbol{d}$ and $\boldsymbol{e}$ are the Burgers vectors of the secondary grain boundary dislocations *b*, *c*, *d* and *e*. The remaining nodal reactions which involve dislocations *f* of $\Sigma27_b$ can be seen from a comparison of figure 6.1(*c*) and

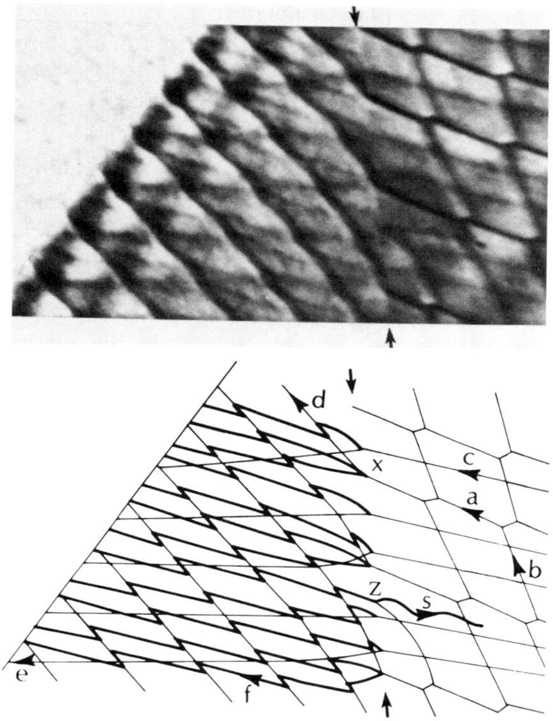

Figure 6.2 Higher magnification image of figure 6.1(d) together with a schematic diagram of the secondary grain boundary dislocation arrays with emphasis given to the array of dislocations f in the $\Sigma 27_b$ boundary.

figure 6.2. Figure 6.2 is a higher magnification image of a portion of figure 6.1(d) together with a schematic drawing in which the dislocations f have been emphasised to illustrate their role in the various nodal balances, since in the experimental image dislocations f are obscured at some of the nodes because they step so markedly where they cross dislocations d. The second reaction involving dislocation b, which is at the region Z, involves a dislocation f of $\Sigma 27_b$ and dislocation s of $\Sigma 3$ and results in a node at the triple line where

$$\boldsymbol{b} = \boldsymbol{s} + \boldsymbol{f} \tag{6.3}$$

where $\boldsymbol{s}$ and $\boldsymbol{f}$ are the Burgers vectors of dislocations s and f. The reactions involving dislocations a of $\Sigma 9$ result in nodes at the triple line where a dislocation a of $\Sigma 9$ meets a dislocation d and two dislocations f of $\Sigma 27_b$ such that

$$\boldsymbol{a} = \boldsymbol{d} + 2\boldsymbol{f} \tag{6.4}$$

where $\boldsymbol{a}$ is the Burgers vector of a dislocation a. Finally, the reactions involving dislocations c of $\Sigma 9$ result in nodes at the triple line where a dislocation c of $\Sigma 9$ meets a dislocation e and a dislocation f of $\Sigma 27_b$ such that

$$\boldsymbol{c} = \boldsymbol{e} + \boldsymbol{f}. \tag{6.5}$$

The Burgers vectors of all the secondary grain boundary dislocations involved in the nodal balances were identified by matching theoretical and experimental images, giving the following results:

$$\boldsymbol{a} = (1/18)[\bar{1}\bar{2}7]_2$$
$$\boldsymbol{b} = (1/9)[122]_2$$
$$\boldsymbol{c} = (1/18)[\bar{4}11]_2$$

which are basis vectors of the DSC lattice corresponding to the $\Sigma 9$ CSL (which when re-indexed in grain 3 are:

$$\boldsymbol{a} = (1/54)[5\ 19\ 10]_3$$
$$\boldsymbol{b} = (1/27)[4\bar{1}8]_3$$
$$\boldsymbol{c} = (1/54)[11\ 4\ \bar{5}]_3)$$

and

$$\boldsymbol{d} = (1/54)[7\ 5\ 14]_3$$
$$\boldsymbol{e} = (1/18)[4\bar{1}\bar{1}]_3$$
$$\boldsymbol{f} = (1/54)[\bar{1}7\bar{2}]_3$$

which are basis vectors of the DSC lattice corresponding to the $\Sigma 27_b$ CSL and

$$\boldsymbol{s} = (1/6)[1\bar{1}2]_3$$

which is a basis vector of the DSC lattice corresponding to the $\Sigma 3$ CSL. These Burgers vectors confirm the nodal balances in equations (6.2)–(6.5). This balance of the Burgers vectors of the secondary grain boundary dislocations at the triple line means that the theory of Bilby and Bollmann applies. All the Burgers vectors of the type identified here have been previously identified in section 5.3.2 by image matching for other cases of $\Sigma 9$, $\Sigma 27_b$ and $\Sigma 3$ boundaries. Therefore, the image matching details for this case will not be presented. However, figure 6.3 illustrates the identification of the sign of the Burgers vectors by a comparison of experimental and matching theoretical images computed for the Burgers vectors given above, with mismatching theoretical images computed for Burgers vectors of the opposite sign.

The only dislocation that could be present along the triple line, which has not been considered so far, would be a partial grain boundary dislocation arising from differences in rigid-body displacements across the three grain boundaries. However, the fact that no dislocation contrast extending along

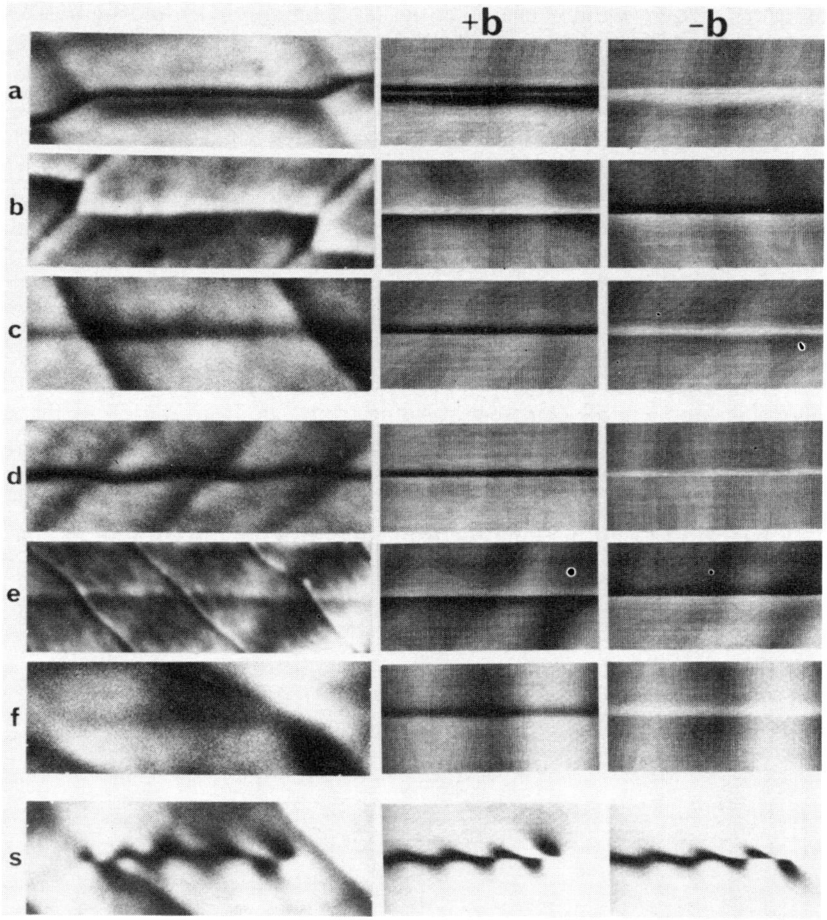

Figure 6.3 Comparison of experimental double two-beam images of grain boundary dislocations *a*, *b*, *c*, *d*, *e*, *f* and *s* with computed images for the Burgers vector $+b$ (see text) and $-b$. The beam directions and diffracting vectors are as follows: for *a*, $[7\ \bar{1}\ 22]_2$, $3\bar{1}\bar{1}_2$, $\bar{2}00_1$; for *b*, $[1\bar{1}5]_2$, $\bar{2}\bar{2}0_2$, $\bar{2}\bar{2}0_1$; for *c*, $[\bar{8}\bar{9}9]_2$, 022_2, $02\bar{2}_1$; for *d*, $[17\bar{1}]_3$, 202_3, $02\bar{2}_1$; for *e*, $[\bar{1}69]_3$, $\bar{3}1\bar{1}_3$, $0\bar{2}0_1$; for *f*, $[\bar{1}69]_3$, $\bar{3}1\bar{1}_3$, $\bar{2}20_1$; for *s*, $[199]_3$, $0\bar{2}2_3$, 220_2.

the length of the triple line was detected in any of the images taken over a wide range of diffracting conditions, suggests, in this case, that the rigid-body displacements also balance at the triple line. To investigate this balance, the rigid-body displacements $\boldsymbol{R}$ between neighbouring grains were analysed from the fringe contrast in images of each boundary with the same diffracting vector $\boldsymbol{g}_c$ operating in neighbouring grains (see section 5.4). For the coherent $\Sigma 3$ boundary, no fringe contrast was observed in a set of non-coplanar same-$\boldsymbol{g}_c$ images (see, for example, figures 6.1(*c*) and (*d*) and figures 6.4(*a*) and (*b*)), so that no rigid-body displacement is present between grains 2 and 3. Therefore any partial dislocation along the triple line must arise from a change in rigid-body displacement of grain 1 across the $\Sigma 9$ and $\Sigma 27_b$ boundaries at the triple line. To determine the rigid-body displacements across these two boundaries, a set of images involving at least three non-coplanar same-$\boldsymbol{g}_c$ diffracting vectors is required for both the $\Sigma 9$ and the $\Sigma 27_b$ boundaries. While such same-$\boldsymbol{g}_c$ images were used for the $\Sigma 9$ boundary, only two different same-$\boldsymbol{g}_c$ images could be used successfully for the $\Sigma 27_b$ boundary (see section 5.7). Figure 6.4 shows these two images where the same-$\boldsymbol{g}_c$ vector is in fact common to grains 1, 2 and 3, so that any significant change in $\boldsymbol{g}_c \cdot \boldsymbol{R}$ across the $\Sigma 9$ and $\Sigma 27_b$ boundaries for these same-$\boldsymbol{g}_c$ vectors should be apparent. For these two images the displacement fringe contrast shows no appreciable change across the triple line from the $\Sigma 9$ to the $\Sigma 27_b$ boundary, so that the $[\bar{1}\bar{1}\bar{3}]_1$ and the $[04\bar{2}]_1$ components of rigid-body displacement of grain 1 with respect to grains 2 and 3 balance at the triple line. The undetermined components of rigid-body displacement could lead to a partial dislocation along the triple line with the direction of its Burgers vector given by $\pm[04\bar{2}]_1 \wedge [\bar{1}\bar{1}\bar{3}]_1 = \pm[\bar{7}12]_1 = \pm[1\bar{7}2]_3$. However, this

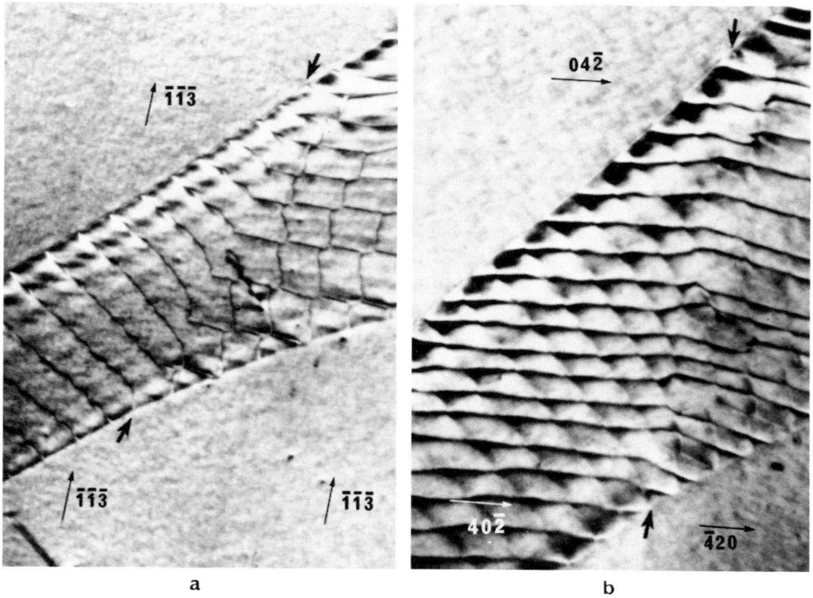

Figure 6.4 Electron micrographs of the triple junction with the same diffracting vector $\boldsymbol{g}_c$ operating in the three grains as indicated. The beam direction is close to $[\bar{2}5\bar{1}]_3$ in (*a*) and $[2\ 13\ 4]_3$ in (*b*).

component, if present, must be very small in magnitude to be compatible with the observed lack of contrast along the triple line.

For the case considered the Burgers vectors of the secondary grain boundary dislocations balance at the triple line in agreement with theory, and the rigid-body displacements also appear to balance at the triple line.

6.3 GRAIN BOUNDARY DISSOCIATION

Electron microscopy of high-angle tilt boundaries in FCC metals and alloys has shown that some boundaries, with misorientations close to particular CSL orientations, dissociate to form intermediate twin-grains. Each new grain resulting from dissociation of a given boundary is bounded by a coherent-$\Sigma3$ boundary and two new boundaries corresponding to CSL orientations which are $\Sigma3$ related to the original near CSL orientation. Clearly, for grain boundary dissociation to occur the total energy of the three new boundaries must be less than that of the original boundary.

The first indication that grain boundaries with near CSL orientations may dissociate was given by the work of Vaughan (1970) on stainless steel, who observed grain boundary configurations for a $\Sigma9$ CSL orientation which could have arisen from dissociation or from the coalescence of $\Sigma3$ twins. Although Vaughan favoured the latter interpretation, subsequent experiments by Goodhew *et al* (1978), Clarebrough and Forwood (1980d) and Forwood and Clarebrough (1984) showed that grain boundaries in FCC metals and alloys with misorientations close to $\Sigma9$ CSL orientations dissociate to form intermediate twinned grains bounded by one coherent and two incoherent $\Sigma3$ boundaries (see figure 6.9)†.

Goodhew *et al* (1978) showed that, for fabricated thin-film bicrystals of gold, dissociation of a $\Sigma9$ boundary and not coalescence was the more likely cause of boundary configurations of the type observed by Vaughan, by using specimens prepared in such a way that intermediate twin-grains were unlikely to have arisen from the coalescence of twins. Proof that $\Sigma9$ boundaries can dissociate into $\Sigma3$ boundaries was given by direct observation of the dissociation of a near-$\Sigma9$ boundary in a polycrystalline specimen of copper, as illustrated in figures 6.5 and 6.6 (Clarebrough and Forwood 1980d). The images in figure 6.5 are all taken under the same diffraction conditions and show the change in inclination of the boundary plane for this near-$\Sigma9$ boundary during annealing. This boundary is that analysed in detail in

† From the determination of the orientation relationships of annealing twins in 99.999% Al by optical microscopy and x-ray diffraction, Fionova *et al* (1981) attributed the formation of the twinned grains to grain boundary dissociation during annealing.

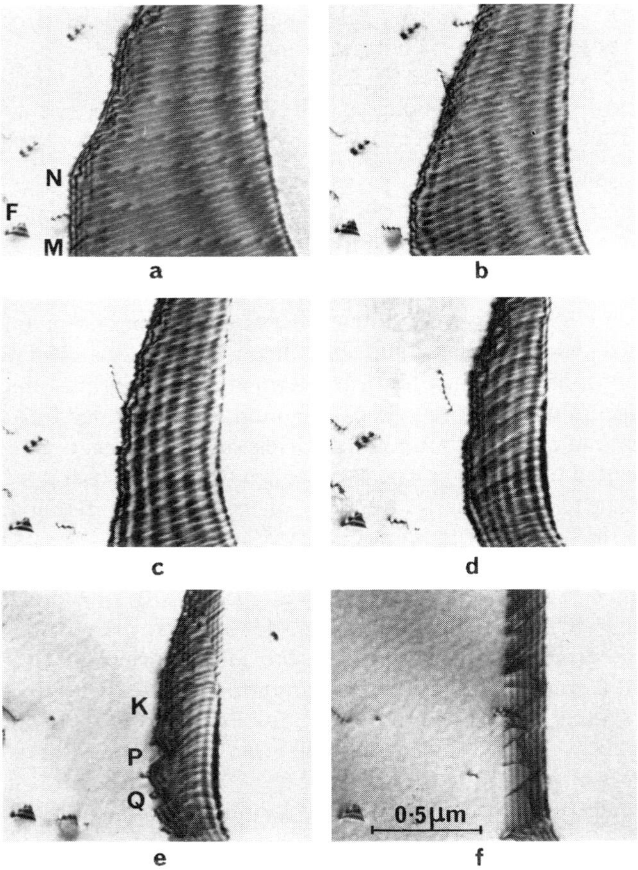

Figure 6.5 Double two-beam images showing the changes in inclination and structure of a Σ9 boundary in copper during annealing. The diffracting conditions for all micrographs are: ***B*** close to $[103]_L$ and $[\bar{1}34]_R$ with diffracting vectors 020_L and $\bar{1}1\bar{1}_R$. The sequence of annealing treatments is: (*a*) room temperature; (*b*) 2 h at 100 °C; (*c*) 0.75 h at 130 °C; (*d*) 1 h at 160 °C; (*e*) 2 h at 160 °C; (*f*) 3.5 h at 220 °C.

section 5.3.2 (i) and the portion of the boundary in figure 6.5(*a*) corresponds to region 1 of figure 5.16. During annealing the boundary develops facets on the $(\bar{1}11)_L$ plane, and after a final annealing treatment at 220 °C the boundary is completely aligned along the $(\bar{1}11)_L$ plane as shown in figure 6.5(*f*). Figure 6.6 shows the boundary of figure 6.5(*f*) with the specimen tilted so as to reveal details of the structure not shown in the projection of figure 6.5(*f*). Figure 6.6 shows that during the final annealing treatment the boundary has dissociated to form the triangular shaped intermediate grains,

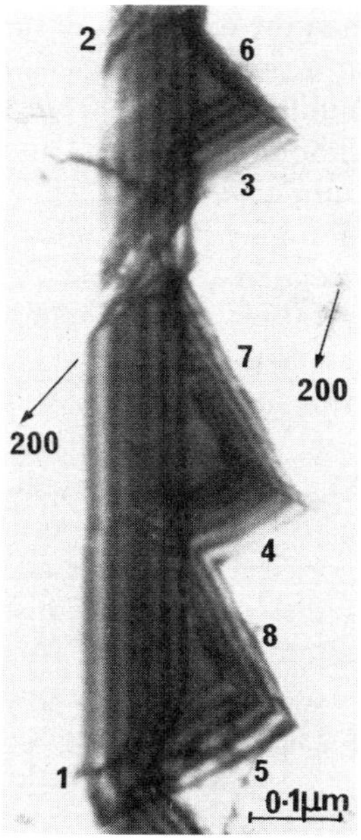

Figure 6.6 Electron micrograph showing the final configuration of the dissociated boundary of figure 6.5 with intermediate grains bounded by coherent twin interfaces along 1 to 2 and 3, 4, 5 with incoherent twin interfaces 6, 7, and 8. The beam direction is close to $[0\bar{1}5]_L$ and $[015]_R$ with diffracting vectors 200_L and 200_R.

and these new grains are Σ3-related to both grains L and R. The boundaries of these intermediate grains are the common boundary 1–2, and the boundaries 6, 3; 7, 4; and 8, 5; where the boundary 1–2 is a Σ3 coherent boundary on $(\bar{1}11)_L$ between the intermediate grains and grain L, the boundaries 3, 4 and 5 are close to being Σ3 coherent boundaries on $(\bar{1}\bar{1}1)_R$ between the intermediate grains and grain R, and the boundaries 6, 7 and 8 are incoherent-Σ3 boundaries, with boundary 6 close to $(6\bar{3}2)_R$ and boundaries 7 and 8 close to $(5\bar{4}3)_R$. These results demonstrate directly that boundary dissociation occurs and is associated with the faceting of the original boundary on to {111} planes.

All cases for which boundary dissociation has been observed involve asymmetric-tilt boundaries with tilt axes along ⟨110⟩ and boundary planes which are {111} planes in one of the grains. In addition to Σ9 boundaries, Goodhew *et al* (1978) report on the dissociation of near-Σ11 and Σ99 boundaries in fabricated thin-film bicrystals of gold. They conclude that the Σ99 boundary dissociates into Σ3 and Σ33 boundaries and that the Σ11 boundary also dissociates into Σ3 and Σ33 boundaries, and on this basis they suggest that Σ33 is a preferred low-energy CSL orientation. Forwood and Clarebrough (1984) studied the dissociation that occurred in several $\Sigma 9$, $\Sigma 27_a$ and $\Sigma 81_d$ boundaries in a polycrystalline Cu – 6 at% Si alloy during annealing at 600 °C. They found that, apart from the $\Sigma 9$ boundaries which always dissociated to lower-Σ boundaries, $\Sigma 27_a$ and $\Sigma 81_d$ boundaries always dissociated to higher-Σ boundaries. That is, $\Sigma 27_a$ boundaries always dissociated to $\Sigma 3$ and $\Sigma 81_d$ boundaries, and not to $\Sigma 3$ and $\Sigma 9$ boundaries; $\Sigma 81_d$ boundaries always dissociated to $\Sigma 3$ and $\Sigma 243_a$ boundaries, and not to $\Sigma 3$ and $\Sigma 27_a$ boundaries. Rather than conclude from these results that $\Sigma 81_d$ and $\Sigma 243_a$ are favoured low-energy CSL orientations an alternative explanation is given in terms of boundary symmetry and kinetics of dissociation (see below).

All the $\Sigma 9$, $\Sigma 27_a$ and $\Sigma 81_d$ boundaries investigated by Forwood and Clarebrough were very close to the exact-CSL orientations, with any small angular departure from these orientations being accommodated by networks of secondary grain boundary dislocations. A consistent form of indexing is used such that each CSL orientation between grains designated 1 and 2 is specified by a right-handed rotation θ of grain 2 with respect to grain 1 around the common [110] axis. Thus a direction $[uvw]_1$, indexed with respect to grain 1, can be expressed as the direction $[u'v'w']_2$, indexed with respect to grain 2, by the matrix equation

$$[uvw]_1 = {}_1(\Sigma N)_2[u'v'w']_2$$

(N taking the values 9, 27_a, 81_d) where

$${}_1(\Sigma 9)_2 = (1/9)\begin{pmatrix} 8 & 1 & -4 \\ 1 & 8 & 4 \\ 4 & -4 & 7 \end{pmatrix} \qquad \theta = -38.94^\circ$$

$${}_1(\Sigma 27_a)_2 = (1/27)\begin{pmatrix} 25 & 2 & 10 \\ 2 & 25 & -10 \\ -10 & 10 & 23 \end{pmatrix} \qquad \theta = 31.58^\circ$$

$${}_1(\Sigma 81_d)_2 = (1/81)\begin{pmatrix} 32 & 49 & 56 \\ 49 & 32 & -56 \\ -56 & 56 & -17 \end{pmatrix} \qquad \theta = 102.11^\circ.$$

$\Sigma 9$ boundaries

Many of the $\Sigma 9$ boundaries studied in the Cu–Si alloy had developed facets during annealing and some of these facets formed tilt boundaries containing the [110] rotation axis. Figure 6.7(*a*) shows an example of boundary faceting where, for most of its length, the $\Sigma 9$ boundary separating grains 1 and 2 is of mixed twist–tilt character and curves between the terminating triple junctions A and B. The pronounced sets of facets that have developed in the

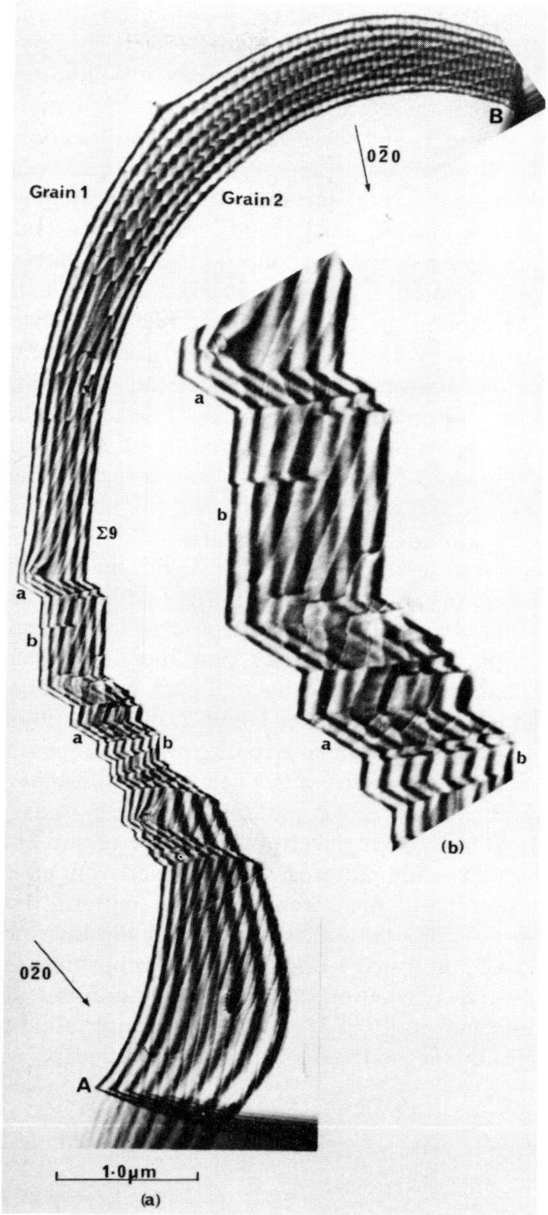

Figure 6.7 Double two-beam image (a) of a near-Σ9 boundary in a Cu − 6 at% Si alloy. The beam direction is close to $[401]_1$ and $[40\bar{1}]_2$ and the diffracting vectors are indicated. In (b) facets a and b are shown at higher magnification.

central region (shown at higher magnification in figure 6.7(*b*)) are asymmetric-tilt boundaries on $(1\bar{1}1)_1/(11\ \overline{11}\ \bar{1})_2$ planes (facets *a*) linked by general boundaries on $(5\bar{1}4)_1/(55\ \overline{19}\ 4)_2$ planes (facets *b*). Figure 6.8(*a*) shows another example where there are two lengths of Σ9 boundary AB and CD separating grains 1 and 2, and these Σ9 boundaries are connected by a Σ3 boundary along CB separating grains 2 and 3. Both Σ9 boundaries have developed distinct facets during annealing. In figure 6.8(*a*) the Σ9 boundary along AB first shows an unfaceted length containing a coarse secondary grain boundary dislocation network (figure 6.8(*b*)), while the remaining length of the boundary is faceted. The facets marked *a* have a mean plane of $(1\bar{1}5)_1/(1\bar{1}1)_2$ corresponding to asymmetric-tilt boundaries. The facets marked *b* and *c* have mean planes of $(1\bar{7}6)_1$ and $(\bar{1}\bar{9}8)_1$ respectively, so that they are not far from $(0\bar{1}1)_1/(1\bar{4}1)_2$. Facet *d* lies on $(\bar{1}14)_1/(1\bar{1}4)_2$ and corresponds to a symmetric-tilt boundary. The Σ9 boundary CD is also strongly faceted. The portion, *e*, of this boundary at the triple junction D is a symmetric-tilt boundary on $(\bar{1}14)_1/(1\bar{1}4)_2$ and is the same as facet *d*; facet *f* lies on $(\bar{1}11)_1/(\bar{1}15)_2$ and is an asymmetric-tilt boundary, while facets *g* lie on $(\bar{1}\bar{1}1)_1/(\bar{5}\ \overline{13}\ 7)_2$ and are not tilt boundaries.

The set of facets *a* in figure 6.8, shown at higher magnification in (*c*), (*d*) and (*e*), contain structure which takes the form of additional facets consisting of triangular prisms. It was found that each prism is bounded by $(1\bar{1}1)_2$, $(\bar{1}11)_1$ and $(1\bar{1}2)_1$ planes with the [110] rotation axis parallel to the axis of the prism, as illustrated schematically in figure 6.9(*a*). These prisms are intermediate grains separating grains 1 and 2 and were found by selected area diffraction to be Σ3 twin-related to both grains 1 and 2 so that boundary dissociation has occurred on these facets. The twin relationship is demonstrated in figure 6.10 which is a diffraction pattern from the boundary taken in the common $[\bar{1}14]_1/[1\bar{1}4]_2$ beam direction from the region shown in figure 6.8(*e*) and similar diffraction patterns were obtained from all *a* facets. Figure 6.10 shows, in addition to the common ⟨114⟩ pattern, diffraction spots corresponding to a ⟨110⟩ pattern, showing that the intermediate grain is Σ3 twin-related to both grains 1 and 2. This relationship can be described by expressing the ${}_1(\Sigma9)_2$ rotation matrix, relating grains 1 and 2, in terms of two Σ3 rotation matrices, ${}_1(\Sigma3)_{\mathrm{INT}}$ and ${}_{\mathrm{INT}}(\Sigma3)_2$, relating grain 1 to the intermediate grain and the intermediate grain to grain 2 respectively, where

$${}_1(\Sigma9)_2 = {}_1(\Sigma3)_{\mathrm{INT}}\ {}_{\mathrm{INT}}(\Sigma3)_2$$

that is

$$(1/9)\begin{pmatrix} 8 & 1 & -4 \\ 1 & 8 & 4 \\ 4 & -4 & 7 \end{pmatrix} = (1/3)\begin{pmatrix} 2 & 1 & 2 \\ 1 & 2 & -2 \\ -2 & 2 & 1 \end{pmatrix}(1/3)\begin{pmatrix} 1 & 2 & -2 \\ 2 & 1 & 2 \\ 2 & -2 & -1 \end{pmatrix}. \tag{6.6}$$

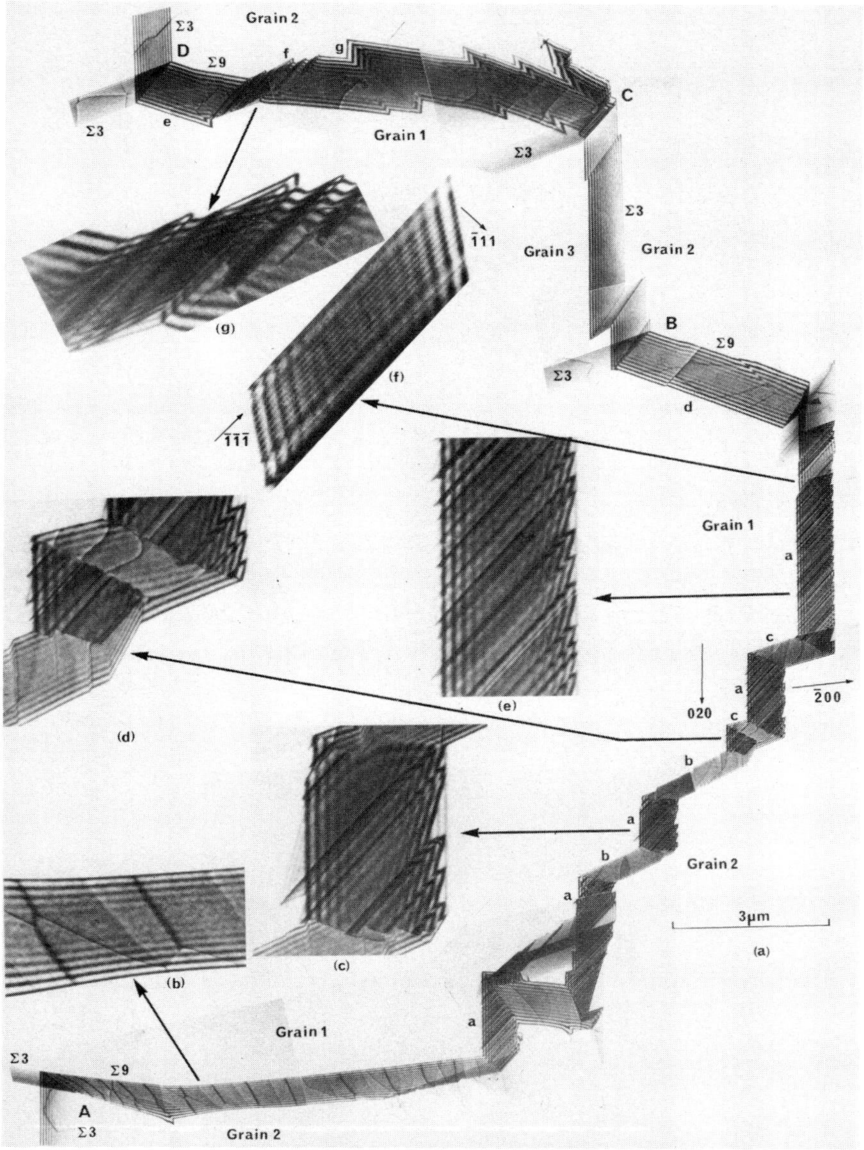

Figure 6.8 Double two-beam image (*a*) of a near-Σ9 boundary in a Cu − 6 at% Si alloy. Parts (*b*)–(*g*) show the indicated portions of the boundary at higher magnification. The beam direction in all images is close to $[\bar{1}02]_1/[0\bar{1}2]_2$ except in (*f*) where it is close to $[\bar{1}\bar{1}2]_1/[0\bar{1}1]_2$. The diffracting vectors are indicated.

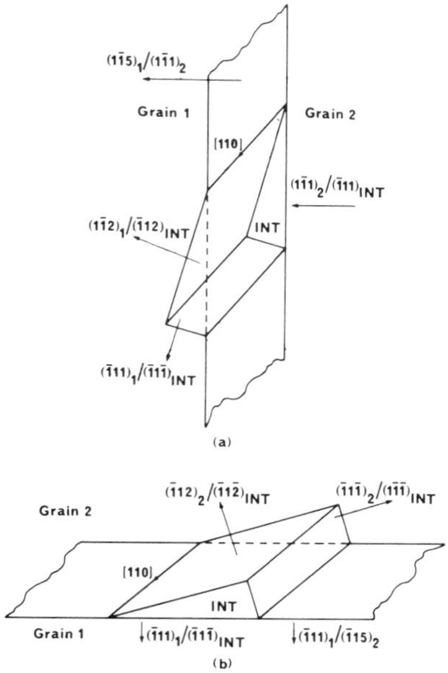

Figure 6.9 Schematic representations (*a*) and (*b*) of boundary dissociation at facets *a* and *f* respectively of figure 6.8.

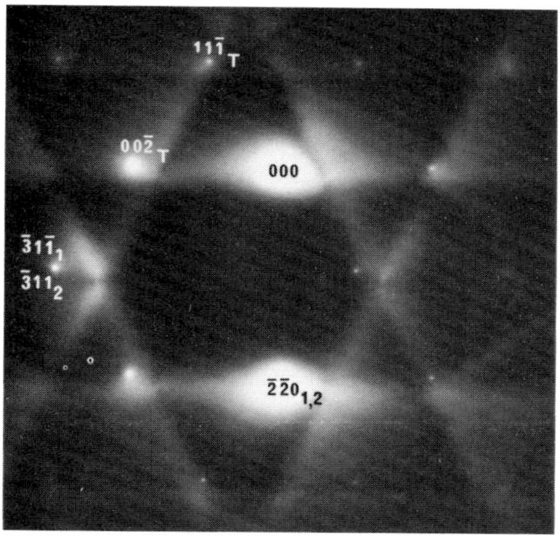

Figure 6.10 Diffraction pattern from facet *a* in the region of figure 6.8(*e*) taken in the common $[\bar{1}14]_1/[1\bar{1}4]_2$ beam direction. Diffraction spots from grains 1 and 2 are indicated and additional twin spots from the intermediate grains are marked T.

For the common [110] rotation axis, the matrices on the right-hand-side of equation (6.6), and all following equations of this type, represent one of two crystallographically equivalent ways of indexing the intermediate grain and this choice does not affect the physical situation. From equation (6.6) it follows that the $[\bar{1}14]_1/[1\bar{1}4]_2$ beam direction of the diffraction pattern in figure 6.10 corresponds to the $[\bar{1}10]_{INT}$ beam direction of the diffraction pattern from the intermediate grain. These results show that the $\Sigma9$ asymmetric-tilt boundaries on $(1\bar{1}5)_1/(1\bar{1}1)_2$ are dissociated to form sets of triangular prisms extending into grain 1, where each prism is bounded by three $\Sigma3$ symmetric-tilt boundaries namely: a $(1\bar{1}1)_2/(\bar{1}11)_{INT}$ coherent twin-boundary separating the intermediate grain from grain 2, together with a $(\bar{1}11)_1/(\bar{1}1\bar{1})_{INT}$ coherent twin-boundary and a $(1\bar{1}2)_1/(\bar{1}12)_{INT}$ incoherent twin-boundary separating the intermediate grain from grain 1. Similarly, the region, *f*, which is shown at higher magnification in figure 6.8(*g*) has dissociated to form triangular prismatic units. However, the dominant plane in this case is $(\bar{1}11)_1/(\bar{1}15)_2$ and the dissociation reaction, which is still described by equation (6.6), has occurred to give sets of three $\Sigma3$ symmetric-tilt boundaries forming prismatic units which extend into grain 2, as shown schematically in figure 6.9(*b*), rather than into grain 1 which was the case for facets *a* of figure 6.8.

In the region of the boundary shown in figure 6.8(*f*) the mean plane is 3° from $(1\bar{1}5)_1/(1\bar{1}1)_2$ and, despite this, the same units of twin-structure shown in figure 6.9(*a*) are present, but on a much finer scale. This suggests that in this region the boundary is stepped with dissociated segments on $(1\bar{1}5)_1/(1\bar{1}1)_2$ separated by short lengths of undissociated $\Sigma9$ boundary, and indicates that boundary dissociation can commence before the $\Sigma9$ boundary forms an extensive facet lying exactly on $(1\bar{1}5)_1/(1\bar{1}1)_2$.

The region shown in figure 6.8(*d*) not only shows a dissociated *a* facet, but also an additional plane. This plane is a $\Sigma9$ asymmetric-tilt boundary lying on $(1\bar{1}1)_1/(11\ \overline{11}\ \bar{1})_2$, i.e. it is the same plane as that of facets *a* in the $\Sigma9$ boundary of figure 6.7 and none of these facets show any dissociation. The only way these facets could dissociate to form a coherent-$\Sigma3$ boundary on $(1\bar{1}1)_1$, between grain 1 and an intermediate twin-grain, would be for the intermediate grain and grain 2 to be separated by two $\Sigma27_a$ boundaries according to the dissociation reaction,

$$_1(\Sigma9)_2 = {}_1(\Sigma3)_{INT}\ {}_{INT}(\Sigma27_a)_2$$

that is

$$(1/9)\begin{pmatrix} 8 & 1 & -4 \\ 1 & 8 & 4 \\ 4 & -4 & 7 \end{pmatrix} = (1/3)\begin{pmatrix} 2 & 1 & -2 \\ 1 & 2 & 2 \\ 2 & -2 & 1 \end{pmatrix}(1/27)\begin{pmatrix} 25 & 2 & 10 \\ 2 & 25 & -10 \\ -10 & 10 & 23 \end{pmatrix}. \tag{6.7}$$

Although such a dissociation would involve replacing a $\Sigma 9$ asymmetric-tilt boundary by three symmetric-tilt boundaries, it is not observed.

The other prominent facets which lie on $\{111\}$ planes are the set, g, but these lie on $(\bar{1}\bar{1}1)_1/(\bar{5}\ \overline{13}\ 7)_2$ and are not tilt boundaries. In this case, the formation of an intermediate twin-grain with a $\Sigma 3$ coherent boundary on $(\bar{1}\bar{1}1)_1/(\bar{1}\bar{1}1)_{INT}$ would involve the formation of two $\Sigma 27_b$ boundaries, which are not tilt boundaries, according to the reaction

$$_1(\Sigma 9)_2 = {_1(\Sigma 3)_{INT}}\ {_{INT}(\Sigma 27_b)_2}$$

that is

$$(1/9)\begin{pmatrix} 8 & 1 & -4 \\ 1 & 8 & 4 \\ 4 & -4 & 7 \end{pmatrix} = (1/3)\begin{pmatrix} 2 & 2 & 1 \\ -1 & 2 & -2 \\ -2 & 1 & 2 \end{pmatrix}(1/27)\begin{pmatrix} 7 & 2 & -26 \\ 22 & 14 & 7 \\ 14 & -23 & 2 \end{pmatrix}. \tag{6.8}$$

However, this dissociation is not observed.

The results show that a $\Sigma 9$ boundary dissociates to form three $\Sigma 3$ symmetric-tilt boundaries only when it lies on a particular $\{111\}$ plane which for the present crystallographic indexing is a $(\bar{1}11)_1$ plane or a $(1\bar{1}1)_2$ plane.

$\Sigma 27_a$ *boundaries*

Like $\Sigma 9$ boundaries $\Sigma 27_a$ boundaries facet during annealing and some of the facets are tilt boundaries containing the [110] rotation axis. Figure 6.11(a) shows several $\Sigma 3$-related boundaries over an extended region of a specimen. The boundaries extend from A to I and were identified as follows: AB, $\Sigma 27_a$; BC, $\Sigma 243_a$; CD, $\Sigma 81_d$; DE, $\Sigma 27_a$; EF, $\Sigma 9$; FG, $\Sigma 27_a$; GH, $\Sigma 9$; and HI, $\Sigma 27_a$. The same-$\boldsymbol{g}_c$ 220 diffracting vector in figure 6.11(a) is common to all grains and under this diffracting condition all the boundaries listed above show fringe contrast due to the presence of rigid-body displacements between the grains. However, the $\Sigma 9$ boundary at the triple junction B and the $\Sigma 3$ boundaries at the remaining six triple junctions C–H are in very weak contrast indicating that the 220 components of rigid-body displacements across these boundaries are small or zero.

The $\Sigma 27_a$ boundary extending from A to B in figure 6.11 is faceted and the most extensive facets, a, lie on $(1\bar{1}1)_1/(13\ \overline{13}\ 43)_2$ and are asymmetric-tilt boundaries. The other facets b and c are on $(7\ 3\ 10)_1/(3\ 7\ 10)_2$ and on the symmetric-tilt boundary plane $(1\bar{1}5)_1/(\bar{1}15)_2$ respectively. The $\Sigma 27_a$ boundary between F and G contains two main facets, a (as above) and d lying on $(535)_1/(355)_2$. The $\Sigma 27_a$ boundary from H to I contains two main facets e and f. Facets e, which constitute the greater part of the boundary, are asymmetric-tilt boundaries on $(\overline{13}\ 13\ 43)_1/(\bar{1}11)_2$ and the small facets f lie on $(425)_1/(245)_2$. All the facets a and e show a fine structure, not present on the other facets, which consists of a set of lines lying along the [110]

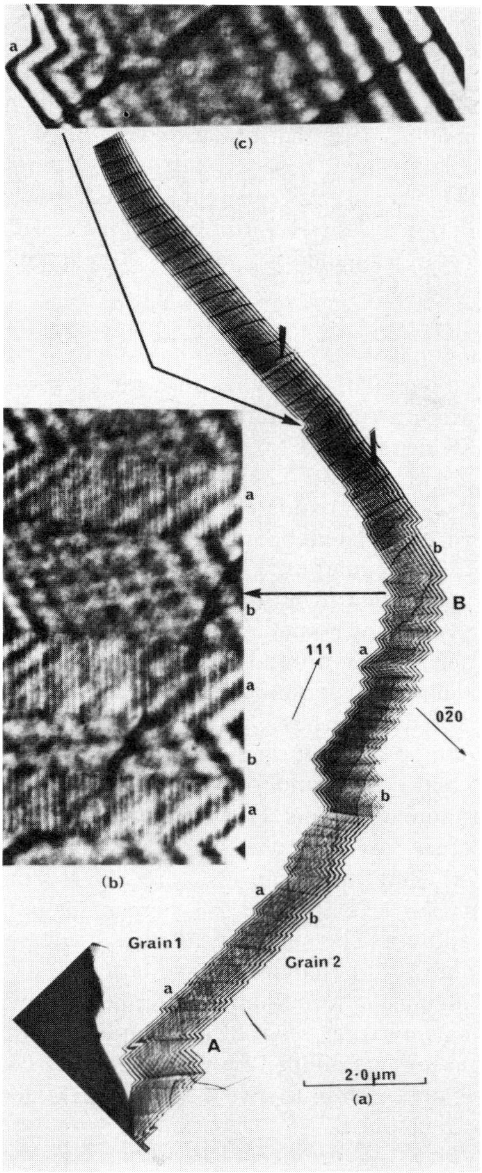

Figure 6.13 Double two-beam image (*a*) of a near-$\Sigma 27_a$ boundary in a Cu – 6 at% Si alloy. In (*b*) and (*c*) the indicated portions of the boundary are shown at higher magnification. The beam direction is close to $[\bar{3}05]_1$ and $[\bar{3}12]_2$ and the diffracting vectors are indicated.

across boundaries. An example of such a small partially developed facet *a* can be seen in figure 6.13(*c*). Clearly dissociation reactions do not depend on the development of extensive facets on {111}. It is of interest that the energy of the multifaceted region of the boundary along AB of figure 6.13(*a*) could have been reduced by the formation of a single facet on $(\bar{1}11)_1$. In principle, such a facet could have dissociated to form intermediate grains bounded by $(\bar{1}11)_1/(\bar{1}1\bar{1})_{INT}$ Σ3 coherent boundaries with grain 1 and Σ9 boundaries with grain 2. However, this type of facet, with its accompanying dissociation to form Σ9 boundaries, was not observed in this case or in any other examples studied.

$\Sigma 81_d$ boundaries

Figure 6.14(*a*) is a region of a specimen showing a series of grain boundaries from A to I separating grain 1 from grains 2, 3, 4, 5 and 6, where grains 3, 4, 5 and 6 are Σ3 twin-related to grain 2. AB, CD, EF and GH are $\Sigma 81_d$ boundaries; BC, DE and HI are Σ243 boundaries and FG is a $\Sigma 27_a$ boundary. This $\Sigma 27_a$ boundary is approximately 3° from the $(1\bar{1}1)_1$ plane and is finely dissociated to form Σ3 coherent boundaries and $\Sigma 81_d$ boundaries as described above for the $\Sigma 27_a$ boundaries. The boundaries BC and DE are $\Sigma 243_f$ boundaries, corresponding to misorientations of 43.08° around $[7\bar{9}1]$ and $[\bar{7}\bar{1}9]$ respectively, and the boundary HI is a $\Sigma 243_a$ boundary corresponding to a misorientation of 7.36° around [110], i.e. the common rotation axis of grains 1 and 2. Figure 6.14(*a*) is taken with the same-$\boldsymbol{g}_c$ 220 diffracting vector in grains 1, 2, 5, and 6 and the $\Sigma 81_d$, $\Sigma 27_a$ and $\Sigma 243_a$ boundaries show fringe contrast due to the presence of rigid-body displacements.

Unlike the Σ9 and $\Sigma 27_a$ boundaries, the $\Sigma 81_d$ boundaries in figure 6.14(*a*) do not show pronounced facets. However, each of these boundaries has a mean boundary plane over most of its length within approximately 3° of $(1\bar{1}1)_1/(\bar{73}\ 73\ 95)_2$ and shows fine structure which is parallel to the [110] rotation axis as for example in the higher magnification images of figures 6.14(*b*) and (*e*). The similarity of this structure to that discussed already for Σ9 and $\Sigma 27_a$ boundaries suggests that it arises from boundary dissociation involving the formation of intermediate twin-grains. The two possibilities for dissociation of a $\Sigma 81_d$ boundary with the formation of intermediate twin-grains bounded by tilt boundaries containing the [110] rotation axis are: dissociation to give a Σ3 with $\Sigma 27_a$ boundaries, or to give a Σ3 with $\Sigma 243_a$ boundaries.

Although the fine structure occurs in all the $\Sigma 81_d$ boundaries of figure 6.14(*a*), detailed examination of the structure was confined to the $\Sigma 81_d$ boundary between G and H which lies in a thinner region of the specimen. In this region, a segment of boundary X lies on $(1\bar{1}1)_1$ (figure 6.14(*c*)) and another segment Y lies on $(\bar{1}11)_2$ (figure 6.14(*d*)). Clearly visible intermediate grains have formed at these segments with the intermediate grains extending into grain 2 at X and into grain 1 at Y. The nature of these intermediate grains was determined by selected area diffraction.

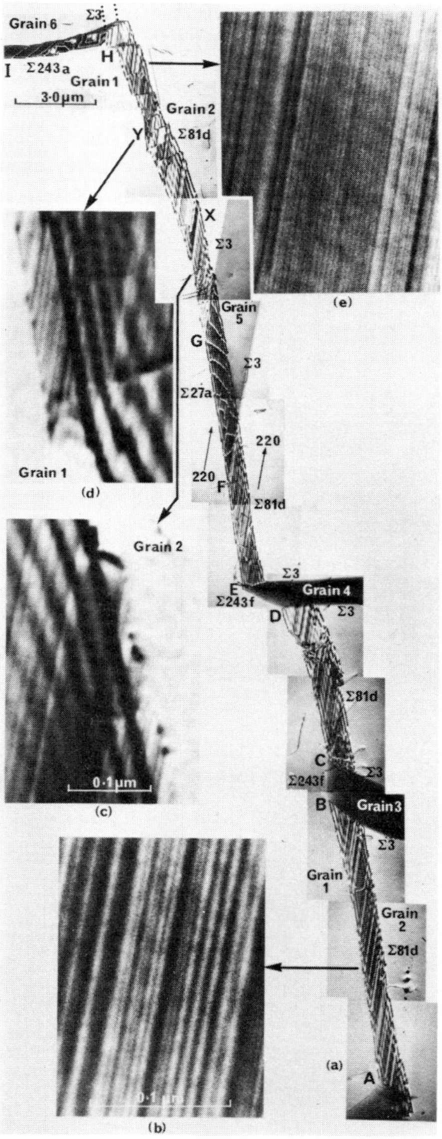

Figure 6.14 Double two-beam image (a) of several $\Sigma 3$-related boundaries as indicated in a Cu – 6 at% Si alloy. Parts (b), (c), (d) and (e) are higher magnification images of the regions indicated for the $\Sigma 81_d$ boundaries. The beam directions are: close to $[7\bar{7}2]_1$ and $[\bar{1}13]_2$ in (a) with $\boldsymbol{g}_1 = \boldsymbol{g}_2 = 220$; close to $[2\bar{1}1]_1$ and $[\bar{1}34]_2$ in (b) with $\boldsymbol{g}_1 = 11\bar{1}_1$ and $\boldsymbol{g}_2 = 1\bar{1}1_2$; close to $[1\bar{2}1]_1$ and $[\bar{3}14]_2$ in (c) and (d) with $\boldsymbol{g}_1 = 111_1$ and $\boldsymbol{g}_2 = 1\bar{1}1_2$. In (e) only grain 2 is diffracting with $\boldsymbol{g}_2 = 200_2$ in the $[012]_2$ beam direction.

Figure 6.15(*a*) is a diffraction pattern from segment X, taken in a beam direction close to $[4\bar{1}1]_1/[23\ 220\ 261]_2$, which contains the $[4\bar{1}1]_1$ pattern of diffraction spots from grain 1 and the $\bar{1}\bar{1}1_2$ spot from grain 2. In addition there is a $\langle 110 \rangle$ pattern of strong spots from the intermediate grains which contains the spots of the $[4\bar{1}1]_1$ pattern. This $\langle 110 \rangle$ pattern is therefore twin-related to the $[4\bar{1}1]_1$ pattern. Thus the intermediate grains at the segment X are twin-related to grain 1, and could only have arisen from dissociation of the $\Sigma 81_d$ boundary to form a $\Sigma 3$ coherent boundary (separating

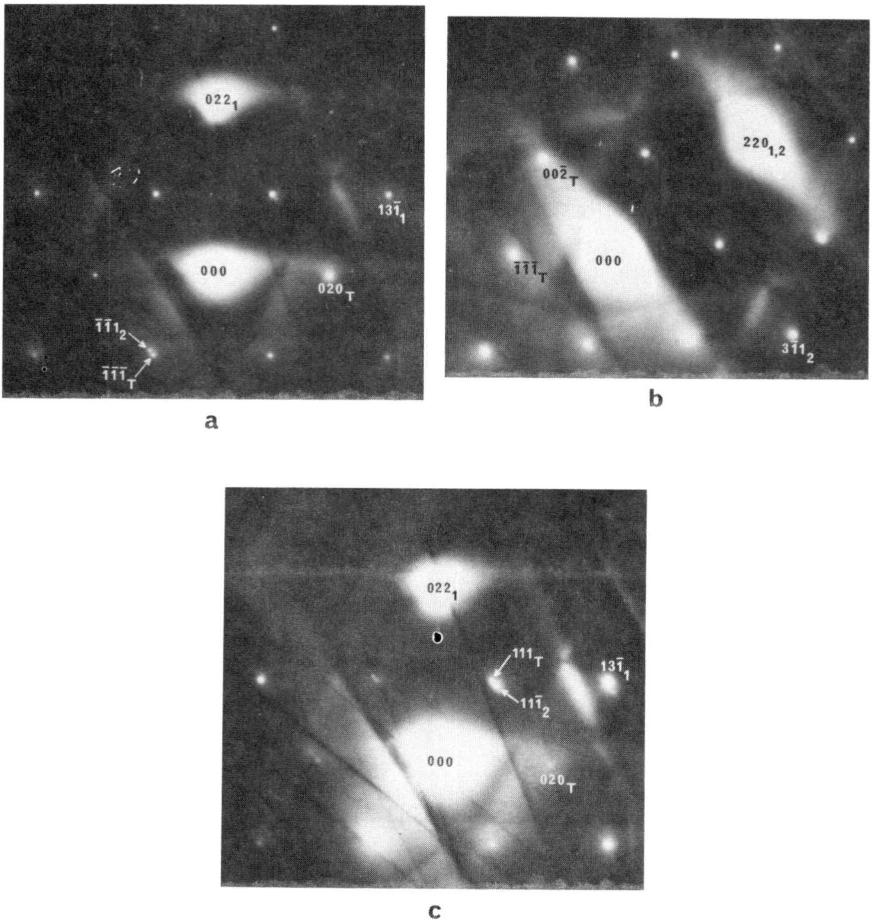

Figure 6.15 Diffraction patterns from: (*a*) region *X* of figure 6.14 in the $[4\bar{1}1]_1/[23\ 220\ 261]_2$ beam direction; (*b*) region *Y* of figure 6.14 in the $[\bar{1}14]_2/[241\ \overline{241}\ 44]_1$ beam direction; (*c*) the region shown in figure 6.14(*e*) in the $[4\bar{1}1]_1/[23\ 220\ 261]_2$ beam direction. Diffraction spots from grains 1 and 2 are indicated and additional twin spots from the intermediate grains are marked T.

grain 1 from each intermediate grain) with $\Sigma 243_a$ boundaries (separating each intermediate grain from grain 2) according to the dissociation reaction

$$_1(\Sigma 81_d)_2 = {}_1(\Sigma 3)_{\mathrm{INT}}\ {}_{\mathrm{INT}}(\Sigma 243_a)_2$$

that is

$$(1/81)\begin{pmatrix} 32 & 49 & 56 \\ 49 & 32 & -56 \\ -56 & 56 & -17 \end{pmatrix} = (1/3)\begin{pmatrix} 2 & 1 & -2 \\ 1 & 2 & 2 \\ 2 & -2 & 1 \end{pmatrix} \times (1/243)\begin{pmatrix} 1 & 242 & 22 \\ 242 & 1 & -22 \\ -22 & 22 & -241 \end{pmatrix} \quad (6.11)$$

which specifies the coherent plane as $(1\bar{1}1)_1/(1\bar{1}\bar{1})_{\mathrm{INT}}$, i.e. the facet plane of segment X.

Figure 6.15(b) is a diffraction pattern from segment Y taken in the $[\bar{1}14]_2/[241\ \overline{241}\ 44]_1$ beam direction which contains the $[\bar{1}14]_2$ pattern of diffraction spots from grain 2 and the spots $\pm 220_1$ from grain 1 (coincident with the spots $\pm 220_2$ from grain 2). In addition there is a $\langle 110 \rangle$ pattern of strong spots from the intermediate grains which contains the spots of the $[\bar{1}14]_2$ pattern and this $\langle 110 \rangle$ pattern is twin-related to the $[\bar{1}14]_2$ pattern. Thus the intermediate grains at segment Y are twin-related to grain 2 and could only have arisen from the dissociation of the $\Sigma 81_d$ boundary to form a $\Sigma 3$ coherent boundary (separating grain 2 from each intermediate grain) with $\Sigma 243_a$ boundaries (separating each intermediate grain from grain 1) according to the dissociation reaction

$$_1(\Sigma 81_d)_2 = {}_1(\Sigma 243_a)_{\mathrm{INT}}\ {}_{\mathrm{INT}}(\Sigma 3)_2$$

that is

$$(1/81)\begin{pmatrix} 32 & 49 & 56 \\ 49 & 32 & -56 \\ -56 & 56 & -17 \end{pmatrix} = (1/243)\begin{pmatrix} 1 & 242 & 22 \\ 242 & 1 & -22 \\ -22 & 22 & -241 \end{pmatrix} \times (1/3)\begin{pmatrix} 2 & 1 & -2 \\ 1 & 2 & 2 \\ 2 & -2 & 1 \end{pmatrix} \quad (6.12)$$

which specifies the coherent plane as $(\bar{1}11)_2/(\bar{1}1\bar{1})_{\mathrm{INT}}$, i.e. the facet plane of segment Y.

For regions of the $\Sigma 81_d$ boundary containing only fine structure, such as that shown in figure 6.14(e), selected area diffraction showed that this fine structure was due to the formation of small intermediate twin-grains of the type identified for segment X, and likewise must have arisen from the dissociation reaction of equation (6.11) giving $\Sigma 3$ coherent boundaries on $(1\bar{1}1)_1/(1\bar{1}\bar{1})_{INT}$. This is demonstrated in figure 6.15(*c*) which is a diffraction pattern, taken under the same conditions as figure 6.15(*a*), but from the region shown in figure 6.14(*e*). In this case, the additional $\langle 110 \rangle$ pattern from the intermediate twin-grains is much weaker due to the smaller volume of twinned material. In regions such as that shown in figure 6.14(*e*) the mean plane of the boundary departs slightly from $(1\bar{1}1)_1$, so that in these regions the boundary will be stepped with dissociated segments separated by undissociated segments of $\Sigma 81_d$ boundary. Therefore, $\Sigma 81_d$ boundaries can also dissociate before an extensive facet is formed on an appropriate $\{111\}$ plane.

In regions X and Y, where the dissociation of the $\Sigma 81_d$ boundary is extensive, each intermediate grain is bounded by two sets of $\Sigma 243_a$ boundary planes in a similar way to the prismatic units formed by the dissociation of the $\Sigma 9$ boundary. In region X, stereographic analysis of the $\Sigma 243_a$ planes was inaccurate due to the short traces available, but showed that the two sets of planes were within 5° of the symmetric-tilt boundaries $(\overline{11}\ 11\ 1)_2/(11\ \overline{11}\ 1)_{INT}$ and $(1\ \bar{1}\ 22)_2/(1\ \bar{1}\ \overline{22})_{INT}$.

In summary, the results show that $\Sigma 81_d$ boundaries dissociate to form $\Sigma 3$ and $\Sigma 243_a$ boundaries, rather than $\Sigma 3$ and $\Sigma 27_a$ boundaries. Further, like the case of $\Sigma 9$ boundaries, the dissociation of $\Sigma 81_d$ asymmetric-tilt boundaries lying close to an approximate $\{111\}$ plane involves the formation of three symmetric-tilt boundaries.

The results described for copper and the copper–silicon alloy show that during annealing, dissociation of $\Sigma 9$, $\Sigma 27_a$ and $\Sigma 81_d$ boundaries occurs when facet planes develop on particular $\{111\}$ planes to form $\langle 110 \rangle$ asymmetric-tilt boundaries. Solely from a geometric point of view, $\Sigma 9$, $\Sigma 27_a$ and $\Sigma 81_d$ boundaries can undergo two possible types of dissociation reaction, i.e. to form intermediate grains with $\{111\}$ $\Sigma 3$ coherent boundaries together with either lower-Σ boundaries or higher-Σ boundaries. For each type of boundary only one of the possible dissociation reactions is favoured, i.e. $\Sigma 9$ boundaries are observed to dissociate only to the lower Σ value of $\Sigma 3$, while $\Sigma 27_a$ and $\Sigma 81_d$ boundaries are observed to dissociate only to the higher Σ values of $\Sigma 81_d$ and $\Sigma 243_a$ respectively. This result means that, since dissociation always involves the $\{111\}$ facet plane forming a $\Sigma 3$ coherent boundary with the intermediate grain, dissociation only takes place on particular $\{111\}$ facets for each type of boundary which, for the crystallographic indexing used here, are $(\bar{1}11)_1$ and $(1\bar{1}1)_2$ for $\Sigma 9$ and $(1\bar{1}1)_1$ and $(\bar{1}11)_2$ for $\Sigma 27_a$ and $\Sigma 81_d$.

For the cases described where the bounding planes of the intermediate grains have been determined (i.e. for $\Sigma 9$ and $\Sigma 81_d$ boundaries), an asymmetric-tilt boundary on a $\{111\}$ plane dissociates to form three symmetric-tilt boundaries. Thus there must be a decrease in energy associated with the transition of asymmetric- to symmetric-tilt boundaries which is large enough to offset the increase in boundary area. Computer calculations by Brokman *et al* (1981) indicate that an asymmetric-tilt boundary can lower its energy by kinking on an atomic scale to form symmetric-tilt boundaries, even though this process involves an increase in boundary area. For example, such computations would suggest that a $\Sigma 9$ asymmetric-tilt boundary on $\{111\}$ would kink to form symmetric $\Sigma 9$ boundaries on $\{114\}$ and $\{221\}$ and such facets have been observed by Sukhomlin and Andreeva (1983) in an extensive study of faceting in $\Sigma 3$ related boundaries. However, what is observed here is the dissociation of a $\Sigma 9$ boundary to form two $\Sigma 3$ coherent boundaries on $\{111\}$ planes and an incoherent-$\Sigma 3$ boundary on $\{112\}$. Since the energies of coherent- and some incoherent-$\Sigma 3$ boundaries are known to be low (see Pond and Vitek 1977, Sutton and Vitek 1983a, b, c), this dissociation reaction could be more favourable than the formation of a simple kinked boundary. Similarly, the dissociation of a $\Sigma 81_d$ boundary, which results in replacing the original asymmetric-$\Sigma 81_d$-tilt boundary on $\{111\}$ by a $\Sigma 3$ coherent boundary and two 7.36° low-angle $\Sigma 243_a$ symmetric-tilt boundaries, could be energetically favoured over kinking of the $\Sigma 81_d$ boundary. However, for the case of a $\Sigma 27_a$ asymmetric-tilt boundary on $\{111\}$, in the absence of any knowledge concerning the relative boundary energies, it is not obvious why the observed $\Sigma 81_d$ boundaries should be preferred to $\Sigma 9$ boundaries in the dissociation reaction, or why dissociation should be preferred to kinking. A possibility to be considered is that boundary dissociation, although requiring a lowering of energy, could well be controlled by kinetics rather than by energy considerations alone, i.e. the ease of atom movements involved in a particular dissociation reaction could result in a stable lower energy state being preferred which is not necessarily the minimum energy state.

Possible atomic movements associated with the dissociation of $\Sigma 9$, $\Sigma 27_a$ and $\Sigma 81_d$ boundaries can be considered by examining models of the atomic structure at $\{111\}$ asymmetric-tilt boundaries. It is found that very simple atom movements can account for the dissociation of a $\Sigma 9$ boundary, and that similar atom movements could account for the type of dissociation observed for the $\Sigma 27_a$ and $\Sigma 81_d$ boundaries.

The atom movements involved in the model for the dissociation of a $\Sigma 9$ boundary will be described in relation to the diagrams of figure 6.16, which are indexed to correspond to the physical situation of figures 6.8(*c*), (*d*) and (*e*). Figure 6.16(*a*) shows the projection along the $[110]$ rotation axis of the FCC lattices of grains 1 and 2, where grain 1 is rotated in a right-handed sense with respect to grain 2 by 38.94° and the two lattices meet at the $(1\bar{1}5)_1/(1\bar{1}1)_2$ boundary plane. In figure 6.16(*a*) the two lattices have been juxtaposed and the lattice sites in grain 1, in the region of the boundary, which are closer than the $(1/2)\langle 110\rangle$ interatomic spacing to the lattice sites in grain 2, have been removed. The resulting structure can be envisaged as having rows of 'vacant lattice sites' in the neighbourhood of the boundary along the $[110]$ direction. This open configuration of atoms at the boundary

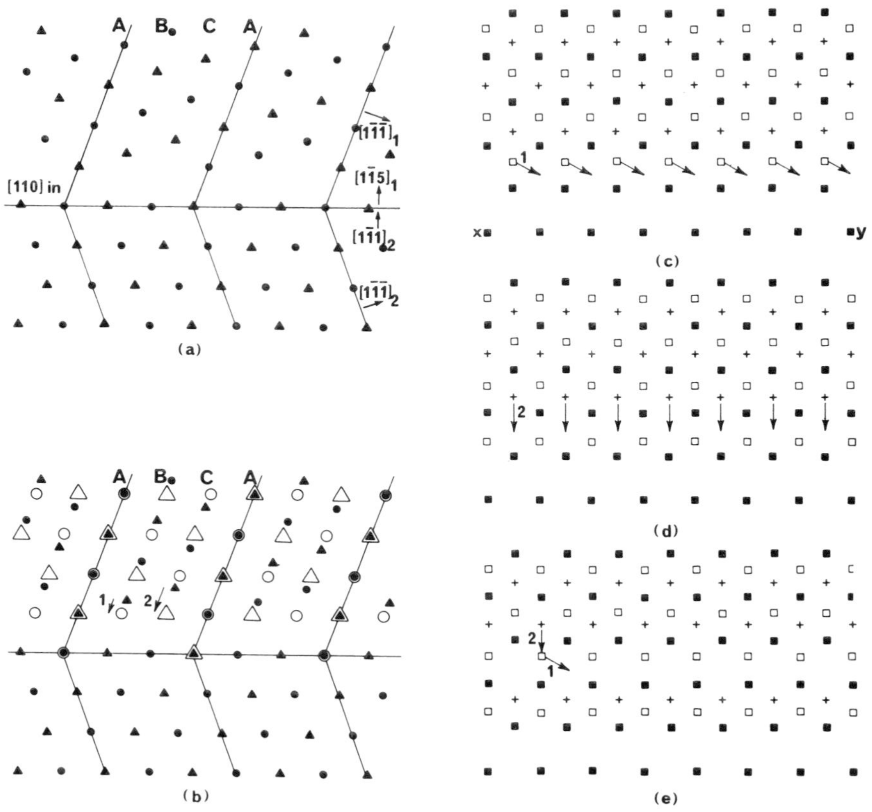

Figure 6.16 Schematic diagrams illustrating the atom movements involved in the model for the dissociation of a $\Sigma 9$ boundary in an FCC structure: (*a*) the [110] projection of the atom positions (solid symbols) at a $(1\bar{1}5)_1/(1\bar{1}1)_2$ $\Sigma 9$ tilt boundary; (*b*) the atom positions (open symbols) in a grain $\Sigma 3$ related to grain 2 are superimposed on the atom positions of grain 1 (● and ○ represent atom positions in the plane of the paper and ▲ and △ represent atom positions displaced $\pm(1/4)[110]$); (*c*), (*d*) and (*e*) have $[\bar{1}11]_1$ into the plane of the paper and are $[\bar{1}11]_1$ projections of the *A*, *B* and *C* planes of grain 1 (the atom positions in the *A*, *B* and *C* planes are represented by ■, □ and + respectively) and illustrate the sequence of atom movements involved in the generation of an intermediate twin-grain.

is similar to the structure determined by Sutton and Vitek (1983b) from computer modelling of a $\{115\}_1/\{111\}_2$ $\Sigma 9$ boundary in aluminium, although their results give a structure in which there are more vacant sites than in figure 6.16(*a*). In figure 6.16(*b*) the projected atom positions for an intermediate grain (marked with open symbols), which is $\Sigma 3$ twin-related to grain 2 with the coherent boundary on $(1\bar{1}1)_2$, have been superimposed on the projected atom positions of grain 1. It can be seen from this diagram that the atom positions in every third $(1\bar{1}\bar{1})_1$ plane in the lattice of grain 1 (planes marked *A*) are exactly coincident with the atom positions in the intermediate twin. Thus the nucleation of an intermediate twin requires only the movement of atoms on the planes marked *B* and *C*, and these movements can take place via the rows of vacant sites along [110] in the neighbourhood of the boundary. The movements involved in the *B* and *C* planes are indicated by the arrows in figure 6.16(*b*) and must take place in the sequence 1, 2. These movements are of the type $(1/6)\langle 211 \rangle_1$ and can be seen more readily in figure 6.16(*c*)–(*e*) which are $[\bar{1}11]_1$ projections of the *A*, *B*, *C* planes of grain 1. Figure 6.16(*c*) corresponds to FCC stacking of the $(1\bar{1}\bar{1})_1$ planes of grain 1 and is to be compared with figure 6.16(*a*). In figure 6.16(*c*) the termination of the atoms in the *A* layer along *xy* is the [110] direction in the boundary corresponding to the termination of the *A* plane in figure 6.16(*a*). The first row of atoms in the *B* layer and the second row of atoms in the *C* layer are vacant, as they would be closer than an interatomic spacing to atoms in grain 2. The first row of atoms in the *C* layer does not exist in grain 1 since it is cut off by the boundary plane. The first atomic movements involved in the development of an intermediate twin-grain are the movements shown by the arrows marked 1 in figure 6.16(*c*), i.e. the movement of the first row of atoms in *B* positions in the *B* layer to *C* positions in the *B* layer by a displacement of $(1/6)[12\bar{1}]_1$ to generate the structure of figure 6.16(*d*). The second set of atom movements shown by the arrows marked 2 in figure 6.16(*d*) involves the movement of the first row of atoms in *C* positions in the *C* layer to *B* positions in the *C* layer by a displacement of $(1/6)[\bar{1}1\bar{2}]_1$ to generate the structure of figure 6.16(*e*). This sequence of movements replaces the *ABC* stacking of grain 1 with the twin stacking *ACB* of the intermediate grain. This twin nucleus can continue to grow by repeating this sequence of atomic displacements which involves the propagation of a row of $\frac{1}{3}$ vacancies in the *B* layer and $\frac{2}{3}$ vacancies in the *C* layer. The only requirement is that there is a structural vacancy content parallel to the rotation axis at the original $\Sigma 9$ boundary and that this vacancy content can be propagated at the boundary between the intermediate twin-grain and grain 1 by suitable atomic displacements. In this model, a minimum vacancy content has been included at the boundary and if a higher vacancy content applies, such as that indicated by the calculations of Sutton and Vitek (1983b) for aluminium, the mechanism would be essentially the same, but would involve a different sequence of atom movements. The mechanism only describes the nucleation of an intermediate twin-grain and is not concerned with the development of the final symmetric-tilt boundary on $(1\bar{1}2)_1/(\bar{1}12)_{INT}$ which would require a change in the vacancy content. This model is simpler than models for the nucleation of annealing twins involving cyclic generation on neighbouring planes of three different $(1/6)\langle 211 \rangle$ dislocations contained in the $\{111\}$ coherent twinning plane (see e.g. Dash and Brown 1963), or growth accidents associated with atoms transferring from one lattice to the other during boundary movement (Gleiter 1969).

Clearly, it is the coincidence of a set of {111} planes containing the [110] rotation axis in grain 1 and in the intermediate twin-grain that allows the easy dissociation process of figure 6.16. In contrast, for the other possible dissociation reaction of a $\Sigma 9$ boundary to give $\Sigma 3$ and $\Sigma 27_a$ boundaries (which is not observed) the smallest angle of misalignment between sets of {111} planes containing the [110] rotation axis both in grain 1 and in the intermediate twin-grain is 31.58°. For this case diagrams of atom positions in grains 1 and 2 and the intermediate twin-grain show that no simple dissociation mechanism of the type shown in figure 6.16 can occur.

For the observed dissociation reaction of a $\Sigma 27_a$ boundary to form $\Sigma 3$ and $\Sigma 81_d$ boundaries, a set of {111} planes containing the [110] rotation axis in grain 1 and another set in the intermediate twin-grain are only misaligned by 7.36°, whereas, for the unobserved dissociation reaction to form $\Sigma 3$ and $\Sigma 9$ boundaries, the smallest angle between {111} planes containing the [110] rotation axis is 31.58°. Similarly, for the observed dissociation reaction of a $\Sigma 81_d$ boundary to form $\Sigma 3$ and $\Sigma 243_a$ boundaries, both sets of {111} planes containing the [110] rotation axis in grain 1 and in the intermediate twin are only misaligned by 7.36°; whereas for the unobserved dissociation reaction to form $\Sigma 3$ and $\Sigma 27_a$ boundaries, the {111} planes containing the rotation axis are misaligned by 31.58°. Since {111} planes are not coincident in these cases, the atom movements involved in boundary dissociation would obviously be more complex than those of figure 6.16, because they do not involve the simple $(1/6)\langle 211 \rangle$ DSC vectors of the $\Sigma 3$ CSL orientation between grain 1 and the intermediate grain, but the DSC vectors corresponding to the CSLs of $\Sigma 81_d$ and $\Sigma 243_a$. However the simple atom modelling of the type used in figure 6.16 suggests that the atom movements required for the observed dissociations, where the {111} planes are only misaligned by 7.36°, would be simpler than those necessary for the unobserved dissociations.

The electron microscopy of boundary dissociation in gold, copper, a copper–silicon alloy and stainless steel shows that it is a general phenomenon in FCC metals and alloys and it is likely that grain boundary dissociation is a source for the nucleation of annealing twins in these materials.

6.4 THE GENERATION OF CRYSTAL DEFECTS BY MOVING GRAIN BOUNDARIES

During recrystallisation and grain growth in deformed polycrystalline metals and alloys, defects such as dislocations, single and overlapping stacking faults, stacking-fault bends and partial and complete stacking-fault tetrahedra are

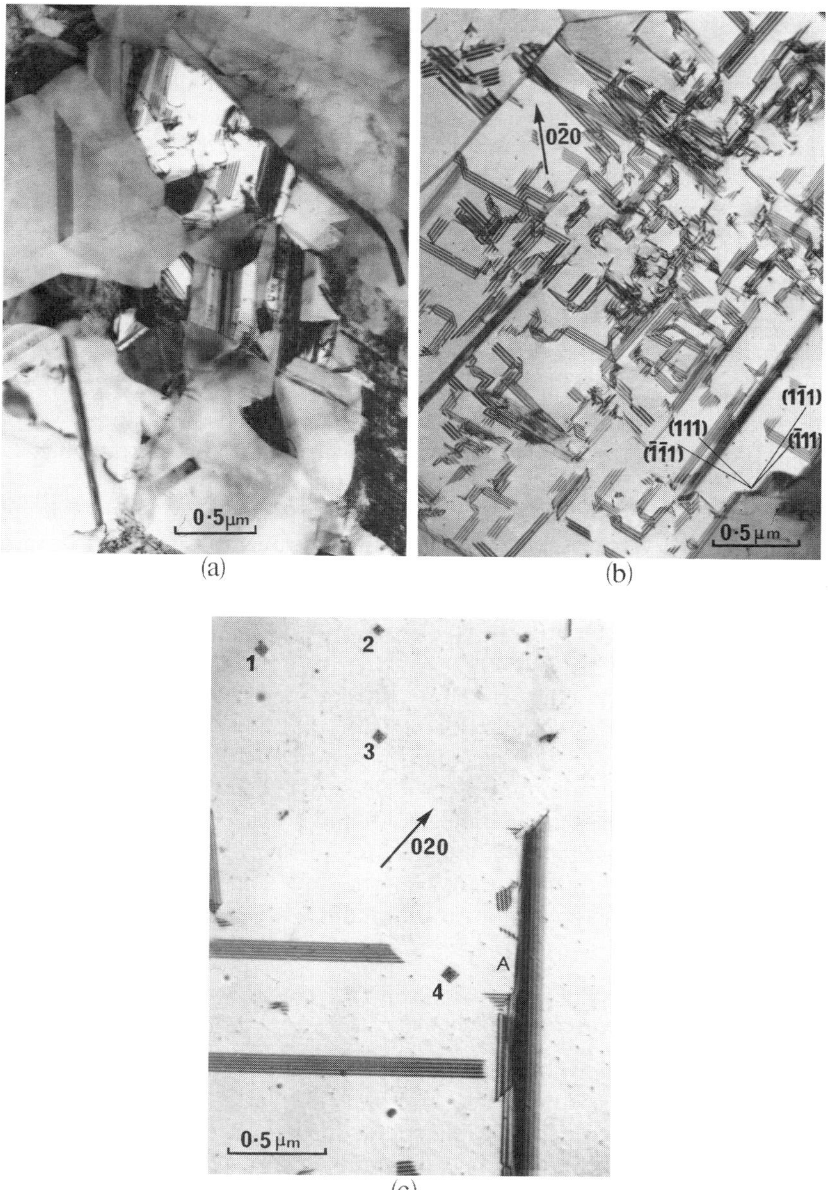

Figure 6.17 Electron micrographs of a Cu – 8 at% Si alloy heavily deformed and then annealed successively for 15 min at 425 °C (*a*), 500 °C (*b*) and 600 °C (*c*). ***B*** is close to [001] in (*b*) and (*c*) and the diffracting vectors are indicated.

generated in the newly formed grains behind the migrating grain boundaries. The generation of these defects has been demonstrated and the defects analysed in silver, copper–silicon alloys and nickel–cobalt alloys (Clarebrough 1974). For example, figure 6.17 shows the defect structure present in new grains after recrystallisation and grain growth in a polycrystalline specimen of a Cu – 8 at% Si alloy which has been heavily deformed and then annealed successively for 15 minutes at 425 °C (*a*), 500 °C (*b*) and 600 °C (*c*), with the specimen being re-thinned after each annealing treatment for examination in the electron microscope. Clearly, defect structure is present in the newly recrystallised grains (figure 6.17(*a*)) and, after extensive grain growth, a high concentration of stacking faults and stacking-fault bends has been generated by the moving boundaries (figure 6.17(*b*)). After annealing at 600 °C most of the extended stacking faults have annealed out and stacking-fault tetrahedra, labelled 1–4, can be clearly seen. Clarebrough and Forwood (1978, 1980e) have observed directly in the electron microscope the generation and growth of stacking faults and stacking-fault bends by moving boundaries in thinned polycrystalline specimens of copper (99.999% Cu). In addition, they have analysed partial stacking-fault tetrahedra attached to grain boundaries in a Cu – 6 at% Si alloy, and have demonstrated that the formation of these defects during grain growth in bulk polycrystalline specimens is related to secondary grain boundary dislocation structure (Clarebrough and Forwood 1987b).

Direct observation of the formation of dislocations during the recrystallisation of evaporated thin-film polycrystals of gold have been made by Kang *et al* (1976). Similar observations of the generation of dislocations by moving boundaries in pure aluminium (99.999% Al), using synchrotron x-ray topography, have been made by Gastaldi and Jourdan (1978) and Gastaldi *et al* (1988). Gleiter *et al* (1980), using etch pitting techniques and electron microscopy, have shown that dislocations are generated behind moving boundaries in indium–phosphorus and indium–arsenic alloys.

6.4.1 Faulted Defects Generated by the Movement of Boundaries in Electron Microscope Specimens

A striking property of high-angle grain boundaries in pure polycrystalline copper (99.999% Cu) is that they are mobile in thin-foil electron microscope specimens at room temperature and rotate during observation, preferentially at the surface intersections, to become more steeply inclined to the plane of the specimen surfaces. This occurs because an inclined boundary traversing a thin-foil specimen experiences a driving force to rotate and become normal to the surfaces of the specimen so as to decrease its area and thus decrease its overall energy. The fact that this takes place preferentially near the surface intersections is to be expected since some of the atom movement required at the boundary can take place by surface diffusion. This rotation occurs for boundaries with misorientations both close to and far from CSL orientations and, during rotation, defects are generated in the newly created portions of the neighbouring grains. The defects observed consist of different types of stacking-fault bend and, in addition, small twinned grains sometimes form by boundary dissociation as discussed in section 6.3 (Clarebrough and Forwood 1978). Stacking-fault bends are sometimes observed to become detached from a boundary during rotation to leave faulted defects in the newly created region of a grain (Clarebrough and Forwood, 1980e).

Figure 6.18 is an electron micrograph of a portion of a grain boundary in copper (99.999% Cu) as first observed in the electron microscope. The misorientation between the two grains corresponds to $\theta = 35.5°$ and $\boldsymbol{u}$ approximately [184]. The abrupt change in spacing across the boundary of the pendellösung fringes along AA indicates a marked change in boundary plane within the thickness of the specimen. In addition, there is a trace along CC displaced from the surface intersection AA of the steeply inclined portion of the boundary. The end points of this trace (not shown in figure 6.18) are those points on the surface intersection of the boundary at which the boundary plane commences to change its inclination to the surface of the specimen. Thus the trace coincides with a former intersection of the boundary plane with the specimen surface. At some locations of the boundary several distinct surface traces are observed marking intermediate positions of the surface intersection of the boundary as it changed its inclination. The marked change in spacing of the pendellösung fringes together with the observed boundary traces in the surface of the specimen show that the boundary has rotated in the thin-foil specimen.

The surface intersection along BB (figure 6.18) is very irregular and this irregularity is not a result of electropolishing, but is associated with the generation of defects. Figure 6.19(*a*) shows another portion of the boundary of figure 6.18. Immediately after recording this image, the microscope was switched off and figure 6.19(*b*) shows the same portion of the boundary after 20 hours at room temperature. Clearly, further rotation of the boundary has

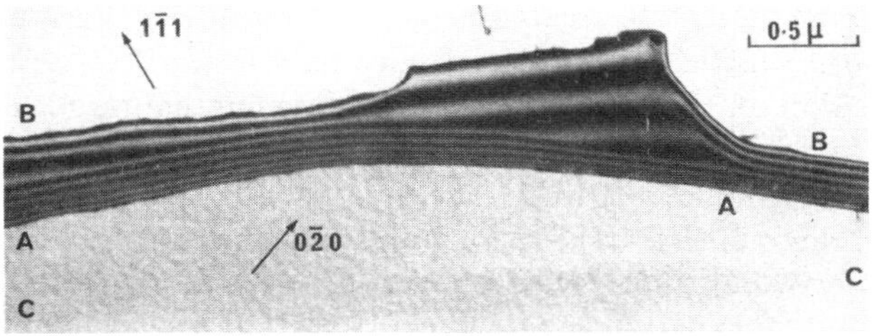

Figure 6.18 Double two-beam image of portion of a high-angle grain boundary in 99.999% copper. $\boldsymbol{B}$ is [$\bar{3}$14] with $\boldsymbol{g}$ = 1$\bar{1}$1 and [$\bar{1}$07] with $\boldsymbol{g}$ = 0$\bar{2}$0.

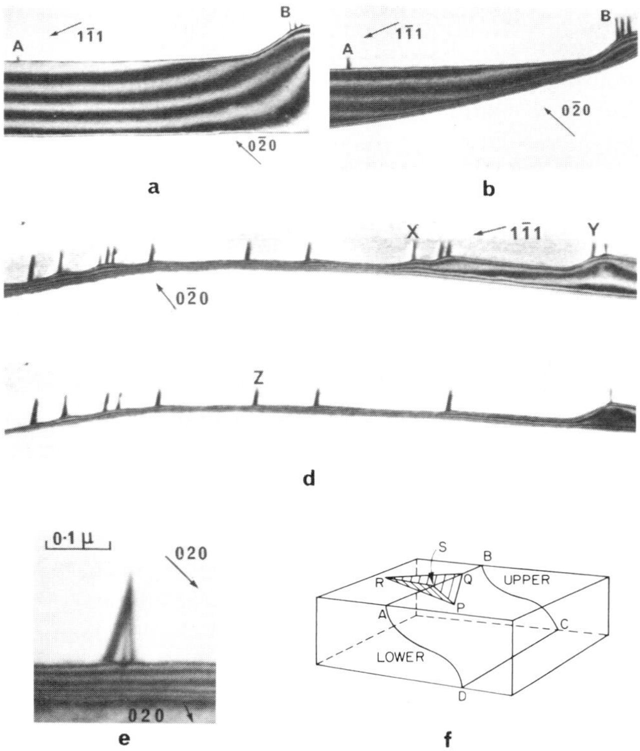

Figure 6.19 Generation of acute intrinsic stacking-fault bends by a moving grain boundary in 99.999% Cu. In (*a*)–(*d*) the diffraction conditions are the same as for figure 6.18. In (*e*), ***B*** is close to [001] in the lower grain and [$\bar{1}$02] in the upper grain. The intrinsic fault bend in (*e*) is illustrated schematically in (*f*).

occurred. In addition the small irregularities at the surface intersections at A and B in figure 6.19(*a*) have been extended by the rotating boundary to form defects in the newly created portion of the lower grain as shown in figure 6.19(*b*). A further example of the formation of defects of this type, at another position on the same boundary, is shown in figures 6.19(*c*) and (*d*). Figure 6.19(*c*) is for the same rest time at room temperature as figure 6.19(*b*) and further development of these defects after 3 days at room temperature is shown in figure 6.19(*d*). The defects always extend from the surface trace to the boundary and some, such as X and Y in figure 6.19(*c*), are no longer present in figure 6.19(*d*). In fact, defects of this type are sometimes observed to 'pop out' of the specimen and others to be generated as the boundary continues to move. Figure 6.19(*e*) is a higher magnification image of defect Z in figure 6.19(*d*), taken in a beam direction which shows that it consists of two faulted planes. Analysis of this and other defects of figure 6.19(d) shows that they are acute intrinsic stacking-fault bends, as illustrated schematically in figure 6.19(f), and involve faults on $(\bar{1}11)$† and $(1\bar{1}1)$ planes (PSR and PQR respectively of figure 6.19(f)) linked by a low-energy stair-rod dislocation along [110] ($\overrightarrow{RP}$ of figure 6.19(f)) with Burgers vector $(1/6)[1\bar{1}0]$. The faults are in the lower grain, intersect the upper surface of the specimen and extend from this surface to the rotated boundary plane as illustrated in figure 6.19(f). Since boundary movement and defect generation occur at room temperature with the electron microscope switched off these processes do not result from electron irradiation or beam heating.

Obtuse fault bends involving extrinsic and intrinsic faults, with a low-energy stair-rod dislocation at the bend, are also formed by moving boundaries as shown in figures 6.20(a)–(f). The misorientation between the grains for the boundary in figure 6.20 corresponds to $\theta = 49.3°$ and $\boldsymbol{u}$ approximately [211]. Figure 6.20(a) shows several defects, labelled 1–5, generated in the upper grain and intersecting the lower surface of the specimen, as illustrated schematically for one defect in figure 6.20(f). Shortly after recording the image in figure 6.20(a), the largest defect, 2, 'popped out' of the specimen. Tilting and contrast experiments showed that defects 1, 3, 4 and 5 are all of the same type and are obtuse bends involving the (111) and $(\bar{1}11)$ planes (planes PSR and PQR respectively of figure 6.20(f)) with a low-energy stair-rod dislocation along $[01\bar{1}]$ ($\overrightarrow{PR}$ of figure 6.20(f)). Figures 6.20(b)–(e) show two pairs of micrographs taken with diffracting vectors 200 and $\bar{2}00$ in the [013] beam direction for defect 1 (figures 6.20(b) and (c)) and defect 3 (figures 6.20(d) and (e)). A comparison of figures 6.20(b) and (c) shows that the outermost fringe for the (111) plane (PSR) changes from black to white on changing the diffracting vector from $\bar{2}00$ to 200, indicating a displacement across the fault plane of $(1/3)[\bar{1}\bar{1}\bar{1}]$ corresponding to an extrinsic fault. The outermost fringe for the $(\bar{1}11)$ plane (PQR) also changes from black to white for the same change in diffracting vector, indicating a displacement across the fault plane of $(1/3)[\bar{1}11]$ corresponding to an intrinsic fault. Similar behaviour is shown by defect 3 in figures 6.20(d) and (e). The edge character of the stair-rod dislocation along $\overrightarrow{PR}$ was demonstrated by the fact that the defects were completely out of contrast for the $02\bar{2}$ diffracting vector.

The observations on both types of fault bend are consistent with such defects being generated in small volumes of grain close to the line of intersection of the boundary with the specimen surface, and then developing by being extended in depth in the new portion of the growing grain as the boundary rotates.

† All indices for planes, directions and Burgers vectors are given with respect to the grain in which the defects are forming or have formed.

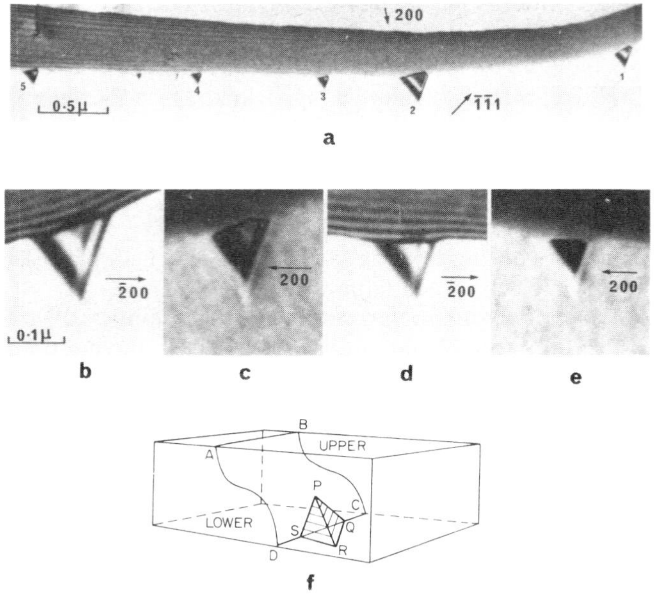

Figure 6.20 Generation of obtuse intrinsic–extrinsic stacking-fault bends by a moving grain boundary in 99.999% Cu. The diffracting vectors are indicated. In (*a*) ***B*** is close to [012] in the lower grain and close to [235] in the upper grain, in (*b*)–(*e*) ***B*** is close to [013] in the upper grain and (*f*) is a schematic representation of an intrinsic–extrinsic obtuse fault bend.

Figure 6.21 shows another portion of the boundary of figure 6.18 where the defects labelled 1, 2 and 3 joining the boundary to the top surface of the specimen were formed during boundary rotation and the trace, linking X, the tips of the defects and X′, marks the intersection of the original position of the boundary with the top surface of the specimen. It is clear from figure 6.21, and from other images of positions on this boundary where similar defects occur, that the formation of the defects results in a local retardation of boundary movement. Defects 1, 2 and 3 are intrinsic–intrinsic acute fault bends which are the same as those analysed previously and illustrated schematically in figure 6.19(*f*), but in the beam direction of figure 6.21 the $(1\bar{1}1)$ plane (PQR) is edge on. Defects 2 and 3 are fully attached to the boundary as illustrated schematically in figure 6.19(*f*), but in the case of defect 1 the $(\bar{1}11)$ plane is partially separated from the boundary as illustrated schematically in figure 6.22. Figure 6.21(*b*) shows a later stage in the movement of the boundary at room temperature. During this movement defects 2 and 3 have unfaulted on both planes and 'popped out' of the

along [011] in the lower grain. Complete detachment of the fault on $(1\bar{1}1)$ occurs when this reaction continues until the $(1/3)[1\bar{1}1]$ dislocation developing along [011] intersects the surface of the specimen leaving the $(1/2)[\bar{1}10]$ dislocation in the boundary.

The identification of the $(1/6)[1\bar{1}2]$ Shockley partial dislocation along $\overrightarrow{ST}$ which first extends the fault on the $(\bar{1}11)$ plane while attached to the boundary and moving with it, but then on leaving the boundary unfaults the $(\bar{1}11)$ plane when moving in the opposite direction, together with the identification of the stair-rod dislocation and the Frank dislocation segments along [110] and [011], provides good support for the proposed mechanism involving dissociation of $(1/2)\langle 110\rangle$ grain boundary dislocations. In a material with a stacking-fault energy as high as that of copper (41 ± 9 mJ m^{-2} (Cockayne *et al* 1971a)) extensive regions of stacking fault such as those generated in the faulted defects would not be expected from normal dislocation dissociation reactions within a grain. However, the mechanism relies on the dissociation of grain boundary dislocations into one of the neighbouring grains with the formation of a stair-rod dislocation, and will therefore necessarily involve the continued development of the stacking fault as long as the partial dislocations bordering the fault remain in the boundary and move with it.

Not all high-angle grain boundaries in thin-foil specimens migrate during observation in the electron microscope. For example, low-energy coherent-$\Sigma3$ twin-boundaries are stable against boundary movement, and it has been demonstrated that $\Sigma9$ symmetric-tilt boundaries on $\{114\}$ planes, and dissociated boundaries, are also stable against movement in electron microscope specimens (Forwood and Clarebrough 1985b, 1984). Figure 6.25 shows an example of the stability of $\Sigma9$ symmetric-tilt boundaries on $\{114\}$ planes

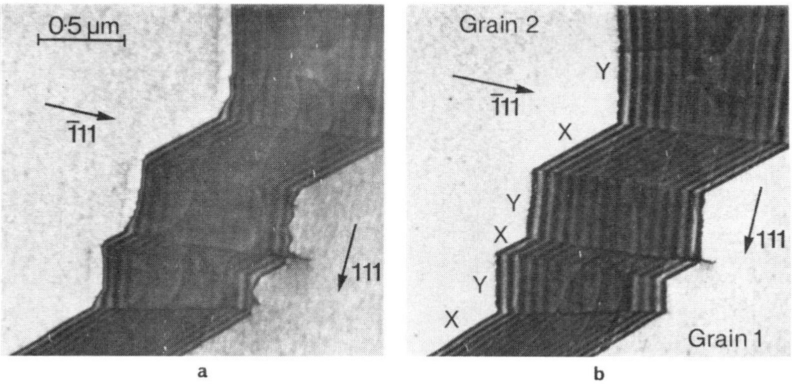

Figure 6.25 Double two-beam images (*a*) and (*b*) taken under the same diffracting conditions with ***B*** close to $[\bar{1}\bar{1}2]_1$; (*a*) was taken one week after (*b*) and the micrographs illustrate the stability of boundaries close to $(\bar{1}14)_1/(1\bar{1}4)_2$ (facets *X*) against migration in the thin-foil specimen.

and this example involves the X and Y facets of figure 5.47. The image in figure 6.25(a) was taken one week after that in (b) and it can be seen from the traces in the surfaces of the specimen in (a) that during this period the facets Y on $(\bar{2}\bar{1}3)_1/(\bar{5}\ \overline{22}\ 25)_2$ have moved and generated defects during

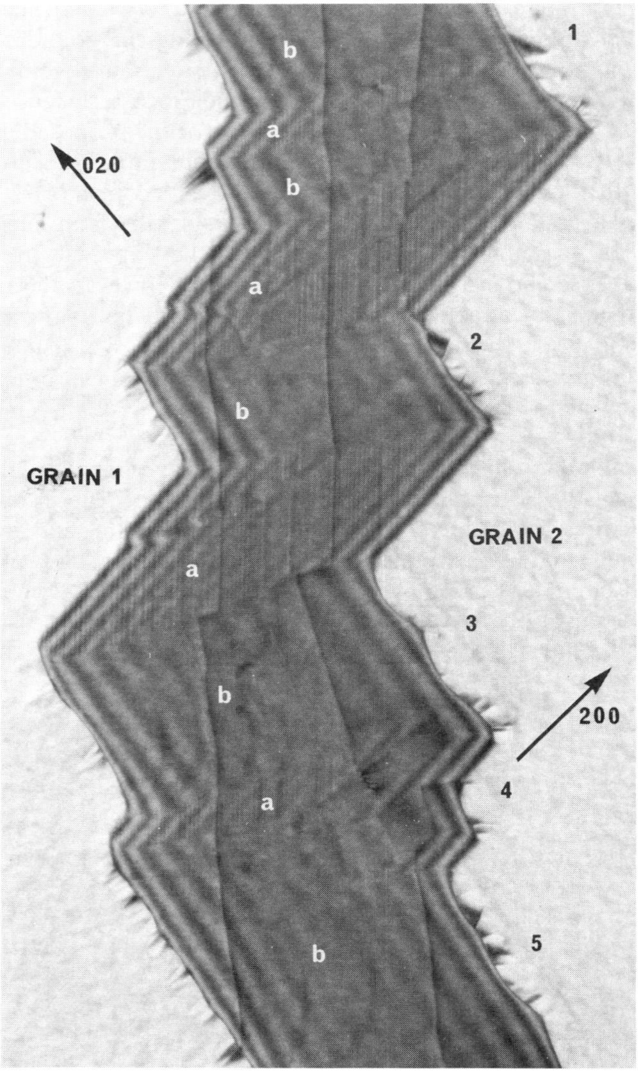

Figure 6.26 Double two-beam image of a near-$\Sigma 27_a$ boundary in a Cu − 6 at% Si alloy illustrating the stability of dissociated facets a against boundary migration. $\boldsymbol{B}$ is close to $[103]_1$ and $[013]_2$ and the diffracting vectors are indicated.

observation, whereas the symmetric-tilt facets X on $(\bar{1}14)_1/(1\bar{1}4)_2$ have not moved. Figure 6.26 is an example demonstrating how boundary dissociation stabilises a boundary against movement by introducing low-energy coherent-$\Sigma 3$ boundaries. The boundary in figure 6.26 is a faceted $\Sigma 27_a$ boundary in a Cu − 6 at% Si alloy with boundary facets a on $(1\bar{1}1)_1/(13\ \overline{13}\ 43)_2$ and b on $(513)_1$. The facets b have not dissociated and have moved at room temperature during observation and generated faulted defects at the surface intersections, while the facets a, which have dissociated to form coherent-$\Sigma 3$ boundaries and symmetric $\Sigma 81_d$ boundaries, have not moved during observation. The original intersections of facets b with one surface of the specimen are indicated by the traces marked 1–5 in grain 2 and similar traces in the other surface of the specimen are observed under different diffracting conditions. These results indicate that faceting of boundaries to form low-energy boundary planes, and that boundary dissociation, which always forms at least one low-energy coherent-$\Sigma 3$ boundary, will stabilise high-angle boundaries against migration during annealing.

6.4.2 Formation of Stacking-fault Tetrahedra by Moving Boundaries

Mechanisms for the formation of faulted defects and stacking-fault tetrahedra by moving boundaries during grain growth in bulk polycrystalline specimens are likely to differ in detail from those just discussed for the generation of defects in thinned electron microscope specimens where the surfaces of the specimen play a dominant role. The formation of stacking-fault tetrahedra during grain growth in bulk polycrystals has been investigated by Clarebrough and Forwood (1987b) in Cu − 6 at% Si, and it is shown that the secondary grain boundary dislocation structure plays an essential role in their formation. Tensile specimens of the Cu–Si alloy, strained 5% and annealed for 1 hour at 600 °C, were used and grain growth in these specimens gave a mean grain size of 20 μm. Following this annealing treatment most grains were free from defects, but some contained very large faulted defects which were sometimes attached and sometimes detached from boundaries.

Figure 6.27 shows three large faulted defects labelled 1, 2 and 3 which have been truncated by intersections with the specimen surfaces and which are attached to different portions of the same grain boundary which separates grains L and R. The misorientation between the grains corresponds to the rotation $\theta = 77.2°$ around an axis $\boldsymbol{u}$ close to $[1\bar{1}1]$ common to both grains and this misorientation is close to a $\Sigma 93_d$ CSL orientation. Image contrast showed that each of the defects consists of three intrinsically faulted planes on $(1\bar{1}1)_L$, $(\bar{1}\bar{1}1)_L$ and $(\bar{1}11)_L$ as indicated, which extend from the boundary into grain L with the junctions of the faulted planes in grain L being the appropriate low-energy $(1/6)\langle 110\rangle_L$ stair-rod dislocations. Defect 3 has been intersected by the specimen surfaces to a lesser extent than defects 1 and 2, and it can be seen that this defect has the character of a stacking-fault

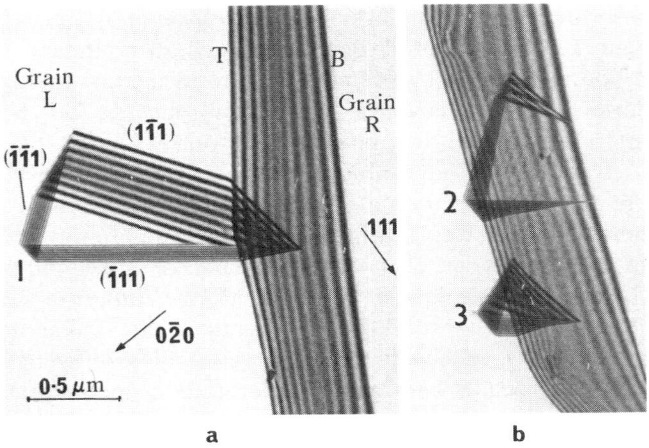

Figure 6.27 Double two-beam images of faulted defects 1, 2, and 3 attached to the same near-$\Sigma 93_d$ boundary. $\boldsymbol{B}$ is close to $[103]_L$ and $[\bar{1}\bar{3}4]_R$ and the diffracting vectors are indicated.

tetrahedron in which the grain boundary replaces the missing fourth faulted $(111)_L$ plane. Defects 1 and 2 have the same character as 3, but due to a difference in their size and position on the boundary, the three faulted planes are intersected to a different extent by the specimen surfaces. These defects have been dragged out in grain L by the movement of the grain boundary during grain growth.

Figure 6.28 shows two large faulted defects, detached from a boundary, involving the intrinsically faulted planes (111), $(1\bar{1}\bar{1})$ and $(1\bar{1}1)$ as indicated.

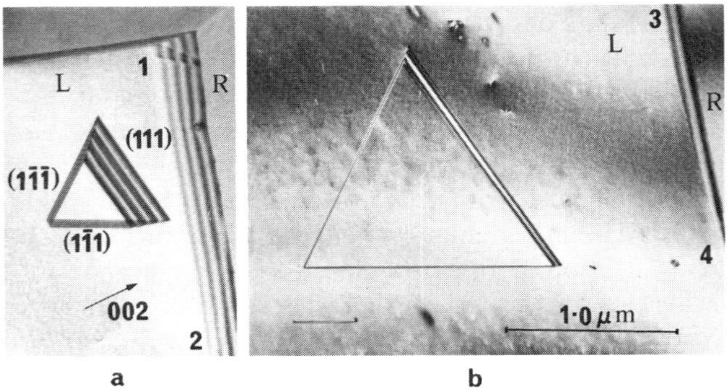

Figure 6.28 Images of large stacking-fault tetrahedra truncated by the surfaces of the thin-foil specimen. Both defects are in the same grain extending from 1, through 2 and 3, to 4. $\boldsymbol{B}$ is close to $[310]_L$ and the diffracting vector is indicated.

Both these defects are in the same grain bordered on the right by the same grain boundary, which starts at the triple junction 1 where the specimen is approximately $4\xi_{002}$ thick, and runs through 2 and 3 until the specimen is approximately $2\xi_{002}$ thick at 4. These defects have the form of stacking-fault tetrahedra, with edge lengths greater than 7800 Å for figure 6.28(*a*) and greater than 16 000 Å for figure 6.28(*b*), which have been truncated by intersection with the specimen surfaces. The formation of such large defects would not be expected from calculations of the equilibrium between Frank dislocation loops and stacking-fault tetrahedra since, for a stacking-fault energy of the alloy of 14.2 mJ m^{-2} (Nordstrom and Barrett 1969), the largest tetrahedron to be expected from such calculations would be only 1840 Å (Humble and Forwood 1968a, b). The observation of such large tetrahedra implies that they have arisen from three-fault-plane defects of the type shown in figure 6.27, which have been extended by grain boundary movement, and have eventually become detached from the boundary by the generation of a fourth faulted plane.

Defects of the type shown in figure 6.27 have been observed attached to many different grain boundaries over a wide range of misorientations between neighbouring grains. In all cases, the misorientations were close to those CSL orientations for which the associated DSC lattices always contained a vector of the type $(1/6)\langle 112\rangle$, so that each of the boundaries could contain a secondary grain boundary dislocation with a Burgers vector of the type $(1/6)\langle 112\rangle$. The near-CSL orientations studied for which defects were observed to be attached to boundaries were $\Sigma 9$, $\Sigma 21_a$, $\Sigma 27_a$, $\Sigma 27_b$ and $\Sigma 93_d$ (the example of figure 6.27).

Figure 6.29 shows two double two-beam images of a near-$\Sigma 27_a$ boundary where grain U is the upper grain with respect to the electron source and grain L is the lower. The misorientation between the grains is such that grain U is rotated in a right-handed sense with respect to grain L around the axis [011] common to both grains by an angle very close to 31.59° which corresponds to the exact-$\Sigma 27_a$ CSL orientation. The three coarsely spaced systems of secondary grain boundary dislocations labelled *a*, *b* and *c* in figure 6.29(*b*) give rise to a very small angular departure from this exact-CSL orientation. The Burgers vectors of the secondary grain boundary dislocations were found by image matching to have the DSC Burgers vectors $\boldsymbol{a} = (1/6)[\bar{1}1\bar{2}]_U$, $\boldsymbol{b} = (1/27)[511]_U$ and $\boldsymbol{c} = (1/54)[2\bar{5}\bar{5}]_U$. The identification of $\boldsymbol{c}$ was not as definite as the identification of $\boldsymbol{a}$ and $\boldsymbol{b}$, because the contrast of the grain boundary dislocations of system *c* was always weak in images taken over a wide range of diffraction conditions. However, the vector $\boldsymbol{c} = (1/54)[2\bar{5}\bar{5}]_U$ was the only DSC vector for which theoretical images gave weak contrast compatible with the experimental images.

In the region X–X′ of the boundary in figure 6.29 the boundary plane normal is close to $(2\bar{2}3)_U$, two faulted defects extend from the boundary into the upper grain and both defects have been truncated by the specimen surfaces.

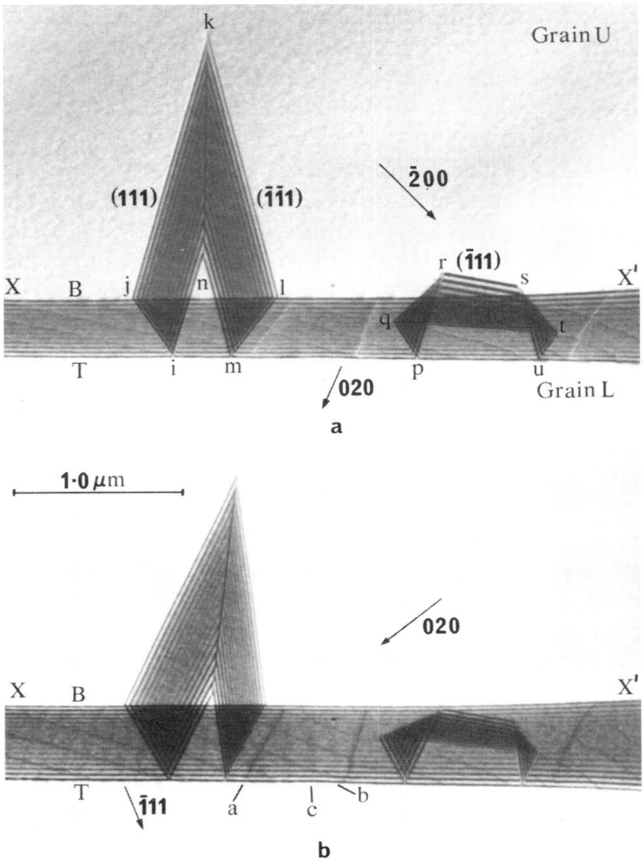

Figure 6.29 Double two-beam images of faulted defects attached to a near-$\Sigma27_a$ boundary XX'. ***B*** in (*a*) is close to $[0\ 1\ 11]_U/[205]_L$ and in (*b*) to $[103]_U/[314]_L$. The diffracting vectors are indicated.

The defect on the left consists of intrinsic faults on $(111)_U$ and $(\bar{1}\bar{1}1)_U$ which intersect the boundary along *ji* and *lm* respectively, and are joined along *kn* by a low-energy stair-rod dislocation with Burgers vector $(1/6)[\bar{1}\bar{1}0]_U$. The defect on the right consists of three intrinsically faulted planes, *pqr* on $(111)_U$, *stu* on $(\bar{1}\bar{1}1)_U$ and *qrst* on $(\bar{1}11)_U$. This defect was chosen to establish the relationship between the secondary grain boundary dislocation structure and the formation of faulted defects by determining the magnitude and sense of the Burgers vectors of the dislocations associated with it, i.e. the stair-rod dislocations along *qr* and *ts* and the dislocations along the lines of intersection *qp*, *tu* and *qt* of the faulted planes with the boundary.

The simplest dislocations in the defect are the stair-rod dislocations at the junctions of the intrinsic faults. Figure 6.30 is an example of the comparison between experimental and theoretical images which identified the stair-rod dislocation along $\overrightarrow{qr}$ as $(1/6)[0\bar{1}\bar{1}]_U$ and that along $\overrightarrow{ts}$ as $(1/6)[\bar{1}01]_U$. To identify the Burgers vectors of the dislocations along the lines of intersection of the faulted planes with the boundary, images for which values of $\boldsymbol{g}_U \cdot \boldsymbol{R}$ were integral, for the fault plane in question, were always used. This was necessary to ensure that the dislocation images were not confused by fault fringes and fringes which occur along that line of intersection by the interaction between stacking-fault fringes and pendellösung fringes from the boundary. Double two-beam images for the line of intersection *qt* of the $(\bar{1}11)_U$ plane with the boundary were taken, for various values of the deviation parameter *w*, with the combinations of diffracting vector shown in table 6.1 for which $\boldsymbol{g}_U \cdot \boldsymbol{R}$ is integral. In none of these images was any diffraction contrast observed along *qt*, which indicates that there was no dislocation strain field along this direction arising from any of the $(1/6)\langle 112 \rangle$ Burgers vectors on $(\bar{1}11)_U$, nor any combination of these vectors with DSC Burgers vectors of the secondary grain boundary dislocation structure. On this basis,

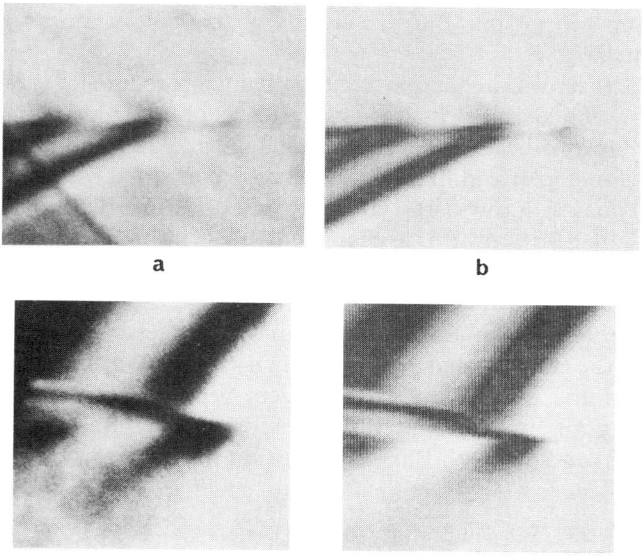

a b

c d

Figure 6.30 Comparison of experimental images (*a*) and (*c*) with corresponding matching computed images (*b*) and (*d*) for the stair-rod dislocations with Burgers vectors $(1/6)[0\bar{1}\bar{1}]_U$ and $(1/6)[\bar{1}01]_U$ along *qr* and *ts* respectively. The diffracting vectors and beam directions are: 220_U and $[\bar{1}\ 1\ 18]_U$ in (*a*) and (*b*) and $13\bar{1}_U$ and $[2\ 7\ 19]_U$ in (*c*) and (*d*).

Table 6.1 Diffraction vectors for images of the line of intersection qt of the $(\bar{1}11)$ plane with the boundary of figure 6.29.

$\boldsymbol{g}_U$	220	220	220	220	$31\bar{1}$	$\bar{3}\bar{1}1$	$31\bar{1}$	$3\bar{1}1$	$\bar{3}1\bar{1}$	$\bar{1}3\bar{1}$	$\bar{1}3\bar{1}$
$\boldsymbol{g}_L$	020	$\bar{3}11$	$1\bar{3}\bar{1}$	$\bar{1}31$	$1\bar{1}\bar{1}$	$\bar{1}31$	$0\bar{2}0$	200	131	$11\bar{1}$	$\bar{1}\bar{1}1$

it was assumed that the faulting on $(\bar{1}11)_U$ was initiated during grain growth by a segment of a secondary grain boundary dislocation, with Burgers vector $\boldsymbol{a} = (1/6)[\bar{1}1\bar{2}]_U$, moving out of the boundary into grain U on the $(\bar{1}11)_U$ slip plane. Such a partial dislocation in grain U could cross-slip on to $(111)_U$ and $(\bar{1}\bar{1}1)_U$ forming the identified stair-rod dislocations along $\overrightarrow{qr}$ and $\overrightarrow{ts}$ and generating the intrinsic faults on the planes qpr and stu. Such cross-slip reactions would involve the formation of partial dislocations with Burgers vectors $(1/6)[\bar{1}2\bar{1}]_U$ and $(1/6)[2\bar{1}1]_U$ intersecting the boundary along $\overrightarrow{qp}$ and $\overrightarrow{tu}$ respectively. However, the diffraction contrast observed along these lines of intersection, for a wide range of diffraction conditions, was always very weak, indicating that the dislocations which have cross-slipped back into the boundary have dissociated into secondary grain boundary dislocations to leave a residual Burgers vector of smaller magnitude along the lines of intersection. All possible Burgers vectors of this type were investigated by image matching.

Figure 6.31 shows an example of the matching procedure for the dislocation along $\overrightarrow{qp}$, which identifies its Burgers vector as $(1/54)[4\bar{1}\bar{1}]_U$. In figure 6.31 three double two-beam experimental images, (*a*), (*b*) and (*c*) in row (i), involving non-coplanar diffracting vectors, are compared with corresponding computed images in rows (ii)–(vi) for different possible Burgers vectors. The computed images in row (vi) are for the $(1/6)[\bar{1}2\bar{1}]_U$ Burgers vector, formed by the cross-slip reaction, and clearly do not match the experimental images. The only theoretical images which do match the experimental images are those in row (ii), which are computed for the Burgers vector $(1/54)[4\bar{1}\bar{1}]_U$. This Burgers vector is that of the residual dislocation resulting from dissociation of the dislocation with Burgers vector $(1/6)[\bar{1}2\bar{1}]_U$ into three secondary grain boundary dislocations, two with Burgers vector $\boldsymbol{c} = (1/54)[\bar{2}55]_U$ and one with Burgers vector $\boldsymbol{a} = (1/6)[\bar{1}1\bar{2}]_U$ according to the reaction

$$(1/6)[\bar{1}2\bar{1}]_U = (1/54)[4\bar{1}\bar{1}]_U + [(1/6)[\bar{1}1\bar{2}]_U + (1/54)[\bar{2}55]_U + (1/54)[\bar{2}55]_U]. \tag{6.13}$$

This residual dislocation has the smallest Burgers vector that can result from the dissociation of the Burgers vector $(1/6)[\bar{1}2\bar{1}]_U$ into secondary grain boundary dislocations. The next smallest Burgers vector, $(1/54)[244]_U$, gives the mismatching images in row (iii), and the next smallest again, $(1/54)[\bar{6}\bar{3}\bar{3}]_U$,

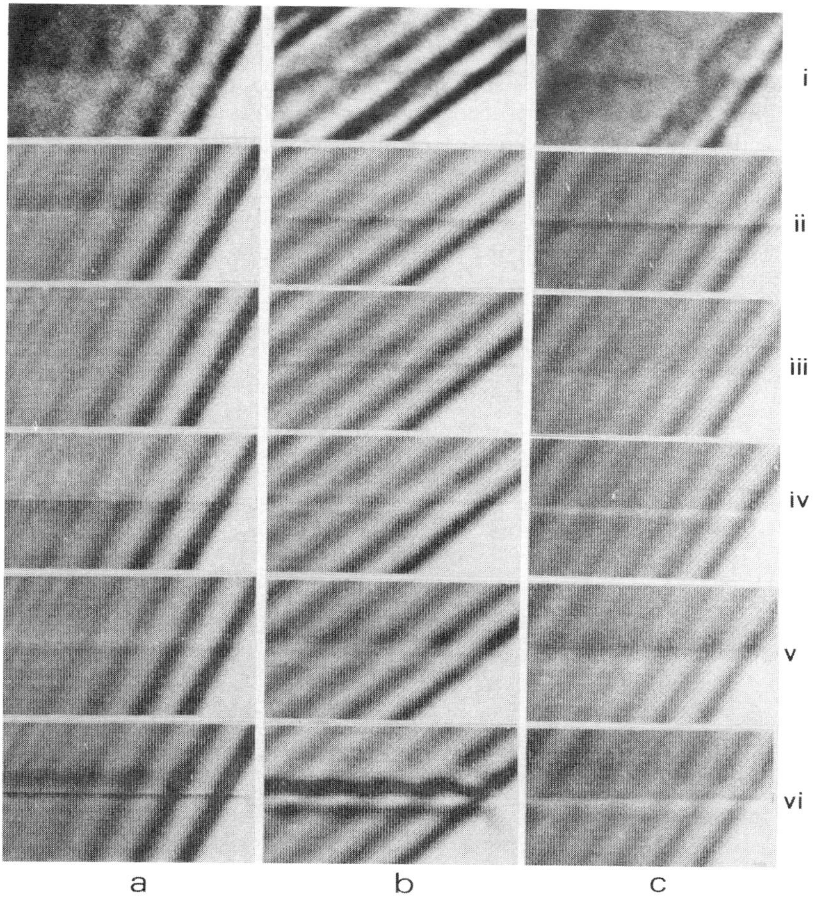

Figure 6.31 Comparison of double two-beam experimental images (*a*), (*b*) and (*c*) in row (i) with corresponding computed images in rows (ii)–(vi) computed for different possible Burgers vectors for the residual dislocations along *qp* of figure 6.29. The diffracting vectors in grains U and L, and beam directions, ***B***, in grain U for the columns (*a*), (*b*) and (*c*) are: $31\bar{1}_U$, $1\bar{1}1_L$, $[5\ 1\ 16]_U$; $13\bar{1}_U$, $\bar{3}11_L$, $[\bar{1}38]_U$; $\bar{3}\bar{1}1_U$, $\bar{1}31_L$, $[217]_U$ respectively.

gives the mismatching images in row (iv). The mismatching images in row (v) are for the Burgers vector $(1/6)[011]_U$ which could have resulted from the simplest dissociation reaction

$$(1/6)[\bar{1}2\bar{1}]_U = (1/6)[011]_U + (1/6)[\bar{1}1\bar{2}]_U. \qquad (6.14)$$

A similar procedure was adopted for the determination of the Burgers vector of the dislocation along $\overrightarrow{tu}$. This involved image matching for the possible residual Burgers vectors left in the boundary along $\overrightarrow{tu}$ after various dissociation reactions into secondary grain boundary dislocations of the dislocation formed by cross-slip with Burgers vector $(1/6)[2\bar{1}1]_U$. Two possibilities for the Burgers vector of the residual dislocation along $\overrightarrow{tu}$ gave equally good matching between experimental and computed images, as shown in figure 6.32. These Burgers vectors were $(1/54)[\bar{5}8\bar{1}]_U$ formed by the dissociation reaction

$$(1/6)[2\bar{1}1]_U = (1/54)[\bar{5}8\bar{1}]_U + [(1/6)[1\bar{1}2]_U + (1/27)[511]_U + (1/54)[2\bar{5}\bar{5}]_U + (1/54)[2\bar{5}\bar{5}]_U] \tag{6.15}$$

and $(1/54)[\bar{3}3\bar{6}]_U$ formed by the dissociation reaction

$$(1/6)[2\bar{1}1]_U = (1/54)[\bar{3}3\bar{6}]_U + [(1/6)[1\bar{1}2]_U + (1/27)[511]_U + (1/54)[2\bar{5}\bar{5}]_U] \tag{6.16}$$

and no distinction could be made between these two possible dissociation reactions.

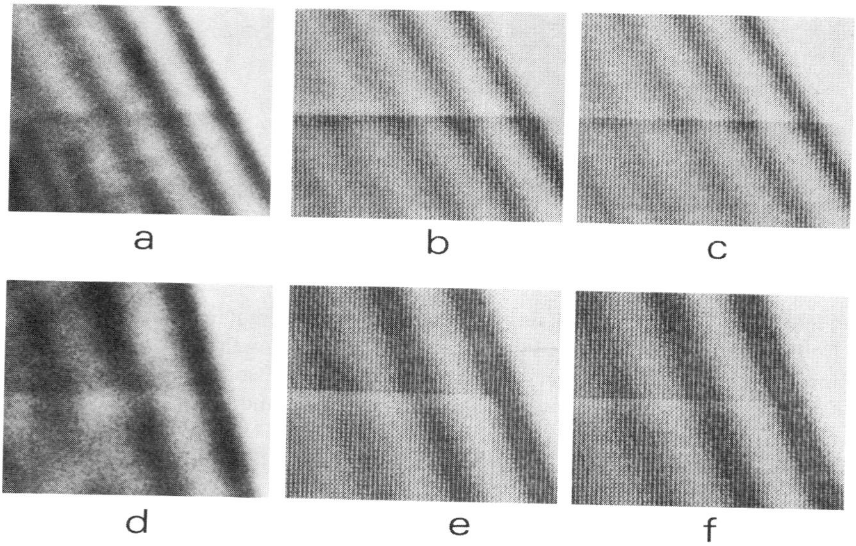

Figure 6.32 Comparison of double two-beam experimental images (*a*) and (*d*) with corresponding computed images (*b*), (*c*) and (*e*), (*f*) respectively, for two possible Burgers vectors of the residual dislocation along *tu* of figure 6.29. The computed images in (*b*) and (*e*) are for the Burgers vector $(1/54)[\bar{5}8\bar{1}]_U$ and in (*c*) and (*f*) for the Burgers vector $(1/54)[\bar{3}3\bar{6}]_U$. The diffracting vectors in grains U and L and the beam directions ***B*** in grain U for (*a*), (*b*) and (*c*) are $2\bar{2}0_U$, $0\bar{2}0_L$ and $[117]_U$ and for (*d*), (*e*) and (*f*) are $13\bar{1}_U$, $11\bar{1}_L$ and $[138]_U$.

The contrast of the dislocations along $\overrightarrow{ji}$ and $\overrightarrow{lm}$ for the defect *ijklmn* in figure 6.29 was always the same as that along $\overrightarrow{qp}$ and $\overrightarrow{tu}$ respectively. Thus the dislocations along $\overrightarrow{ji}$ and $\overrightarrow{lm}$ have the same Burgers vectors as those along $\overrightarrow{qp}$ (i.e. $(1/54)[4\bar{1}\bar{1}]_U$) and $\overrightarrow{tu}$ (i.e. $(1/54)[\bar{5}8\bar{1}]_U$ or $(1/54)[\bar{3}3\bar{6}]_U$) respectively. The defect *ijklmn* can therefore be considered as a larger example of the defect *pqrstu*, but in this case the $(\bar{1}11)_U$ plane has been removed by intersection of the defect with the lower surface of the specimen during thinning.

The dissociation reactions along $\overrightarrow{qp}$ and $\overrightarrow{tu}$ will be discussed with reference to the schematic diagram of figure 6.33, which has been drawn for reactions (6.13) and (6.15) rather than (6.13) and (6.16). Figure 6.33 shows the defect *pqrstu*, where the light lines indicate the intersections of the boundary and the faulted planes with the specimen surfaces as they appear in figure 6.29. If the defect were not truncated by the specimen surfaces, the faulted planes would extend as indicated by the dashed lines to include the stair-rod dislocation with Burgers vector $(1/6)[\bar{1}\bar{1}0]_U$ along $\overrightarrow{wv}$. There is no dislocation joining *q* and *t* and the dotted line merely indicates the termination at the boundary of the fault on $(\bar{1}11)_U$, which has been generated by the movement of a secondary grain boundary dislocation with Burgers vector $\boldsymbol{a} = (1/6)[\bar{1}1\bar{2}]_U$ into grain U. With reference to dissociation reactions (6.13) and (6.15) along $\overrightarrow{qp}$ and $\overrightarrow{tu}$ respectively, the common dissociation products involving two grain boundary dislocations with Burgers vector $(1/54)[\bar{2}55]_U$ and one grain boundary dislocation with Burgers vector $(1/6)[\bar{1}1\bar{2}]_U$ form loops in the boundary between nodal points *q* and *t*, and the additional grain boundary dislocation with Burgers vector $(1/27)[511]_U$, resulting from reaction (6.15),

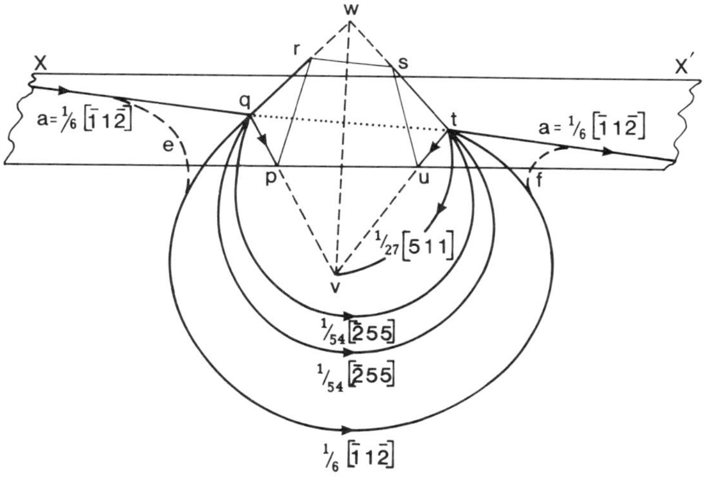

Figure 6.33 Schematic diagram illustrating the dissociation reactions in the boundary along *qp* and *tu* of figure 6.29.

terminates in the boundary at the nodal points t and v. Since the $(1/6)[\bar{1}1\bar{2}]_U$ Burgers vector of one of the dissociation products is the same as that of the original secondary grain boundary dislocation with Burgers vector $\boldsymbol{a}$, dislocations with the same Burgers vector enter and leave the nodal points at q and t. Therefore a continuous length of grain boundary dislocation with Burgers vector $\boldsymbol{a} = (1/6)[\bar{1}1\bar{2}]_U$ is free to move away in the boundary during grain growth, as indicated by the dashed lines e and f in figure 6.33. Thus, this dislocation would not be expected to be observed at the nodal points q and t in figure 6.27. However, the other grain boundary dislocations resulting from the dissociation reactions should be observed at the nodal points q and t, namely an additional grain boundary dislocation of system b with $\boldsymbol{b} = (1/27)[511]_U$ terminating at t, and two additional grain boundary dislocations of system c with $\boldsymbol{c} = (1/54)[2\bar{5}\bar{5}]_U$ terminating at q and t. The image in figure 6.34 shows the termination at t of an additional secondary grain boundary dislocation of system b as expected, but no definitive evidence was obtained for the termination of two grain boundary dislocations of system c at either q or t. The failure to observe the additional grain boundary dislocations of system c may be due to the weak contrast which is always associated with this system. If the operative dissociation reaction along $\overrightarrow{tu}$ had been reaction (6.16) rather than reaction (6.15) a similar arrangement of dissociation products would result, but in this case only one grain boundary dislocation of system c would be expected to terminate at the nodal point t.

This analysis of the defect $pqrstu$ of figure 6.29 shows that the residual grain boundary dislocations in the boundary along $\overrightarrow{qp}$ and $\overrightarrow{tu}$ have Burgers vectors which are not DSC vectors. However, for the mechanism proposed, these dislocations cannot be the usual partial grain boundary dislocations (which separate regions of different rigid-body displacement between the grains at a boundary), because the only displacements that occur in the boundary in the formation of these residual dislocations, are associated with dissociation products which all have DSC Burgers vectors. Thus if images were taken with same-$\boldsymbol{g}_c$ diffracting vectors in both grains no change in fringe

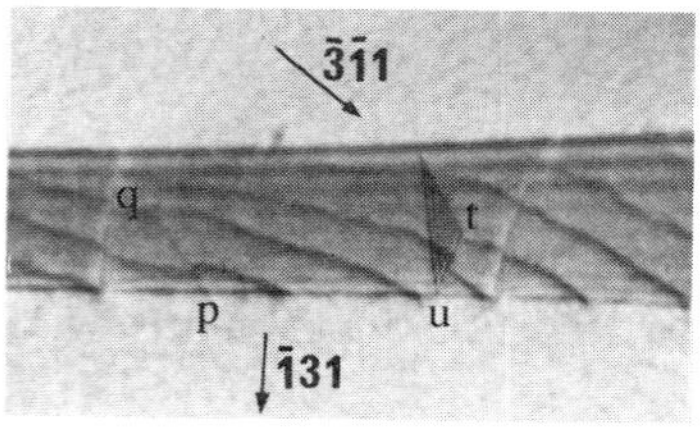

Figure 6.34 Double two-beam image of a portion of the boundary XX' of figure 6.29 showing an additional grain boundary dislocation of system b terminating at t. The diffracting vectors are indicated and $\boldsymbol{B}$ is close to $[217]_U/[10\ \bar{1}\ 13]_L$.

contrast should occur across $\overrightarrow{pq}$ and $\overrightarrow{tu}$. It was not possible to check this because three suitable non-coplanar same-$\boldsymbol{g}_c$ are not available for a $\Sigma 27_a$ boundary. However, for a near-Σ9 boundary three non-coplanar same-$\boldsymbol{g}_c$ are available, and it was shown, for the same type of faulted defect attached to such a boundary, that there was no difference in fringe contrast across residual dislocations similar to those at *pq* and *tu*. The defect examined is labelled *jklmn* in figure 6.35(*a*), where the planes *jkl*, *nlm* and *klm* are $(\bar{1}\bar{1}1)_R$, $(111)_R$ and $(1\bar{1}1)_R$ respectively. The misorientation between the grains is close to the exact-Σ9 CSL orientation corresponding to a rotation of 67.11° around the $[3\bar{1}1]$ axis common to both grains. The same-$\boldsymbol{g}_c$ diffracting vectors used were $3\bar{1}1_L/3\bar{1}1_R$, $024_L/240_R$ and $\bar{2}04_L/042_R$. For none of these same-$\boldsymbol{g}_c$ vectors was any significant difference in displacement fringe contrast found across *kj* or *mn*, and the result for the $3\bar{1}1_L/3\bar{1}1_R$ same-$\boldsymbol{g}_c$ diffracting vector is shown in figure 6.35(*b*). This observation, therefore, supports the mechanism given for the formation of faulted defects at grain boundaries during grain growth.

During grain growth the secondary grain boundary dislocation structure in high-angle grain boundaries will be undergoing continuous rearrangement. For grain boundaries containing secondary grain boundary dislocations with Burgers vectors of the type $(1/6)\langle 112\rangle$, the rearrangement of the secondary grain boundary dislocation structure, in the model, involves the alignment of one of these grain boundary dislocations so that it moves out of the boundary on a $\{111\}$ slip plane into the growing grain. This dislocation then cross-slips on to two other $\{111\}$ planes, moves back into the boundary, and dissociates to leave low-energy residual dislocations in the boundary. The

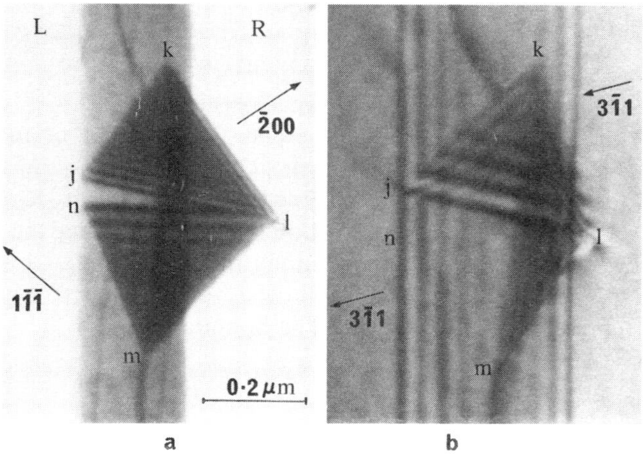

Figure 6.35 Double two-beam images of a faulted defect attached to a near-Σ9 boundary. The diffracting vectors are indicated and $\boldsymbol{B}$ is close to $[\bar{1}43]_L/[015]_R$ in (*a*) and $[\bar{4}75]_L/[\bar{1}14]_R$ in (*b*).

presence of such dislocations implies that the defect as a whole would be stable and would continue to be dragged out during grain growth by climb of the residual dislocations in the boundary plane and extension of the three stacking faults in the growing grain. However, dissociation of $(1/6)\langle 112\rangle$ dislocations to form low-energy residual dislocations in the boundary is not a necessary condition for the dragging-out of faulted defects. Such a defect can be dragged out by a moving boundary as soon as a portion of the grain boundary dislocation which has left the boundary cross-slips to form a stair-rod dislocation so as to lock the defect in the growing grain.

In order for a defect involving three faulted $\{111\}$ planes intersecting a boundary to become detached to form a stacking-fault tetrahedron, the boundary plane during grain growth would first need to become oriented parallel to the fourth $\{111\}$ plane required to close the tetrahedron. If this happens before the $(1/6)\langle 112\rangle$ dislocations formed by cross-slip have dissociated in the boundary, then detachment can take place by one of these $(1/6)\langle 112\rangle$ dislocations dissociating to form a $(1/6)\langle 110\rangle$ dislocation and a $(1/3)\langle 111\rangle$ dislocation. For example, for the defect *wqvt* illustrated in figure 6.33, if the dislocation along $\overrightarrow{qv}$ undergoes the dissociation

$$(1/6)[\bar{1}2\bar{1}]_U = (1/3)[\bar{1}1\bar{1}]_U + (1/6)[101]_U \tag{6.17}$$

then the dislocation with the Burgers vector $(1/3)[\bar{1}1\bar{1}]_U$ could climb in the boundary plane, react along tv to give the $(1/6)[01\bar{1}]_U$ stair-rod dislocation, that is

$$(1/3)[\bar{1}1\bar{1}]_U + (1/6)[2\bar{1}1]_U = (1/6)[01\bar{1}]_U \tag{6.18}$$

and dissociate along $\overrightarrow{qt}$ to give the $(1/6)[\bar{1}10]_U$ stair-rod dislocation, and reconstitute the secondary grain boundary dislocation $(1/6)[\bar{1}1\bar{2}]_U$ according to the reaction

$$(1/3)[\bar{1}1\bar{1}]_U = (1/6)[\bar{1}10]_U + (1/6)[\bar{1}1\bar{2}]_U \tag{6.19}$$

thereby detaching the tetrahedron from the boundary.

It is possible for detachment to take place without the boundary plane becoming oriented parallel to the fourth $\{111\}$ plane. This can occur by a $(1/6)\langle 112\rangle$ dislocation, formed by cross-slip, expanding to reach a nodal point such as q or t (figure 6.33) without fully intersecting the boundary, and a portion of it becoming aligned along the appropriate $\langle 110\rangle$, so that a reaction such as (6.17) can occur, followed by reactions (6.18) and (6.19), to leave a tetrahedron detached from the boundary.

6.5 INTERACTION OF SLIP DISLOCATIONS WITH GRAIN BOUNDARIES AND TRANSFER OF SLIP ACROSS BOUNDARIES

High-angle grain boundaries present strong barriers to the movement of slip dislocations from one grain to another during plastic deformation of polycrystals, and this leads to the well known dependence in metals and alloys of yield stress on grain size, i.e. the yield stress is proportional to (grain size)$^{-1/2}$ (see, for example, McLean 1962). When a polycrystal responds plastically to an applied stress, the plastic strain in each grain is usually accommodated at the grain boundaries without loss of cohesion. In order to understand how this occurs, it is necessary to consider (i) how the boundary itself responds to an incoming slip dislocation and (ii) the mechanisms which enable slip to take place between neighbouring grains across the boundary. When a slip dislocation enters a high-angle boundary it can interact with and modify the secondary grain boundary structure. For slip to transfer across a boundary from one grain to the other a number of different mechanisms can be involved, and these depend on the misorientation between the grains and the plane of the boundary. In this section, the interaction of slip dislocations with secondary grain boundary dislocations will be treated first and this will be followed by a discussion of mechanisms for slip transfer.†

6.5.1 Interaction of Slip Dislocations with Secondary Grain Boundary Dislocations

On the CSL model which describes the dislocation structure of high-angle grain boundaries in terms of secondary grain boundary dislocations with DSC Burgers vectors, slip dislocations with Burgers vectors which are lattice vectors (and therefore also vectors of the DSC lattice) can be considered to interact with a boundary structure in the following way. A slip dislocation from one of the grains on forming a line of intersection with the grain boundary can lower its energy by dissociating into secondary grain boundary dislocations with smaller DSC Burgers vectors. Depending on their Burgers vectors, these new dislocations can interact with the existing secondary grain boundary dislocation structure to add dislocations to it, or remove dislocations from it. These processes will involve glide and climb of grain boundary dislocations depending on whether their Burgers vectors lie in or out of the plane of the boundary respectively.

Schober and Balluffi (1971c) were the first to observe interactions between lattice dislocations and secondary grain boundary dislocation structure using (001) twist boundaries in fabricated thin-film bicrystals of gold and, using the same type of specimens, Darby *et al* (1978) extended these observations to a wider range of twist boundaries and to several tilt boundaries. In these experiments not all the dislocations that were expected from the dissociation of lattice dislocations in the boundaries were observed, this being attributed to some of the DSC Burgers vectors involved being too small for dislocations

† In addition to work on metals and alloys, interaction of slip dislocations with grain boundaries has been studied in $\Sigma 9$ symmetric-tilt boundaries in silicon by Baillin *et al* (1987) using two-beam microscopy and by Elkajbaji and Thibault-Desseaux (1988) using *n*-beam lattice images.

with these Burgers vectors to be detected by strain contrast in the electron microscope.

Following the early work of Schober and Balluffi, Bollmann *et al* (1972) published their now famous electron micrograph showing the dissociation of a $(1/2)\langle 110\rangle$ lattice dislocation into five grain boundary dislocations in a near-$\Sigma 29_a$ grain boundary in a polycrystalline specimen of stainless steel. Although the Burgers vectors of these dislocations were not determined, the observation was consistent with the proposed dissociation reaction

$$(1/2)[110] = 3[(1/58)[370]] + 2[(1/58)[10\ 4\ 0]]$$

where $(1/58)[370]$ and $(1/58)[10\ 4\ 0]$ are DSC vectors corresponding to the $\Sigma 29_a$ CSL orientation.

Kegg *et al* (1973) and Horton *et al* (1974) studied the interaction of slip dislocations with $\Sigma 9$ and $\Sigma 27_a$ boundaries during creep of specially prepared bicrystals of aluminium. They reported dissociation reactions of the type

$$(1/2)[011] = 2[(1/4)[011]]$$

but positive identification of the non-DSC vector $(1/4)[011]$ was not made.

Pond and Smith (1977) and Dingley and Pond (1979) examined the interaction of a number of slip dislocations with a near-$\Sigma 41_a$ boundary in polycrystalline aluminium. They reported dissociation reactions of the type

$$(1/2)[0\bar{1}1] = (1/82)[4\ \bar{5}\ 41] + 4[(1/82)[\bar{1}\bar{9}0]]$$

and

$$(1/2)[110] = (1/82)[41\ \bar{4}\ \bar{5}] + 5[(1/82)[091]]$$

but not all the components formed by these proposed dissociation reactions were detected. Pond and Smith (1977) reported the dissociation of $(1/2)\langle 110\rangle$ dislocations, in an incoherent-$\Sigma 3$ boundary in polycrystalline aluminium on a $(1\bar{2}1)$ plane, into pairs of $(1/6)\langle 112\rangle$ dislocations according to the reaction

$$(1/2)[10\bar{1}] = (1/6)[2\bar{1}\bar{1}] + (1/6)[11\bar{2}].$$

This dissociation process involves climb of the $(1/6)\langle 112\rangle$ dislocations in the boundary, and was observed to have taken place in the thin-foil specimen, after one month at room temperature, to give a separation of the $(1/6)\langle 112\rangle$ dislocations of approximately 2000 Å at one of the intersections of the boundary plane with the surface of the specimen.

As discussed in section 5.4.1, Pond (1977) identified two different states of rigid-body displacement, α_1 and α_2, on $\{112\}$ incoherent-$\Sigma 3$ boundaries in aluminium, which have the same energy and are separated by a partial grain boundary dislocation. During hot-stage microscopy, he observed a $(1/2)\langle 110\rangle$ lattice dislocation on a $(11\bar{1})$ slip plane move into a $(1\bar{2}1)$

incoherent-$\Sigma 3$ twin-boundary and, after this event, he observed partial dislocations separating α_1 and α_2 states. He attributed the formation of these partial dislocations to the following specific reactions. Firstly he considered the $(1/2)\langle 110\rangle$ slip dislocation to have dissociated into DSC vectors according to the reaction

$$(1/2)[101] = (1/3)[111] + (1/6)[1\bar{2}1]$$

and then into partial dislocations, which have non-DSC vectors, according to the reaction

$$(1/3)[111] = (1/9)[111] + (2/9)[111].$$

As he pointed out, these two partial dislocations are glissile in the plane of the boundary and can move apart unheeded under the influence of the repulsive force between them, since the α_1 and α_2 states have the same grain boundary energy.

The interaction of a slip dislocation with the secondary grain boundary dislocation structure of a near-$\Sigma 27_a$ boundary in a Cu – 6 at% Si alloy has been studied in detail by the authors. The secondary grain boundary dislocation structure for this boundary was analysed in section 5.3.2 (i) and is shown in figure 5.27. The changes in this boundary structure resulting from interaction with a slip dislocation are shown in figures 6.36(*a*)–(*f*). The region of the boundary in figure 6.36 almost borders that in figure 5.27 and the unperturbed array of secondary grain boundary dislocations in this part of the boundary is illustrated schematically in figure 6.36(*b*) where the secondary grain boundary dislocations are labelled *A*, *B*, *C*, *D* as in figure 5.27(*b*). In this region of the boundary dislocations *D* run parallel and close to dislocations *A* over short lengths where the two dislocations cross. Figure 6.36(*a*) shows a slip dislocation *S* in the lower grain, grain 2, on the $(\bar{1}\bar{1}1)_2$ slip plane for which the Burgers vector was determined by image matching as $\boldsymbol{b}_S = (1/2)[0\bar{1}\bar{1}]_2$. This dislocation extends from the bottom surface of the specimen and intersects the boundary near the line of intersection of the boundary with the top surface of the specimen. It can be seen from figure 6.36(*a*) that the regular secondary grain boundary dislocation structure is altered by the presence of the slip dislocation. The details of the changes in the secondary grain boundary dislocation structure are illustrated schematically in figure 6.36(*c*) and shown in the higher magnification images in figures 6.36(*d*)–(*f*). It can be seen from figures 6.36(*d*)–(*f*) that the segments of the secondary grain boundary dislocation network labelled *A*, *B* and *C* in figure 6.36(*c*) and the slip dislocation terminate at the node *n*, and that a segment *D* also runs into the node *n* in combination with a segment *A*. The magnitude and sense of all the Burgers vectors of the dislocations terminating at the node have been determined by image matching (see section 5.3.2 (i)) and are specified for the line directions from the bottom to the top

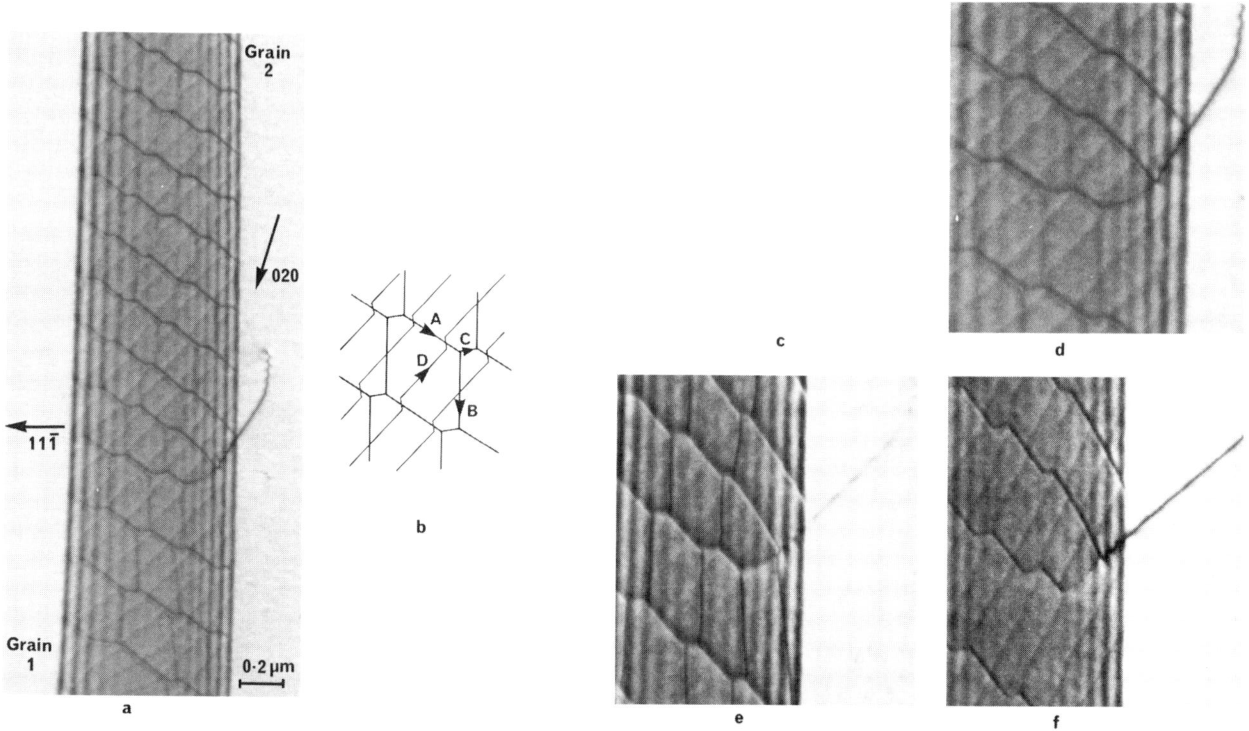

Figure 6.36 Double two-beam images (*a*), (*d*), (*e*) and (*f*) and schematic diagrams (*b*) and (*c*) illustrating the interaction of a slip dislocation with the secondary grain boundary dislocation structure of a near-$\Sigma 27_a$ boundary in a Cu – 6 at% Si alloy. The diffracting conditions in (*a*) and (*d*) are the same as those in figure 5.27. For (*e*) $\boldsymbol{B}$ is $[102]_1$, $[001]_2$ with $\boldsymbol{g}_1 = 020_1$ and $\boldsymbol{g}_2 = \bar{2}00_2$ and for (*f*) $\boldsymbol{B}$ is $[105]_1$, $[\bar{1}05]_2$ with $\boldsymbol{g}_1 = 020_1$ and $\boldsymbol{g}_2 = 020_2$.

of the specimen indicated in figures 6.36(b) and (c) as

$$\boldsymbol{b}_S = (1/2)[0\bar{1}\bar{1}]_2$$

$$\boldsymbol{b}_A = (1/6)[\bar{1}12]_1 = (1/6)[\overline{11}\ 13\ 14]_2$$

$$\boldsymbol{b}_B = (1/27)[\bar{5}\bar{1}1]_1 = (1/27)[\bar{5}1\bar{1}]_2$$

$$\boldsymbol{b}_C = (1/54)[1\ 11\ 16]_1 = (1/54)[\bar{1}\ 11\ 16]_2$$

and

$$\boldsymbol{b}_D = (1/54)[2\bar{5}5]_1 = (1/54)[\bar{2}\bar{5}5]_2.$$

These Burgers vectors satisfy the nodal balance, namely

$$\boldsymbol{b}_S + \boldsymbol{b}_A + \boldsymbol{b}_C = \boldsymbol{b}_B + \boldsymbol{b}_D$$

indicating that the slip dislocation S has dissociated in the boundary in accordance with the reaction

$$\boldsymbol{b}_S = \boldsymbol{b}_B + \boldsymbol{b}_D - \boldsymbol{b}_A - \boldsymbol{b}_C.$$

Liu and Balluffi (1985) examined the interaction of $(1/2)\langle 110\rangle$ slip dislocations with a near-$\Sigma 13$ (510) symmetric-tilt boundary with a [001] tilt axis in an evaporated thin-film bicrystal of aluminium. The bicrystal was grown so that the deviation from the exact-$\Sigma 13$ CSL orientation was also mainly a tilt deviation around an axis close to [001]. Therefore, for this boundary, the secondary grain boundary dislocation structure consisted mainly of a single array of edge dislocations in which the individual dislocations had the DSC Burgers vector, $(1/26)[510]_1/(1/26)[5\bar{1}0]_2$, and in the electron micrograph of figure 6.37(a), this array is the inclined set of dislocation lines in strong contrast. The schematic diagram in figure 6.37(b)

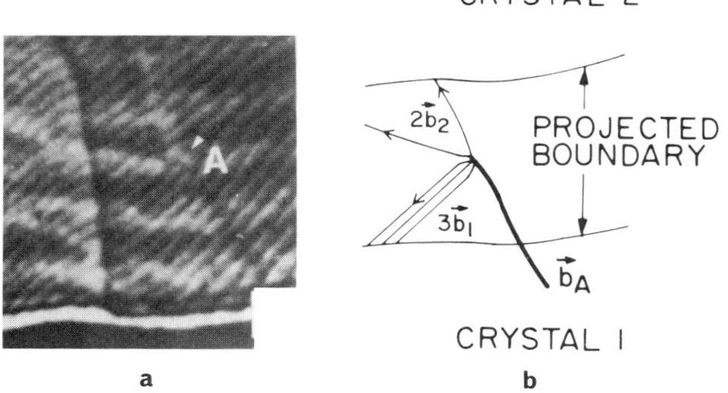

a b

Figure 6.37 Interaction of a slip dislocation with the secondary grain boundary dislocation structure in a near-$\Sigma 13$ symmetric-tilt boundary (Liu and Balluffi 1985).

represents the observed interaction at the nodal point A in the electron micrograph of figure 6.37(a) and shows a slip dislocation in crystal 1 with Burgers vector $\boldsymbol{b}_A$ impinging on the boundary at A, where it joins three grain boundary dislocations with Burgers vector $\boldsymbol{b}_1$ and two grain boundary dislocations with Burgers vector $\boldsymbol{b}_2$. From $\boldsymbol{g}\cdot\boldsymbol{b}$ criteria the magnitude of the Burgers vectors $\boldsymbol{b}_A$, $\boldsymbol{b}_1$ and $\boldsymbol{b}_2$ were determined and, with an assumed sense, these Burgers vectors are given as;

$$\boldsymbol{b}_A = (1/2)[110]_1, \qquad \boldsymbol{b}_1 = (1/26)[510]_1/(1/26)[5\bar{1}0]_2$$

and

$$\boldsymbol{b}_2 = (1/26)[\bar{1}50]_1/(1/26)[150]_2.$$

For the line directions of the dislocations given in figure 6.37(b), these Burgers vectors balance at the node A and therefore are consistent with the slip dislocation having dissociated in the boundary according to the reaction

$$\boldsymbol{b}_A = 3\boldsymbol{b}_1 + 2\boldsymbol{b}_2.$$

The dissociation of four other slip dislocations into reaction products in the boundary was also observed. Two of these slip dislocations, with Burgers vectors $\boldsymbol{b}_D$ and $\boldsymbol{b}_E$, entered the boundary from crystal 1 and the other two, with Burgers vectors $\boldsymbol{b}_B$ and $\boldsymbol{b}_C$, from crystal 2. The dissociation reactions were obtained using $\boldsymbol{g}\cdot\boldsymbol{b}$ criteria with the same assumptions concerning senses as for the dissociation reaction of $\boldsymbol{b}_A$, as

$$\boldsymbol{b}_B = 2\boldsymbol{b}_1 + 3\boldsymbol{b}_2$$

$$\boldsymbol{b}_C = 3\boldsymbol{b}_1 - 2\boldsymbol{b}_2$$

and

$$\boldsymbol{b}_D = \boldsymbol{b}_E = -2\boldsymbol{b}_1 + \boldsymbol{b}_2 + \boldsymbol{b}_3$$

where $\boldsymbol{b}_B = (1/2)[110]_2$, $\boldsymbol{b}_C = (1/2)[1\bar{1}0]_2$, $\boldsymbol{b}_D = \boldsymbol{b}_E = (1/2)[\bar{1}01]_2$ and $\boldsymbol{b}_3 = (1/26)[\bar{2}\ \bar{3}\ 13]_1/(1/26)[\bar{3}\ \bar{2}\ 13]_2$.

The modification of the secondary grain boundary dislocation structure in a near-Σ9 boundary in a polycrystalline specimen of copper by the interaction of two $(1/2)\langle 110\rangle$ slip dislocations, one from each of the neighbouring grains, has been analysed by Clarebrough and Forwood (1980b). The analysis is for the same Σ9 boundary as that shown in figure 5.16 and the position on the boundary at which the interaction of the slip dislocations occurred is position 3 in figure 5.16. In section 5.3.2 (i) the secondary grain boundary dislocation structure of this near-Σ9 boundary at position 1 was analysed in detail and the magnitude and sense of the Burgers vectors of the segments A, B and C (labelled in figure 5.21 (d)) were identified as

$$\boldsymbol{A} = (1/18)[217]_L/(1/6)[112]_R$$

$$\boldsymbol{B} = (1/9)[\bar{2}\bar{1}2]_L/(1/9)[\bar{2}12]_R$$

tions, are interpreted directly in terms of dislocation reactions. For example, the periodic structure shown for the $\Sigma 57_a$ boundary in the electron micrographs of figure 5.37 consists of two periodic sets of fringes in the boundary, both of which are due to periodic secondary grain boundary dislocation structure, but neither of which are direct images of secondary grain boundary dislocations. It was pointed out in section 5.3.2 (ii) that the two sets of fringes result from interference diffraction contrast from two periodic arrays of grain boundary dislocations, the fine set representing a mean and the coarse set a moiré difference of the two periodic arrays. Figure 6.42 shows the interaction of a slip dislocation with this boundary (Forwood and Clarebrough 1978). A $(1/2)\langle 110\rangle$ dislocation, which has dissociated on a $\{111\}$ plane in one of the grains into two $(1/6)\langle 112\rangle$ partial dislocations, has intersected the boundary with one partial terminating at A and the other at B with the stacking-fault fringes on the $\{111\}$ plane lying parallel to AB. At A a dark fringe in the coarse periodic system of fringes in the boundary extending from the top of the micrograph terminates, and at B there is a similar termination of a dark fringe extending from the bottom of the micrograph. Neither of these terminating fringes directly represent terminating secondary grain boundary dislocations, and therefore the analysis of such images is more complex than for those cases where the secondary grain boundary dislocation structure can be imaged directly.

For high-angle grain boundaries where there is no resolvable secondary grain boundary dislocation structure, it has been observed that the image contrast associated with an impinging slip dislocation along the line of intersection of the slip plane with the boundary gradually disappears on

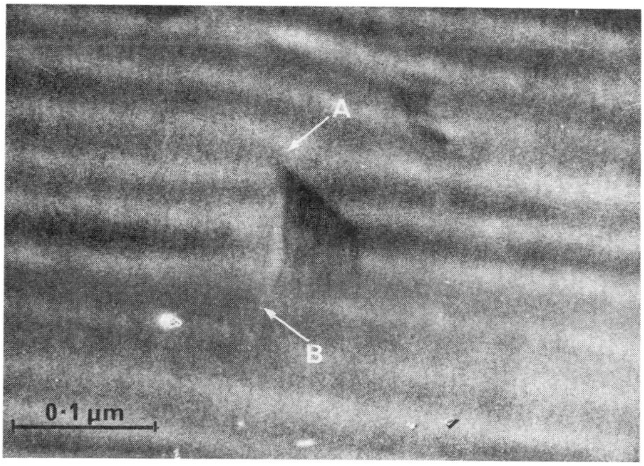

Figure 6.42 Double two-beam image showing the interaction of a slip dislocation with the coarse system of interference fringes in the near-$\Sigma 57_a$ boundary of figure 5.37.

annealing (Pumphrey and Gleiter 1974). Observations of this type led Pumphrey and Gleiter to suggest that the displacement field of a slip dislocation becomes delocalised by a spreading out of the dislocation core in the boundary. Some evidence for delocalisation of dislocation cores in a grain boundary was obtained by Pumphrey *et al* (1977) using weak-beam imaging to determine dislocation core widths. An alternative explanation would be that the slip dislocation dissociates into grain boundary dislocations with sufficiently small Burgers vectors for the strain contrast of these dislocations to escape detection by electron microscopy. Darby *et al* (1978) point out that, in the limit, these two explanations are equivalent because the uniform spreading of the displacement field can be represented by a distribution of an infinite number of grain boundary dislocations with infinitesimally small Burgers vectors.

6.5.2 Slip Transfer Across High-angle Grain Boundaries

There are some special grain boundaries which do not act as strong barriers to slip and allow easy transfer of slip from one grain to another. For example, at symmetric-tilt boundaries the slip planes in the neighbouring grains are symmetrically disposed about the boundary plane and can have a common line of intersection in the boundary, so that a slip dislocation should be able to transfer readily from one grain to the other. Examples of such symmetric-tilt boundaries are coherent-$\Sigma 3$ twin-boundaries in FCC metals, where slip can transfer quite readily, because the $\{111\}$ slip planes in both grains have common lines of intersection in the boundary and dislocations with $(1/2)\langle 110 \rangle$ Burgers vectors can transfer from one grain to the other either directly, or by the generation of residual $(1/6)\langle 112 \rangle$ dislocations in the boundary (see, for example, Hirth and Lothe 1968, p 759). However, for general high-angle grain boundaries, the condition that slip planes from neighbouring grains have a common line of intersection with the boundary is not generally satisfied. For such boundaries the transfer of slip requires that the displacement field of an incoming slip dislocation, along the line of intersection of its slip plane with the boundary, can be redistributed in the boundary interface so as to coincide with the line of intersection of the slip plane for an outgoing slip dislocation. Such a redistribution could take place via a rearrangement of the secondary grain boundary dislocation structure (see, for example, Das and Dwarakadasa 1974), but, in general, this would require diffusion-controlled climb of grain boundary dislocations. For this reason direct transfer of slip is inhibited and grain boundaries generally present strong barriers to the movement of dislocations, giving rise to dislocation pile-ups.

Studies of the mechanisms of slip transfer across high-angle grain boundaries during plastic deformation of polycrystalline metals and alloys date back to the early observations of Ogilvie (1952). He used optical microscopy to study the continuity of slip lines across grain boundaries in polycrystalline specimens of α-brass and aluminium, and found that continuity of slip was favoured if the slip planes in the neighbouring grains had a common line of intersection in the boundary. This result was confirmed by Davis *et al* (1966) using bicrystals of aluminium, and they also showed that slip continuity was still favoured provided the lines of intersection of the slip planes in the neighbouring grains with the boundary were misaligned at an angle less than 15°. Another study of slip transfer using optical microscopy was made by Lim and Raj (1985) on $\langle 110 \rangle$ symmetric-tilt boundaries in specially prepared bicrystals of nickel covering a wide range of misorientation. In all these bicrystals each slip plane in one grain showed a common $\langle 110 \rangle$ line of intersection in the boundary with a symmetrically disposed slip plane in the neighbouring grain. They found a high incidence of slip continuity when the geometry of the bicrystals was such that screw dislocations would have impinged on the boundary along the common line of intersection, and a low incidence of slip continuity when the geometry was such that mixed dislocations would have impinged on the boundary along the common line of intersection. They concluded that a high incidence of slip transfer between neighbouring grains occurs when screw dislocations are involved because no residual dislocations are left in the boundary, whereas when mixed dislocations are involved a lower incidence of slip transfer occurs because residual dislocations must be left in the boundary.

Electron microscopy has been used to study the individual dislocation processes involved in slip transfer across grain boundaries both statically, where the specimens are first deformed and then thinned for examination, and dynamically, where thin-foil specimens are either strained in the electron microscope or heated in a hot stage to induce dislocation movement in the neighbouring grains. The dynamic methods are the most direct way of observing the sequences of events involved in slip transfer, but unfortunately the data usually obtained are not adequate to allow detailed crystallographic analysis of the geometry of the process, or the determination of the Burgers vectors of the dislocations taking part. In addition, the deformation of thin-foil specimens is not representative of the deformation of bulk polycrystals. The static method, in which a bulk polycrystalline specimen is first deformed and then thinned, indicates the processes of slip transfer that have occurred in the bulk specimen, and this method also allows the collection of all the data necessary for the determination of the geometry of the processes and the Burgers vectors of the dislocations involved. However, the disadvantage of this method is that only the final state of slip transfer is observed and the sequence of events involved must be inferred.

Kurzydlowski *et al* (1984) have made a study of slip transfer across a high-angle boundary in a polycrystalline specimen of stainless steel which departed from an exact-$\Sigma 9$ CSL orientation by approximately 2°. After determination of the misorientation between the grains and the plane of the boundary, the specimen was strained *in-situ* during observation in the electron microscope in four stages, and the dislocation activity recorded and slip planes determined for each stage. Slip activity on a (111) plane in one grain was seen to give a pile-up of dislocations at the $\Sigma 9$ boundary and to generate dislocation loops on $(1\bar{1}1)$ planes in the neighbouring grain. The lines of intersection of the boundary with the (111) and $(1\bar{1}1)$ slip planes from the neighbouring grains departed from coincidence by several degrees, and the dislocation loops on $(1\bar{1}1)$ were considered to have been generated at the crossing point of these lines of intersection in the boundary. Although slip transfer was observed directly, complete analysis of the process was not possible because none of the Burgers vectors of the dislocations generated by straining were determined.

Bamford *et al* (1986) examined slip transfer across a high-angle grain boundary separating grains designated as 1 and 2 in a polycrystalline specimen of α-brass, which was 3.82° from an exact-$\Sigma 55$ CSL orientation. Although no controlled deformation was applied, it was assumed that the slip dislocation configurations observed at the boundary were generated in the bulk polycrystalline specimen before thinning. A pile-up of six dislocations in grain 1 on a $(1\bar{1}1)_1$ slip plane, each with a Burgers vector of $\pm(1/2)[110]_1$ $(\equiv \pm(1/110)[12\ 65\ \overline{41}]_2)$ determined by $\boldsymbol{g}\cdot\boldsymbol{b}$ criteria, was observed at the boundary. A group containing four dislocations on a $(111)_2$ slip plane was also observed in grain 2 adjacent to the leading dislocation in the pile-up in grain 1. The Burgers vector of each of these dislocations was identified by $\boldsymbol{g}\cdot\boldsymbol{b}$ criteria as $\pm(1/2)[01\bar{1}]_2$. In addition, two other dislocations were present on a parallel $(111)_2$ slip plane in grain 2 each with a Burgers vector of $\pm(1/2)[\bar{1}10]_2$, but the origin of these dislocations was not discussed. Considering only the arrays containing six and four dislocations in grains 1 and 2 respectively on the $(1\bar{1}1)_1$ and $(111)_2$ slip planes, whose lines of intersection in the boundary were misaligned by 4.4°, the following explanation was given for the origin of the dislocations in grain 2. It was assumed that the dislocation nearest the boundary in grain 1 was the leading dislocation of an incoming array and that this dislocation first dissociated in the boundary into dislocations with Burgers vectors which were basis DSC vectors corresponding to the $\Sigma 55$ CSL. Then, the dislocations with these basis vectors combined in a different way to give a $(1/2)[01\bar{1}]_2$ slip dislocation, along the line of intersection of $(111)_2$ with the boundary, which was then emitted into grain 2 leaving a residual dislocation in the boundary. The assumed reactions were

$$(1/2)[110]_1 = (1/110)[12\ 65\ \overline{41}]_2 = (1/110)[657]_2 + 3[(1/110)[2\ 20\ \overline{16}]_2]$$

for the initial dissociation, and

$$(1/110)[657]_2 + 3[(1/110)[2\ 20\ \overline{16}]_2] = (1/2)[01\bar{1}]_2 + 2[(1/110)[657]_2]$$

for the recombination. While this explanation of the observations is possible and of the type that has been proposed theoretically, the results could be equally well interpreted as being consistent with two dislocation pile-ups, one in grain 1 and the other in grain 2. It is not possible to distinguish between these two interpretations without a determination of the sense of the Burgers vectors of the slip dislocations.

Shen *et al* (1988) recorded the slip dislocation activity on video tape during *in-situ* straining of polycrystalline specimens of stainless steel. They reported an unexpected slip transfer process in which a pile-up of dislocations at a boundary on a $\{111\}$ slip plane in one grain caused the transfer of slip to the neighbouring grain together with the emission of dislocations from the boundary back into the grain containing the original pile-up, but on a different slip plane. Unfortunately the quality of the images was such that the process could not be verified by quantitative analysis.

Slip transfer processes at a high-angle boundary in stainless steel have been studied by Forwood and Clarebrough (1981) using the static method. A specimen of stainless steel (18% Cr, 8% Ni, 74% Fe) 75 μm thick was annealed at 990 °C, quenched to room temperature, given a small plastic strain and then thinned for examination in the electron microscope. Figure 6.43 shows a composite electron micrograph of the high-angle boundary where the upper and lower grains are labelled U and L and the intersection of the boundary with the top surface of the specimen is indicated. The misorientation corresponds to a rotation of grain L in a right-handed sense

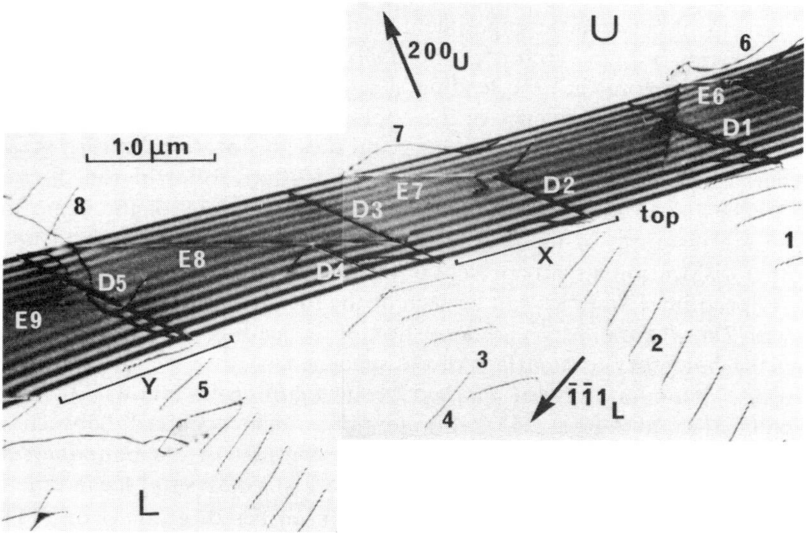

Figure 6.43 Double two-beam image of a high-angle boundary in stainless steel with boundary plane normal close to $[6\ 18\ \bar{7}]_U$. The diffracting vectors and the top surface of the specimen are indicated and ***B*** is close to $[01\bar{1}]_U$ and $[3\bar{1}2]_L$.

with respect to grain U by $\theta = 158.27°$ around an axis $\boldsymbol{u}$ common to both grains of [cos 41.42°, cos 48.59°, cos 89.03°] to an accuracy of $\pm 0.09°$. This rotation axis makes an angle of 3.17° with [110]. No secondary grain boundary dislocation structure was observed in this boundary because the misorientation is not close to any low-Σ CSL orientation.

The slip dislocations in grain L occur as pile-ups on the $(1\bar{1}1)_L$ slip planes, marked 1–5 in figure 6.43, and the Burgers vectors of these dislocations were determined by image matching. All the dislocations in arrays 1 and 3 have the Burgers vector $(1/2)[10\bar{1}]_L$, while those in arrays 2, 4 and 5 have the Burgers vector $(1/2)[110]_L$. Similarly, image matching showed that the slip dislocations 6–8 in grain U were on $(1\bar{1}1)_U$ slip planes with the Burgers vector $(1/2)[\bar{1}\bar{1}0]_U$. The sense of the dislocation line directions is in all cases from the bottom to the top of the specimen, i.e. from left to right in the micrograph of figure 6.43 for both grains. The slip dislocations marked *D1–D5* in figure 6.43 have entered the boundary and lie along the lines of intersection of the boundary with the $(1\bar{1}1)_L$ slip planes 1–5 operative in grain L. Similarly, the slip dislocations marked *E6–E8* are along the lines of intersection of the boundary with the $(1\bar{1}1)_U$ slip planes 6–8 operative in grain U. Image matching confirmed that these slip dislocations in the boundary have the same Burgers vectors, in sense as well as magnitude, as the slip dislocations on the operative slip planes in grains U and L. For example, figure 2.10 shows the agreement between experimental and computed images for the dislocation *D2* which identifies its Burgers vector as $(1/2)[110]_L$. Similarly, the agreement between the experimental and computed images in figure 6.44 identifies the Burgers vector of the dislocation *E7* as $(1/2)[\bar{1}\bar{1}0]_U$. It is concluded therefore that the leading slip dislocations have moved into the boundary under the influence of the applied stress. The boundary then acts as a barrier to their further movement, and these dislocations in turn present barriers to the further movement of slip dislocations following on the same slip plane due to mutual repulsion between dislocations of the same sign. There are instances in figure 6.43 where the slip dislocation is held up by another dislocation already present in the boundary due to slip activity from the other grain. Examples can be seen at the intersection of *D4* with *E8* and *E8* with *D3*, where in each case a portion of the leading slip dislocation runs from the boundary to a surface of the specimen.

In the boundary region marked X in figure 6.43 (shown at higher magnification in figure 6.45) an intersection has occurred between the dislocation *E7*, which shows mainly white contrast in the boundary, and dislocation *D2*, which shows mainly dark contrast. Analysis of the interaction in figure 6.45 shows that the slip dislocation $\overrightarrow{de}$ lies in grain U on $(1\bar{1}1)_U$, has the Burgers vector $(1/2)[\bar{1}\bar{1}0]_U$ and extends from the boundary to the top surface of the specimen. Thus dislocation $\overrightarrow{de}$ corresponds to the leading dislocation that has given rise to the dislocation *E7* (identified in figure 6.44). However, $\overrightarrow{de}$ is linked to the termination of *D2*. A similar situation exists

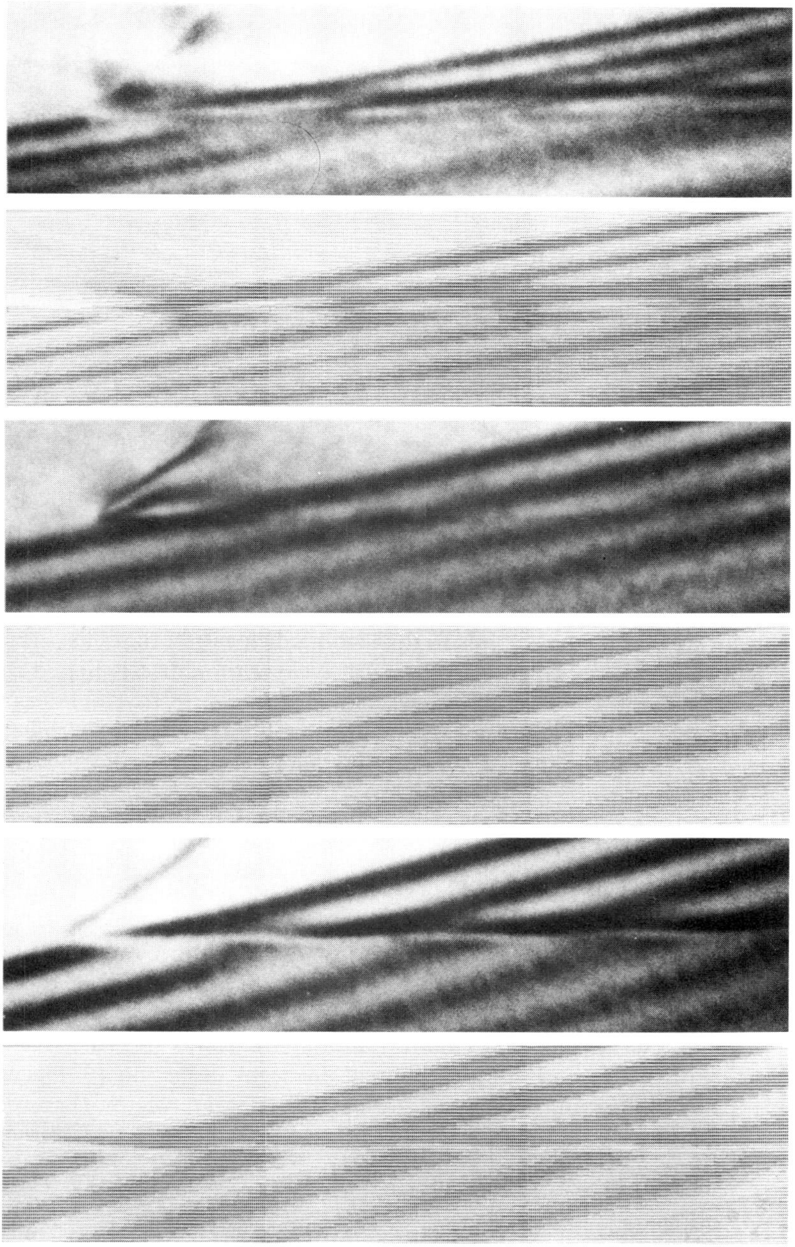

Figure 6.44 Three double two-beam experimental images of dislocation *E7* with the matching set of computed images for a Burgers vector $(1/2)[\bar{1}\bar{1}0]_U$; in (*a*) $\boldsymbol{B}$ is close to $[\bar{6}9\bar{5}]_U$ with $\boldsymbol{g}_U = 1\bar{1}\bar{3}_U$ and $[1\bar{1}0]_L$ with $\boldsymbol{g}_L = \bar{1}\bar{1}1_L$, in (*b*) $\boldsymbol{B}$ is close to $[\bar{1}2\bar{1}]_U$ with $\boldsymbol{g}_U = 31\bar{1}_U$ and $[3\bar{2}0]_L$ with $\boldsymbol{g}_L = 002_L$ and in (*c*) $\boldsymbol{B}$ is close to $[01\bar{1}]_U$ with $\boldsymbol{g}_U = \bar{3}11_U$ and $[5\bar{2}3]_L$ with $\boldsymbol{g}_L = \bar{1}\bar{1}1_L$.

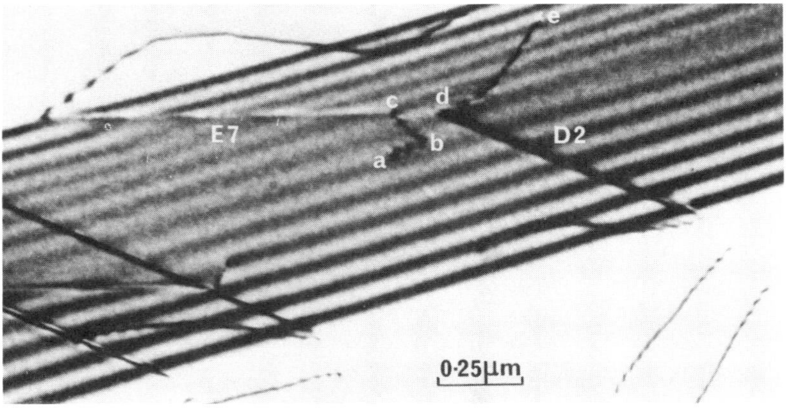

Figure 6.45 Higher magnification image of region X of figure 6.43.

for dislocation $\overrightarrow{abc}$ which lies in grain L, has the Burgers vector $(1/2)[110]_L$ and extends from the bottom surface of the specimen to the boundary. Thus the dislocation $\overrightarrow{abc}$ corresponds to the leading dislocation that has given rise to the dislocation *D2* (identified in figure 2.10), although it is linked to the termination of *E7*. This overall geometry implies, therefore, that an interaction has taken place resulting in an interchange of the leading slip dislocations between *D2* and *E7*. The nature of this interaction is indicated by the configuration of the leading $(1/2)[110]_L$ dislocation $\overrightarrow{abc}$. Analysis of this configuration showed that only the segment $\overrightarrow{ab}$ lies on the $(1\bar{1}1)_L$ slip plane, whereas the segment $\overrightarrow{bc}$ lies on a prismatic glide plane in grain L close to $(3\bar{3}1)_L$ containing the $(1/2)[110]_L$ Burgers vector and the line direction of *E7*. This interaction can be described with the aid of the schematic diagrams of figure 6.46. Figure 6.46(*a*) shows the proposed initial dislocation configuration arising from the intersection at the point (*c*, *d*) of *E7* and its leading slip dislocation *de*, with dislocation *D2* and its leading slip dislocation *abc*. In figure 6.46(*b*) the observed configuration of figure 6.45 is shown as resulting from the prismatic glide of the dislocation segment $\overrightarrow{bc}$ on $(3\bar{3}1)_L$ which generates a dislocation segment along $\overrightarrow{cd}$ with Burgers vector f, according to the reaction

$$f = (1/2)[\bar{1}\bar{1}0]_U + (1/2)[110]_L = [-0.059\,26, 0.067\,35, 0.006\,65]_L. \tag{6.20}$$

This interaction represents a mechanism which enables plastic response to the applied stress in the neighbourhood of the boundary, and is an energy-lowering reaction since a length of dislocation along $\overrightarrow{cd}$ with a $(1/2)\langle 110\rangle$ Burgers vector has been replaced by a residual dislocation with a very small Burgers vector. As might be expected from the small Burgers

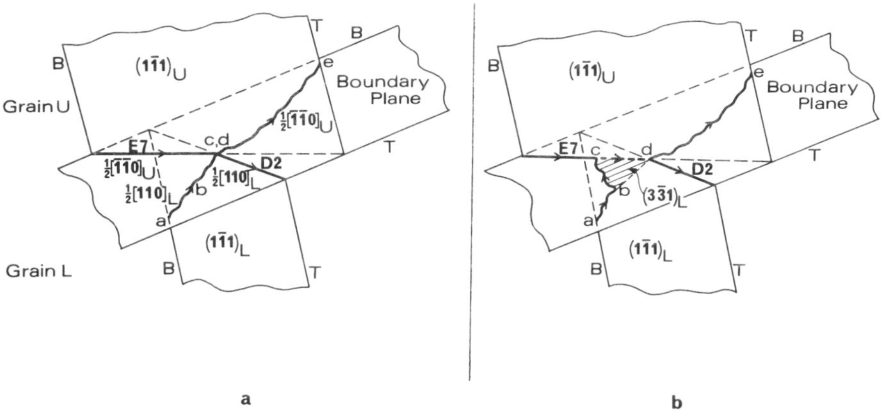

Figure 6.46 Schematic illustration of the interaction at the region X of figure 6.43. T and B indicate intersections with the top and bottom of the specimen respectively and the arrows indicate the sense of the line directions for the different dislocation segments.

vector involved, the contrast associated with the residual dislocation along $\vec{cd}$, although detectable, was always extremely weak under all diffracting conditions, and because of this and its short length its Burgers vector could not be confirmed by image computation.

In the boundary region marked Y in figure 6.43 (shown at higher magnification in figure 6.47 and schematically in figure 6.48(*c*)), a dislocation *D5* has been introduced into the boundary by slip activity on $(1\bar{1}1)_L$ and this dislocation has provided a source for prismatic glide in grain U giving slip transfer from grain L to grain U. The analysis of this dislocation configuration will be discussed with reference to figures 6.47 and 6.48(*c*) (in figure 6.48 the dislocation *E9* of figure 6.43 is not shown since *E9* is not involved in the slip transfer process). The dislocation *D5* has the Burgers vector $(1/2)[110]_L$ (i.e. identical with that of the slip dislocations in array 5 in grain L) and extends in the boundary completely across the specimen, so that the leading slip dislocation that has given rise to it does not appear in the micrograph. Dislocation *S2* lies in grain L on $(1\bar{1}1)_L$ and only contacts the boundary at the point *a* on dislocation *D5*. It has the same Burgers vector, $(1/2)[110]_L$, as *S3* and the other slip dislocations on the $(1\bar{1}1)_L$ slip plane of array 5. Dislocation *S3′* also lies in grain L on $(1\bar{1}1)_L$, extends from the point *e* on dislocation *D5* to the bottom surface of the specimen and has the Burgers vector $(1/2)[110]_L$ for this sense of line direction, which is opposite to that of *S3*. It can be seen from figure 6.47 that *S3* and *S3′* are, in fact, linked in the bottom surface of the specimen by a faint slip trace. In the schematic drawing of figure 6.48(*c*), *S3* and *S3′* are shown as one continuous dislocation line containing a segment close to the bottom surface

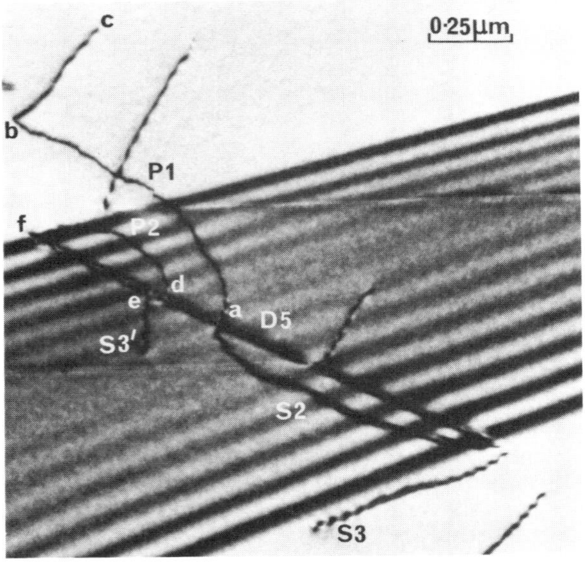

Figure 6.47 Higher magnification image of region *Y* of figure 6.43.

of the specimen. After thinning the bulk specimen, this segment must have slipped out of the thin-foil specimen to generate the slip trace in figure 6.47. The dislocations *P1* and *P2* along $\overrightarrow{ab}$ and $\overrightarrow{df}$ lie in grain U and have the Burgers vector $(1/2)[\bar{1}\bar{1}0]_U$. The segment $\overrightarrow{bc}$ of *P1* lies on the $(1\bar{1}1)_U$ slip plane; however, the remainder of *P1* and the entire length of *P2* are not on the $(1\bar{1}1)_U$ slip plane, but lie on the prismatic glide plane close to $(3\bar{3}7)_U$, containing the $(1/2)[\bar{1}\bar{1}0]_U$ Burgers vector and the line direction of *D5*. This information, which is summarised in figure 6.48(*c*), will now be used in conjunction with figures 6.48(*a*) and (*b*) to describe the mechanism proposed for the slip transfer process.

With reference to figure 6.48(*a*) dislocation *D5* has been originally introduced into the boundary by a leading slip dislocation on $(1\bar{1}1)_L$ as mentioned above. The second slip dislocation *S2* is prevented from entering the boundary owing to mutual repulsion with *D5*. Figure 6.48(*b*) shows schematically how this repulsive force is reduced when a segment of *D5*, extending from the point *a* in the boundary towards the bottom of the bicrystal, undergoes prismatic glide into grain U generating the dislocation *P1* under the action of the applied stress. This involves a dislocation reaction in the boundary given by equation (6.20) which leaves a residual dislocation with the small Burgers vector ***f*** along the segment of *D5* involved. Over the length of this segment the repulsive force on *S2* has been markedly reduced and it can now enter the boundary to reconstitute dislocation *D5* which will

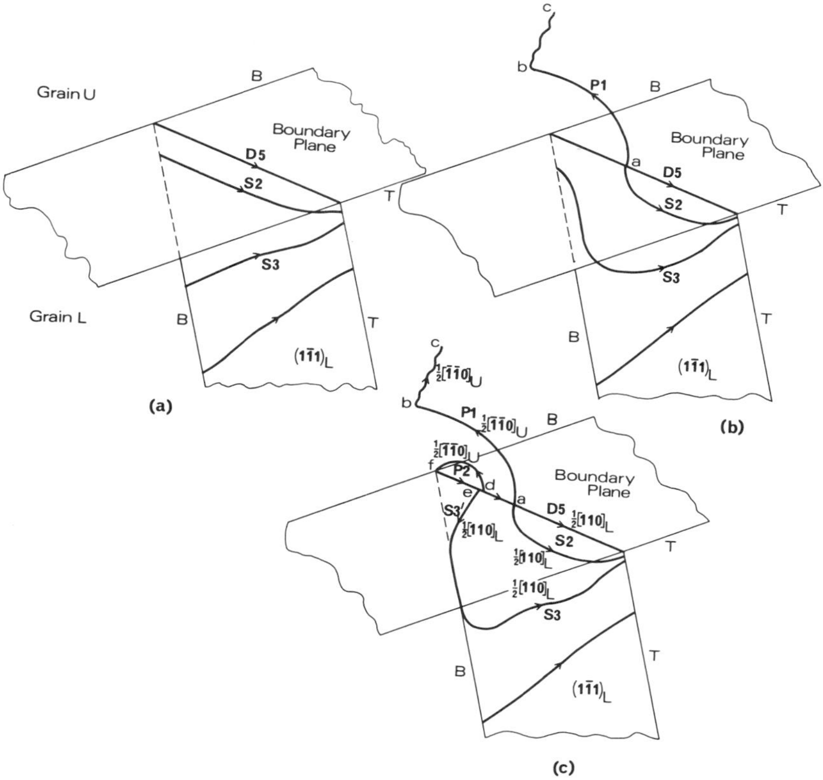

Figure 6.48 Schematic illustration of the sequence of events leading to slip transfer in the region *Y* of figure 6.43. T and B indicate intersections with the top and bottom of the specimen respectively and the arrows indicate the sense of the line directions for the different dislocation segments.

now have the Burgers vector $(1/2)[110]_L + f$. This process can now be repeated as *S3* approaches and then enters the boundary following the formation of a second dislocation *P2* on the same $(3\bar{3}7)_U$ prismatic glide plane, as shown schematically in figure 6.48(*c*). At this stage, the length of *D5*, from which prismatic glide has been generated, consists of the segment $\overrightarrow{fd}$ with Burgers vector $(1/2)[110]_L + 2f$, the very short segment $\overrightarrow{ed}$ with Burgers vector $2f$ and the segment $\overrightarrow{da}$ with Burgers vector $(1/2)[110]_L + f$. The segment $\overrightarrow{bc}$ of dislocation *P1* lies on the $(1\bar{1}1)_U$ slip plane so that it has arisen from a segment of *P1* in screw orientation along the $[110]_U$ direction transferring from the prismatic glide plane to the normal slip plane.

Two processes have been identified which occur at the grain boundary to give a plastic response to the applied stress. In one case (region X of figure

6.43) the plastic response takes place by normal slip changing to prismatic glide within one grain close to the boundary. In the other case (region Y of figure 6.43) the plastic response takes place by normal slip in one grain producing prismatic glide in the other. In this analysis, the senses as well as the magnitude of the Burgers vectors of the dislocations involved in the processes of plastic response have been identified. Thus, any alternative explanation of the observations, based on the premise that the configuration of dislocations on the slip planes arises from pre-existing secondary grain boundary dislocations acting as dislocation sources (see, for example, Murr and Hecker 1979), cannot be substantiated. This follows because such a model would require, for the senses of Burgers vectors identified here, an unrealistic lack of continuity for the applied stress, namely a reversal of sign across the boundary.

The occurrence of prismatic glide close to the boundary takes place by reactions involving $(1/2)[110]$ dislocations in grain L and $(1/2)[\bar{1}\bar{1}0]$ dislocations in grain U. For example, prismatic glide is only observed in association with slip arrays 2 and 5 of figure 6.43 and not for slip arrays 1 and 3, where the Burgers vectors are $(1/2)[10\bar{1}]_L$. This behaviour is a direct consequence of the fact that the common axis of rotation between the two grains is close to $[110]$. Thus, the Burgers vectors $(1/2)[110]_L$ and $(1/2)[\bar{1}\bar{1}0]_U$ are approximately equal and opposite, and the residual dislocation with Burgers vector $\boldsymbol{f}$ resulting from their reaction in the boundary is always small. In the mechanism of slip transfer, the number of slip dislocations from a single slip plane in grain L, that can be transferred to prismatic dislocations in grain U, will be determined by the residual Burgers vector $\boldsymbol{f}$ formed by each transfer. For example, the transfer of the nth slip dislocation from grain L to grain U leaves a residue in the boundary with Burgers vector $n\boldsymbol{f}$. Clearly, there will be a value of n for which the applied stress will be unable to force the $(n+1)$th slip dislocation into the boundary against the repulsion of the residue Burgers vector $n\boldsymbol{f}$. For the example of figure 6.47, $n = 2$, in that two prismatic glide dislocations have been generated in grain U, and over the length $\overrightarrow{fd}$ there is a residual dislocation with Burgers vector $2\boldsymbol{f}$ together with a third slip dislocation which has entered the boundary over the length $\overrightarrow{fe}$. Assuming that the applied stress σ has forced the third dislocation with Burgers vector $\boldsymbol{b} = (1/2)[110]_L$ to approach within a distance $|\boldsymbol{b}|$ of the residue $2\boldsymbol{f}$ in the boundary, then the repulsive force which has been overcome must be less than or equal to $\sigma|\boldsymbol{b}|$. Thus, from Hirth and Lothe (1968, p 110), using isotropic elasticity

$$\sigma|\boldsymbol{b}| \geqq \left(\frac{\mu}{2\pi|\boldsymbol{b}|}\right)\left\{[(\boldsymbol{b}\cdot\boldsymbol{r})(2\boldsymbol{f}\cdot\boldsymbol{r})] + \left[\left(\frac{1}{1-\nu}\right)[(\boldsymbol{b}\wedge\boldsymbol{r})\cdot(2\boldsymbol{f}\wedge\boldsymbol{r})]\right]\right\}$$

where μ is the shear modulus, $\nu \approx 0.3$ is Poisson's ratio and $\boldsymbol{r}$ is the unit vector along dislocation $D5$. This gives $\sigma \gtrsim 1.7 \times 10^{-3}\mu$ which is of the order of magnitude of the yield stress of stainless steel, i.e. the stress used in the light deformation of the specimen.

For the present example the common axis of rotation between the two grains deviates by 3.17° from [110]. Therefore, in the early stages of plastic deformation it is expected that slip transfer by the same mechanism would certainly take place for misoriented grains where the common axis of rotation deviates from [110] by less than 3.17°. On this basis, if grains are randomly oriented in a polycrystal, such that all directions in the standard stereographic projection are equally likely as common axes of rotation, then this mechanism would operate across at least 1% of boundaries. However, it is likely that this mechanism of slip transfer can operate for misoriented grains where the common axis of rotation deviates from $\langle 110 \rangle$ by an angle ρ greater than 3.17°. It is possible, for a given order of applied stress, to obtain a rough estimate of the maximum value of ρ that will allow a single slip dislocation from one grain to generate a prismatic loop in the other, and for this loop to be stabilised by the entry of a second slip dislocation into the boundary. For example, for an initial slip dislocation with Burgers vector $\boldsymbol{b}$, the residue dislocation after prismatic glide will have a magnitude of $|\boldsymbol{b}|(1 - \cos \rho)$ and a direction approximately normal to $\boldsymbol{b}$. The repulsive force that must be overcome by the applied stress σ, to allow the second slip dislocation to approach within $|\boldsymbol{b}|$ of this residue, will be of the order $\mu|\boldsymbol{b}|(1 - \cos \rho)(\lambda/2\pi)$, where λ is a geometric factor less than unity determined by the detail of the misorientation between the grains and the boundary plane. Thus,

$$\sigma|\boldsymbol{b}| \gtrsim \frac{\mu|\boldsymbol{b}|}{2\pi}(1 - \cos \rho)\lambda$$

that is

$$\rho \lesssim \cos^{-1}\left(1 - \frac{2\pi\sigma}{\mu\lambda}\right).$$

Taking $2\pi\sigma/\lambda \sim 10^{-2}\mu$, this gives $\rho \lesssim 8°$. Again assuming randomly oriented grains in the polycrystal, a value of ρ of 8° corresponds to approximately 6% of misoriented grains for which the slip transfer mechanism involving prismatic glide could operate. Of course, preferred orientation in polycrystals and higher applied stresses than considered here will result in an increase in the percentage of misoriented grains for which this mechanism could operate.

Although slip transfer occurs initially at the boundary by the generation of a prismatic loop, continued dislocation movement need not be confined to prismatic glide because segments of the prismatic dislocation in screw orientation can slip on the normal $\{111\}$ slip planes, as illustrated by the segment *bc* of loop *P1* in figure 6.47. Further, any screw segment of a prismatic loop transferring to a $\{111\}$ slip plane has the potential, under the influence of a suitable stress, to act as a source of slip dislocations operating between the points where it joins the prismatic parts of the loop.

It is clear from the results of section 6.5 that in using transmission electron microscopy to investigate the interaction between slip dislocations and secondary grain boundary dislocation structure, or dislocation processes such as slip transfer across grain boundaries, it is essential to be able to determine the sense and magnitude of Burgers vectors involved in order to obtain an understanding of the operative dislocation mechanisms.

7

Interphase Interfaces

7.1 INTRODUCTION

In the chapters on low- and high-angle grain boundaries the concern has been with metals and single-phase alloys where the misorientation between neighbouring grains in a polycrystal has been described by a simple rotation. In this chapter the emphasis will be on the dislocation structure of general interphase interfaces separating different crystal structures in alloys consisting of more than one phase. In this case, the orientation relationship between the phases cannot be described by a simple rotation, but requires a general deformation. As discussed in section 1.3, a general deformation can be represented by arrays of interfacial dislocations, with a net Burgers vector content in the interface specified by equation (1.17), in an analogous way to the accommodation of a rotation by grain boundary dislocations. In addition, it will be seen in this chapter that the CSL model for the structure of high-angle grain boundaries can be adapted to give a description of the structure of interphase interfaces, by appropriate choices of near-coincident cells in each structure when the two crystal structures are considered to interpenetrate.

In the discussion of the structure of interphase interfaces the main concern will be with dislocation structures so that only a very small part of the large volume of material available in the literature will be reviewed. For example, a considerable amount of work on interfaces in two-phase alloys is concerned with the details of the process of precipitation from solid solution and the nature of the interfaces associated with the various stages in the growth of a precipitate leading to the development of the final equilibrium structure (for a review of early work see Kelly and Nicholson 1963). However, this type of work will not be discussed in detail, the emphasis being on the dislocation structures associated with interfaces between a matrix and a well-established second phase at or close to equilibrium. Similarly, the large amount of work on the epitaxy of evaporated thin films will not be reviewed, and only those thin-film studies which are directed towards the dislocation structure of the interface between two different phases will be discussed. The electron microscopy of martensitic interfaces will not be treated, but the recent high-resolution n-beam lattice imaging of martensitic interfaces in the classical transformation in steel, i.e. austenite (FCC) → martensite (body-centred tetragonal (BCT)), will be mentioned briefly.

7.2 ADAPTION OF THE CSL MODEL TO INTERPHASE INTERFACES

In section 5.6 it was pointed out that the CSL model for high-angle grain boundaries in metals and alloys with cubic crystal structure can be extended to grain boundaries in non-cubic structures, if the secondary grain boundary dislocations are taken as accommodating not only the small angular rotation, but also the strain necessary to bring near-coincident lattice sites in the two interpenetrating lattices into a constrained coincidence. It is this type of modification of the CSL model, first developed by Bonnet and Durand (1975a, b) and later reviewed and discussed for particular cases by Balluffi *et al* (1982), which is required to allow a description of the structure of interphase interfaces in terms of interfacial dislocations with DSC Burgers vectors.

This extension of the CSL model to two crystal structures will be illustrated using the two-dimensional lattices 1 and 2 in the schematic diagram of figure 7.1. These lattices are drawn so that they are misoriented relative to one another at the orientation corresponding to that which they would have at some interphase interface. Cells are chosen in the two misoriented lattices such as M_1 in lattice 1 and M_2 in lattice 2 (outlined in figure 7.1 by dashed lines) so that they are very similar in size, shape and orientation. The two lattices at the given misorientation are then considered to interpenetrate in such a way that a site at a corner of cell M_1 and a similar site at a corner of cell M_2 form a common origin for the two lattices as illustrated in figure 7.2. Since the cells M_1 and M_2 are not identical this origin is the only site in exact coincidence, but the other corners of the two cells are in near coincidence. All the corner sites of the two cells can be brought into exact constrained coincidence by either straining lattice 2, so that the near-coincident sites of the cell M_2 are brought into exact coincidence with those of cell M_1, or by the inverse process of straining lattice 1, so that the near-coincident sites of the cell M_1 are brought into exact coincidence with those of cell M_2. Alternatively, both lattices can be strained to form an exact coincidence cell $\bar{M}$ which is intermediate between M_1 and M_2. Figure 7.3 shows the exact coincidence obtained when the sites of cell M_2 are made coincident with the sites of cell M_1 by straining lattice 2 to give a new lattice, namely lattice 2′, indicated by the primitive cell labelled 2′ in figure 7.3. Figure 7.3 also shows the coarsest lattice which contains all the lattice sites

Burgers vectors of the interfacial dislocations were found to be lattice vectors of either phase. This means, on the constrained coincidence model, that for each case the difference in lattice parameter between the phases is such that the M_1 and M_2 cells are cubic cells and the DSC$_1$ and DSC$_2$ lattices correspond to the crystal lattices. Interfaces of this type have Σ_1 equal to Σ_2 equal to 1 and are analogous to low-angle grain boundaries.

7.4 INTERFACES BETWEEN PHASES WITH DIFFERENT CRYSTAL STRUCTURES

7.4.1 Face-centred-cubic Structures/Non-cubic Structures

7.4.1 (i) FCC/hexagonal

For interfaces involving FCC structures with non-cubic structures the simplest is probably that between the two close-packed structures FCC and HCP and an example of such an interface will be considered first. The most widely studied interfaces of this type are probably the interfaces between an FCC aluminium matrix (α) and plate-like HCP precipitates of Ag_2Al (γ') which in equilibrium has an ordered structure (γ). The orientation relationship for this Al/Ag system is $\{0001\}_{HCP} \| \{111\}_{FCC}$, $\langle 11\bar{2}0\rangle_{HCP} \| \langle 1\bar{1}0\rangle_{FCC}$. Early electron microscope studies of this system were concerned with the nucleation of the precipitate and the processes which lead to the formation of ordered Ag_2Al (see, for example, Nicholson and Nutting 1961, Hren and Thomas 1963, Laird and Aaronson 1967). These investigations showed that the broad $\{0001\}_{HCP}/\{111\}_{FCC}$ interfaces of the plate-like precipitates contained either one, two or three arrays of interfacial dislocations with Burgers vectors of the type $(1/6)\langle 112\rangle_{FCC}$ lying in the $\{111\}_{FCC}$ plane of the interface. These dislocations were associated with steps which acted as growth ledges for extension and thickening of the plates.

The conclusions concerning the dislocation structure of the Ag/Ag_2Al interfaces from two-beam microscopy have been confirmed recently by Howe *et al* (1987) and Howe and Mahon (1989) using high-resolution *n*-beam lattice images. Figure 7.6 shows an *n*-beam lattice image of a portion of a $\{0001\}_{HCP}/\{111\}_{FCC}$ interface viewed edge-on along a beam direction parallel to $\langle 11\bar{2}0\rangle_{HCP}/\langle 110\rangle_{FCC}$ in an Al – 15 wt% Ag alloy. Included in the image is the computer-simulated image for the α/γ' interface where the ABAB HCP stacking of the γ' phase has A planes that are nearly pure silver and B planes that are one third silver and two thirds aluminium. Clearly the image in figure 7.6 is of a planar interface between the α and γ' phases, but Howe *et al* have detected steps in such interfaces associated with $(1/6)\langle 112\rangle_{FCC}$ dislocations. Figure 7.7 is an *n*-beam lattice image of a narrow edge of a γ' plate taken in the same beam direction as figure 7.6. This edge of the

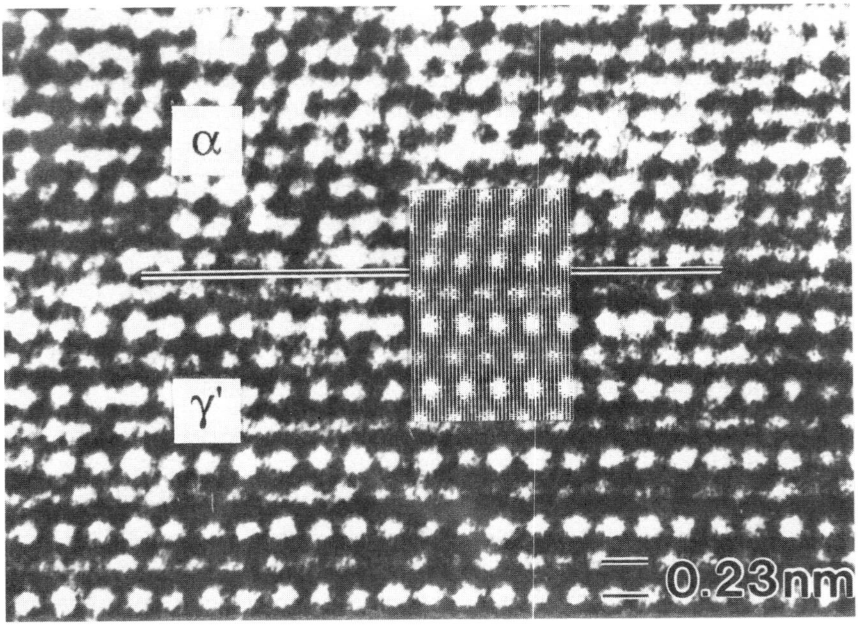

Figure 7.6 Experimental n-beam lattice image of an α/γ' interface in an Al – 15 wt% Ag alloy, with a corresponding computed image in the inset (Howe and Mahon 1989).

precipitate plate is made up of an array of $(1/6)\langle 112\rangle_{FCC}$ dislocations, but because of the close spacing of the dislocations and the large changes in the composition between the α and γ' phases the individual dislocation cores cannot be distinguished in the image. However, by constructing two different Burgers circuits around the edge of the precipitate, as shown in figure 7.7, Howe *et al* found that there were fifteen 30° and eight 90° (edge) $(1/6)\langle 112\rangle_{FCC}$ dislocations present for a plate thickness of forty-six $(0001)_{HCP}$ planes. That is, one $(1/6)\langle 112\rangle_{FCC}$ dislocation was present every other plane of the γ' plate so that edge growth of the plate would transform α to γ' by glide of these dislocations accompanied by the appropriate change in composition. Moreover, the fact that there were two 30° dislocations for every 90° dislocation indicated that there were equal numbers of all three types of $(1/6)\langle 112\rangle_{FCC}$ dislocation, thus minimising the net Burgers vector content and energy of the interface. This dislocation structure is that which would be expected for a transformation of FCC to HCP in a situation where the HCP structure has the ideal c/a ratio for close packing of atoms. In other words the constrained coincidence model has no application in this idealised case because if the two lattices are considered to interpenetrate, the atoms in $\{0001\}_{HCP}$ planes will be in exact coincidence with atoms in the $\{111\}_{FCC}$

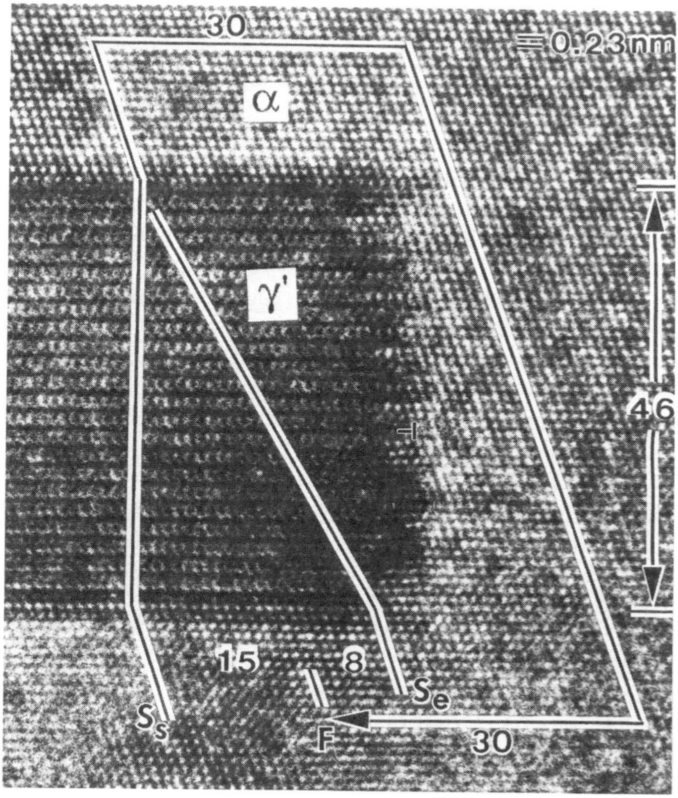

Figure 7.7 n-beam lattice image of an α/γ' interface in the Al–Ag alloy of figure 7.6 showing the structure at the edge of the precipitate. The symbols S_s and S_e indicate the start of the 30° and 90° Burgers circuits respectively and F indicates the finish of the circuits (Howe *et al* 1987).

planes in two adjacent planes every six layers. This can be seen from a comparison of the ABC... stacking sequence for FCC and the ABAB... stacking sequence for HCP, i.e.

ABCABCABC...

ABABABABA....

Howe and Mahon (1989) have also used n-beam lattice images to study the growth of ordered HCP precipitates of Ti_3Al in an ordered face-centred-tetragonal (FCT) matrix of TiAl in near equiatomic Ti–Al alloys, and have shown that the structure of the interfaces and the mechanism of growth of the precipitates are very similar to those found in their work on the Al – 15 wt% Ag alloy.

7.4.1 (ii) FCC/tetragonal

Interfaces between FCC and tetragonal structures are another common group of interphase interfaces and examples are found in the age-hardening Al–Cu alloys and in alloys which undergo martensitic transformations.

A large part of the electron microscopy on Al–Cu alloys has been concerned with determining the complex precipitation sequence; supersaturated α matrix → Guinier–Preston zone → θ'' → θ' → θ ($CuAl_2$) and identifying the structures involved (see, for example, Kelly and Nicholson 1963, Phillips 1973, 1975). The θ' and the equilibrium $CuAl_2$ precipitate θ both have tetragonal structures, but with different c/a ratios and for both the θ' and θ precipitates the orientation relationship with the FCC α matrix is $\{001\}_\alpha \| \{001\}_{\theta'/\theta}$ and $\langle 100\rangle_\alpha \| \langle 100\rangle_{\theta'/\theta}$. In an early investigation of $(001)_\alpha/(001)_{\theta'}$ interfaces, Weatherly and Nicholson (1968) used $\boldsymbol{g}\cdot\boldsymbol{b}$ criteria to identify the Burgers vectors of the interfacial dislocations, which occurred in irregular arrays, as being of the type $\pm[100]_\alpha$ and $\pm[010]_\alpha$. In addition to interfacial dislocations, large growth and dissolution ledges surrounding regions of the interface at different levels, with ledge heights in the range 15–65 Å, are found to be a common feature of the interfacial structure of α/θ' and α/θ interfaces and such ledges have been studied extensively by Weatherly and Sargent (1970) and Weatherly (1971).

Kang and Laird (1974) have also studied the structure of α/θ' and α/θ interfaces, but for precipitates formed in thin-foil electron microscope specimens following heating in a hot stage. Some aspects of the morphology and orientation relationship of these precipitates differed from those formed in bulk specimens. However, they found both ledges and interfacial dislocations for both types of interface, but did not determine the Burgers vectors of the interfacial dislocations. In their experiments, they were able to distinguish readily between ledges and interfacial dislocations because, in their direct observations at the annealing temperature, the ledges were mobile and could be seen to be responsible for growth or dissolution of the precipitates.

Similar interfacial structures to those observed at θ precipitate particles, namely interfacial dislocations and ledges, were found by Garmong and Rhodes (1974) for interfaces between extensive volumes of α and θ phases in a directionally solidified eutectic of $\alpha/CuAl_2$, but again the Burgers vectors of the interfacial dislocations were not identified.

Weatherly and Sargent (1970) and Weatherly and Mok (1972) have treated the contrast effects associated with dislocation-free ledges and have discussed the problem of distinguishing between ledges and interfacial dislocations. This topic has also been addressed by Pond (1984) in a review of contrast effects associated with interphase interfaces. Growth and dissolution ledges give diffraction contrast in electron microscope images which is similar, in some respects, to the contrast arising from the strain fields of interfacial dislocations. It is thought to arise from a strain field associated with the difference in misfit between the phases at the riser and the broad flat face of the ledge. However, the type of strain field associated with ledges has not been quantified, so detailed contrast calculations of the type made for interfacial dislocations cannot be made for ledges. On the assumption that the strain field of a ledge has a dislocation-like character, the best agreement between observed and calculated ledge contrast has been obtained by treating ledges as prismatic dislocation loops in the plane of the interface with some effective Burgers vector normal to the interface with a magnitude which is related to the step height and the misfit at the step riser. For example, calculations by Weatherly and Sargent suggest that the contrast associated with a 35 Å ledge in the θ' phase in a Cu–Al alloy is equivalent to the strain contrast from a dislocation with a Burgers vector of $(1/2)[001]_{\theta'}$.

So far the discussion of the structure of FCC/tetragonal interfaces has been concerned with conventional two-beam microscopy, but *n*-beam high-resolution electron microscopy has recently been applied to FCC/BCT phases at some martensite interfaces. For example Mahon *et al* (1989) have been able to study the martensite interface between the FCC matrix (γ) and a BCT martensite plate (α) in an Fe – 7.9 at% Cr – 1 at% C alloy. The orientation relationship between the phases was $(11\bar{1})_\gamma \| (10\bar{1})_\alpha$ with $[101]_\gamma \| [111]_\alpha$, i.e. the Kurdjumow–Sachs orientation relationship. The habit plane of the interface was $(25\bar{2})_\gamma$ so that this plane was edge-on in a specimen oriented with the beam direction along $[101]_\gamma/[111]_\alpha$. Figure 7.8 shows an *n*-beam lattice image of a γ/α interface in this beam direction, on which a computer-simulated image is superimposed, which identifies the columns of atoms in the experimental image as bright spots on a dark background. In figure 7.8

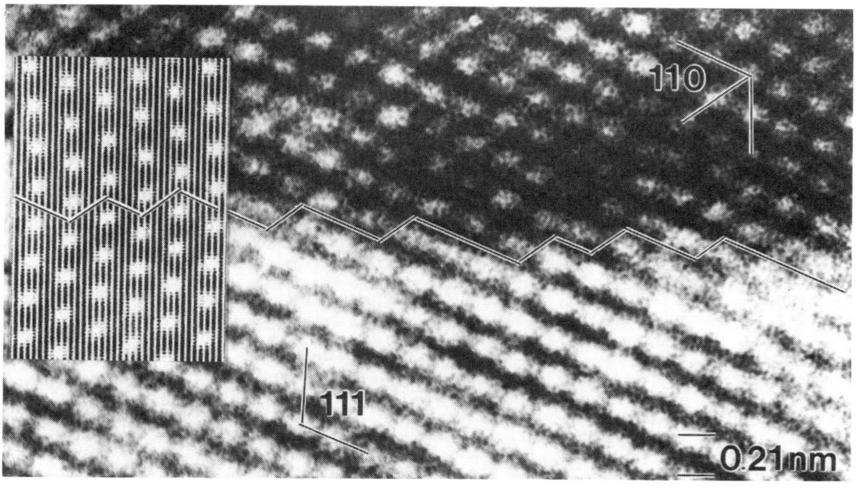

Figure 7.8 Experimental *n*-beam lattice image and corresponding computed image in the inset of a γ/α interface in an Fe–Cr–C alloy (Mahon *et al* 1989).

the interface can be seen to be faceted on an atomic scale, as indicated by the black–white line, and these facets are on the $(11\bar{1})_\gamma$ plane, which is the nearest low-index plane to the mean plane of the interface. Further, in the image of figure 7.8 there is good matching across the interface of the planes of atoms in each phase and in fact the interface is coherent and no dilatation is present. From the angle between corresponding $\{111\}_\gamma$ and $\{110\}_\alpha$ planes across the interface, Mahon *et al* obtained the magnitude of the shear on each $(11\bar{1})_\gamma$ facet plane as either $(1/12)[112]_\gamma$ or $(1/12)[2\bar{1}1]_\gamma$ and interpreted these shears as being consistent with the growth of the martensite plate by the movement of transformation dislocations with these Burgers vectors. However, in common with all work using n-beam lattice images, they were unable to detect the component of the transformation parallel to the beam direction.

7.4.1 (iii) FCC/orthorhombic

The structure of interphase interfaces between FCC and orthorhombic phases has been investigated for several alloy systems using directionally solidified eutectics.

The interfaces between ordered FCC Ni_3Al (γ'), with a lattice parameter $a = 3.59$ Å, and ordered orthorhombic Ni_3Nb (δ), with lattice parameters $a = 5.11$ Å, $b = 4.25$ Å and $c = 4.54$ Å, have been studied by Nakagawa and Weatherly (1972)†. In this directionally solidified eutectic the growth direction was parallel to $[1\bar{1}0]_{\gamma'} \| [100]_\delta$. They investigated two different orientation relationships: (i) $(111)_{\gamma'} \| (010)_\delta$ with $[1\bar{1}0]_{\gamma'} \| [100]_\delta$ and (ii) $(113)_{\gamma'} \| (031)_\delta$ with $[110]_{\gamma'} \| [100]_\delta$. For the orientation relationship (i) they found two orthogonal arrays of interfacial dislocations on the $(111)_{\gamma'}/(010)_\delta$ plane of the interface, with the dislocations in one array (array 1) parallel to the $[1\bar{1}0]_{\gamma'}/[100]_\delta$ growth direction spaced at 150 Å, and those in the second orthogonal array (array 2) spaced at 400 Å. Using $\boldsymbol{g} \cdot \boldsymbol{b}$ criteria they showed that both arrays of interfacial dislocations were composed of edge dislocations with Burgers vectors which were lattice vectors of the FCC phase, $(1/2)[112]_{\gamma'}$ for array 1 and $(1/2)[1\bar{1}0]_{\gamma'}$ for array 2. The dislocations in array 1 were interpreted as accommodating the atomic misfit between the γ' and δ phases of approximately 4% in the $[11\bar{2}]_{\gamma'}/[001]_\delta$ direction and those in array 2 the misfit of approximately 1% in the $[1\bar{1}0]_{\gamma'}/[100]_\delta$ direction. For orientation relationship (ii) only a single array of edge dislocations was observed on the $(113)_{\gamma'}/(031)_\delta$ interface plane, and these dislocations were found to have a Burgers vector of $(1/2)[1\bar{1}0]_{\gamma'}$. Garmong and Rhodes (1975) have studied interface structure in the closely related Ni_3Al/Ni_3Cb system and found two arrays of interfacial dislocations with the same Burgers vectors as those

† This system was one example used by Bonnet and Cousineau (1977) to demonstrate the computation of M_1, M_2 cells and DSC_1, DSC_2 vectors for the constrained coincidence model.

determined by Nakagawa and Weatherly for the Ni_3Al/Ni_3Nb system. However, in addition to interfacial dislocations they observed large ledges and facets.

A more complex example of interfaces between FCC and orthorhombic phases in a directionally solidified eutectic is provided by the work of Spiller and Smith (1980). They examined interfaces between fibres of an orthorhombic carbide Cr_3C_2 phase, which they denoted f, and a two-phase matrix, which they denoted m, consisting of an ordered FCC Ni_3Al (γ') phase and an FCC solid solution of aluminium in nickel (γ). The matrix structure was that already discussed for γ/γ' interfaces in section 7.3 and, since the lattice parameters at room temperature of the two phases forming the matrix are very similar ($a_\gamma = 3.52$ Å, $a_{\gamma'} = 3.56$ Å), no distinction was made between the Cr_3C_2/γ' and Cr_3C_2/γ interfaces. The lattice parameters of the orthorhombic carbide phase at room temperature are $a = 11.46$ Å, $b = 5.52$ Å and $c = 2.82$ Å. Two orientation relationships between the carbide fibres (f) and the matrix (m) were observed: (i) $(100)_f$ approximately parallel to $(331)_m$ with $[001]_f \| [1\bar{1}0]_m$ (the growth direction) and (ii) $(100)_f$ approximately parallel to $(111)_m$ with $[001]_f \| [1\bar{1}0]_m$. Interfaces corresponding to the orientation relationship (i) were selected for study and the types of interface plane examined were $(110)_f/(33\bar{5})_m$, $(1\bar{1}0)_f/(117)_m$ and $(100)_f/(331)_m$, where the indices given for the planes in the matrix are approximations to the true high-index planes involved. Only one array of interfacial dislocations, parallel to the growth direction, was observed by Spiller and Smith for the $(110)_f/(33\bar{5})_m$ and $(1\bar{1}0)_f/(117)_m$ interfaces, but the Burgers vectors were not determined, and no interfacial dislocations were observed on the $(100)_f/(331)_m$ interfaces. They gave a qualitative explanation of the occurrence of the preferred interface planes in terms of the mismatch (measured normal to the growth direction) between rows of atoms in the matrix and in the fibre parallel to the growth direction at a given interface plane. This mismatch was considered to be small (4.6–6.8%) for the interfaces most frequently observed, namely $(111)_f/(33\bar{5})_m$ and $(1\bar{1}0)_f/(117)_m$, but was large (23.9%) for the less frequently observed interface plane, $(100)_f/(331)_m$. The observed presence or absence of interfacial dislocations correlated qualitatively with these estimated mismatches, in that coarsely spaced interfacial dislocations parallel to the growth direction were observed for small mismatches, but no dislocations were resolved for large mismatches.

Knowles (1982) carried out a detailed analysis of the interfaces studied by Spiller and Smith using the constrained coincidence model for M_1 cells in the matrix and M_2 cells in the Cr_3C_2 fibres. Applying the general theory of interface structure (see expressions (1.28) and (1.29)), he considered three arrays of interfacial dislocations for the interface planes observed by Spiller and Smith and tested four different combinations of three non-coplanar Burgers vectors for these dislocations' arrays. The Burgers vectors tested were basis vectors and their linear combinations for the DSC_1 lattice and for the

DSC_2 lattice. Each combination of Burgers vectors predicted two arrays of dislocations with line directions parallel to the growth direction and one array of dislocations with a line direction perpendicular to the growth direction. The spacing of the predicted array perpendicular to the growth direction was 11 Å in all cases; such an array would not have been resolved by Spiller and Smith. However, none of the predicted arrays parallel to the growth direction had the dislocation spacing of 125 Å, which was the spacing of the single array of dislocations observed by Spiller and Smith for $(110)_f/(33\bar{5})_m$ and $(1\bar{1}0)_f/(117)_m$ interfaces. Knowles also predicted arrays of dislocations parallel to the growth direction with readily resolvable spacings on the $(100)_f/(331)_m$ interface plane for which no dislocations were observed by Spiller and Smith. Knowles considered that the disagreement between his calculations and the limited experimental observations available may be due in part to his use of room temperature lattice parameters in a situation where lattice parameters corresponding to some higher temperature would have been more appropriate.

Since the limited experimental data of Spiller and Smith did not allow an adequate test of the constrained coincidence model, Knowles and Goodhew (1983a and b) made a detailed experimental and theoretical investigation of the structure of interfaces between FCC and orthorhombic phases in a directionally solidified Al–Al_3Ni eutectic, following on earlier work by Breinan *et al* (1972) and Cantor and Chadwick (1975). The preparation of this alloy has been described by Cantor and Chadwick (1974). The orthorhombic Al_3Ni fibres (f) have the room temperature lattice parameters $a = 6.6115$ Å, $b = 7.3664$ Å and $c = 4.8118$ Å and the aluminium matrix (m) has the room temperature lattice parameter $a = 4.0496$ Å, i.e. there is virtually no solid solubility of nickel in aluminium. The orientation relationship between the fibres and the matrix is $(102)_f$ approximately parallel to $(111)_m$ with $[010]_f \| [1\bar{1}0]_m$ (the growth direction) to an accuracy of $\pm 0.5°$. The $(102)_f$ and $(111)_m$ planes are only approximately parallel because the $(102)_f$ plane is rotated slightly from the $(111)_m$ plane towards $(11\bar{1})_m$ by $0.4° \pm 0.3°$.

In the directionally solidified eutectic the fibres facet to form discrete interface planes lying in the zone $[010]_f/[1\bar{1}0]_m$ and the interface planes investigated for each fibre were $(104)_f/(332)_m$, $(\bar{1}02)_f/(66\bar{1})_m$ and $(504)_f/(6\ 6\ 13)_m$. Three arrays of interfacial dislocations (A, B and C) were observed in these interface planes, with different spacings and directions in each plane. Figure 7.9(a) is a dark-field electron micrograph showing how the dislocations in array A change with interface plane and this change is illustrated schematically in figure 7.9(b) where the array in the $(104)_f/(332)_m$ plane is labelled A_1, that on the $(\bar{1}02)_f/(66\bar{1})_m$ plane, A_2 and that on the $(504)_f/(6\ 6\ 13)_m$ plane, A_3. In figure 7.9(a) the array A_3 can be seen more clearly in the insert. The dislocations in the arrays B and C are not in contrast in figure 7.9(a), but they show similar variations with interface plane in micrographs in which they are in good contrast.

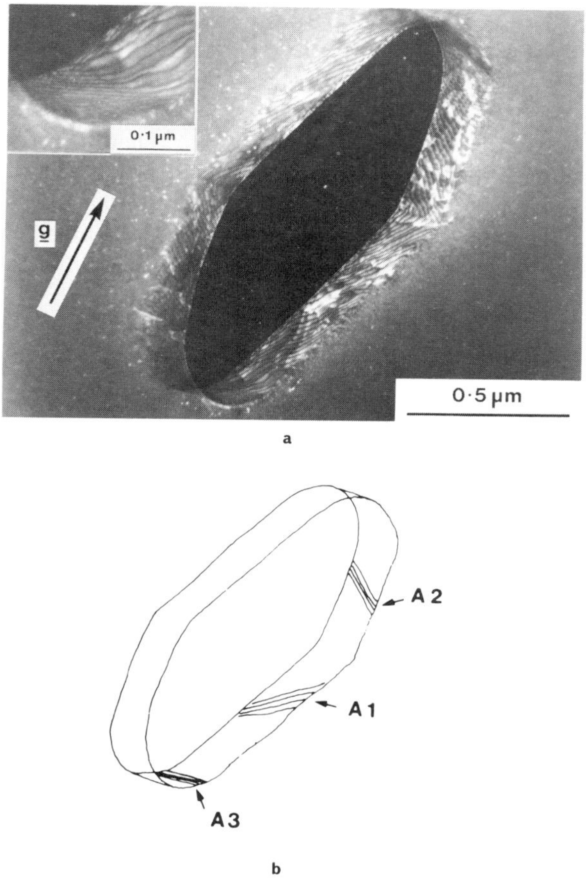

Figure 7.9 Dark-field electron micrograph (*a*) and schematic illustration (*b*) showing the change in spacing and direction of the interfacial dislocations A_1, A_2 and A_3 of array A with change in interface plane for an Al–Al_3Ni interface. The dislocations A_3 are shown at higher magnification in the inset in (*a*) (Knowles and Goodhew 1983a).

The electron microscopy of many fibre–matrix interfaces in this alloy system strongly indicated that the structure of these interfaces always consisted of three arrays of dislocations *A*, *B* and *C* with Burgers vectors $\boldsymbol{b}_1$, $\boldsymbol{b}_2$ and $\boldsymbol{b}_3$ respectively, which remain constant for all interface planes, but whose spacings and directions vary with the interface as a function of the interface normal $\boldsymbol{v}$.

Knowles and Goodhew state that, for their weak-beam images in which only the matrix phase was diffracting, no reliable determination of Burgers vectors could be made using $\boldsymbol{g} \cdot \boldsymbol{b}$ criteria. Therefore, they analysed the dislocation structure of the interfaces using a geometric method of the type already discussed for low-angle boundaries in section 4.3 and for high-angle boundaries in section 5.3.2 (i). In this case their analysis was based on the treatment by Knowles (1982) of the dislocation structure of a general interphase interface given already in section 1.3. There it was shown for an interphase interface that if a general deformation $\mathbf{S}$ relating the two phases is accommodated by three independent arrays of interfacial dislocations with non-coplanar Burgers vectors $\boldsymbol{b}_1$, $\boldsymbol{b}_2$ and $\boldsymbol{b}_3$, line directions $\boldsymbol{r}_1$, $\boldsymbol{r}_2$ and $\boldsymbol{r}_3$ and spacings d_1, d_2 and d_3, then it follows that

$$\boldsymbol{r}_1 \parallel \tilde{\mathbf{T}}\left(\frac{\boldsymbol{b}_2 \wedge \boldsymbol{b}_3}{\boldsymbol{b}_1 \cdot (\boldsymbol{b}_2 \wedge \boldsymbol{b}_3)}\right) \wedge \boldsymbol{v} \tag{7.2}$$

and

$$d_1 = \left(\left|\tilde{\mathbf{T}}\left(\frac{\boldsymbol{b}_2 \wedge \boldsymbol{b}_3}{\boldsymbol{b}_1 \cdot (\boldsymbol{b}_2 \wedge \boldsymbol{b}_3)}\right) \wedge \boldsymbol{v}\right|\right)^{-1} \tag{7.3}$$

with similar expressions for $\boldsymbol{r}_2$, d_2 and $\boldsymbol{r}_3$, d_3, where

$$\mathbf{T} = (\mathbf{I} - \mathbf{S}^{-1}) \tag{7.4}$$

and $\tilde{\mathbf{T}}$ is the transpose of $\mathbf{T}$. Expressions (7.2) and (7.3) enable a geometric analysis of the interfacial dislocation structure in the following way. For an interface with a variation in interface plane (i.e. a variation in $\boldsymbol{v}$) expression (7.2) shows that for constant Burgers vectors the different directions $\boldsymbol{r}_1$ of the interfacial dislocations with Burgers vector $\boldsymbol{b}_1$ must always be normal to

$$\boldsymbol{n}_1 = \tilde{\mathbf{T}}\left(\frac{\boldsymbol{b}_2 \wedge \boldsymbol{b}_3}{\boldsymbol{b}_1 \cdot (\boldsymbol{b}_2 \wedge \boldsymbol{b}_3)}\right) \tag{7.5}$$

that is, they form parts of planar parallel loops with a common normal parallel to $\boldsymbol{n}_1$. Similarly, interfacial dislocations with Burgers vectors $\boldsymbol{b}_2$ and $\boldsymbol{b}_3$ form planar parallel loops with normals parallel to $\boldsymbol{n}_2$ and $\boldsymbol{n}_3$ given by equations similar to equation (7.5). Then, from equation (7.3), the spacings of the three independent arrays of interfacial dislocation loops, measured normal to their planes, are

$$|\boldsymbol{n}_1|^{-1}, |\boldsymbol{n}_2|^{-1} \quad \text{and} \quad |\boldsymbol{n}_3|^{-1}.$$

Knowles and Goodhew showed by stereographic analysis that each of the dislocation arrays *A*, *B* and *C* on the different interface planes did form parts of parallel planar loops and determined these loop plane normals, i.e. the directions $\boldsymbol{n}_1$, $\boldsymbol{n}_2$ and $\boldsymbol{n}_3$. The magnitudes of $\boldsymbol{n}_1$, $\boldsymbol{n}_2$ and $\boldsymbol{n}_3$ were determined from the measured spacings of the dislocations in the arrays *A*, *B* and *C* and gave values (in units of Å^{-1}), indexed with respect to the matrix, of

$$\boldsymbol{n}_1 = 0.0131[\cos 63.1^\circ \; \cos 31.6^\circ \; \cos 74.7^\circ]_{\mathrm{m}}$$

$$\boldsymbol{n}_2 = 0.0625[\cos 138.7^\circ \; \cos 130.8^\circ \; \cos 84.8^\circ]_{\mathrm{m}}$$

$$\boldsymbol{n}_3 = 0.0151[\cos 44.7^\circ \; \cos 53.4^\circ \; \cos 68.0^\circ]_{\mathrm{m}}.$$

Figure 7.10 is a [001] projection, indexed with respect to the matrix, illustrating the determination of the direction $\boldsymbol{n}_1$ normal to the loops formed by the interfacial dislocations in array A. The zones drawn with dashed lines correspond to the three interface planes considered and the bold lines, marked with error bars on these zones, indicate the three different determined directions of $\boldsymbol{r}_1$ on these planes. Within the limits of experimental error, these directions of $\boldsymbol{r}_1$ define the plane of the interfacial dislocation loops in array A and this is indicated by the zone drawn with a continuous line in figure 7.10, with the direction $\boldsymbol{n}_1$, the loop plane normal, indicated by the triangle containing a cross. The same type of stereographic analysis was used to determine $\boldsymbol{n}_2$ and $\boldsymbol{n}_3$.

In order to determine the Burgers vectors $\boldsymbol{b}_1$, $\boldsymbol{b}_2$ and $\boldsymbol{b}_3$ of the three arrays of dislocations A, B and C, Knowles and Goodhew considered different possible matrices $\tilde{\mathbf{T}}$ in equations for $\boldsymbol{n}_1$, $\boldsymbol{n}_2$ and $\boldsymbol{n}_3$ (i.e. equations of the type (7.5)). Each of these matrices $\tilde{\mathbf{T}}$ was obtained on the basis of the constrained coincidence model for the misoriented lattices of the two phases. They examined eight possible pairs of near-coincident cells M_m and M_f in the matrix and fibre respectively. For each pair of near-coincident cells they calculated the deformation $\boldsymbol{S}$ required to bring the near-coincident lattice sites at the corners of the cells M_m and M_f into exact coincidence, and thus

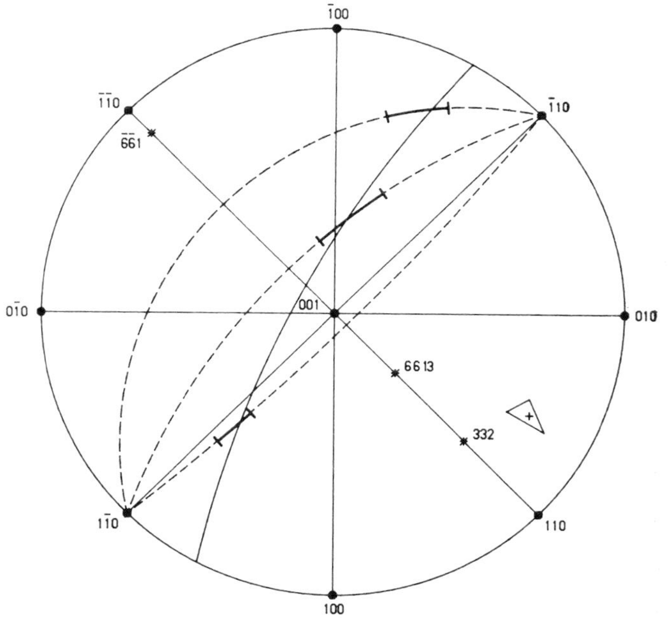

Figure 7.10 Stereographic projection showing the determination of the direction $\boldsymbol{n}_1$ (indicated by +) normal to the loops formed by the interfacial dislocations in array A (Knowles and Goodhew 1983a).

obtained a matrix **T** for each case from equation (7.4). By substituting the measured values of $\boldsymbol{n}_1$, $\boldsymbol{n}_2$ and $\boldsymbol{n}_3$ in equations of the type (7.5), they determined sets of values for the Burgers vectors $\boldsymbol{b}_1$, $\boldsymbol{b}_2$ and $\boldsymbol{b}_3$, where each set corresponded to each constrained coincident cell considered. Then, for each constrained coincident cell, they compared the calculated set of Burgers vectors with the set of DSC vectors appropriate to that cell. From these comparisons only one constrained coincident cell was found which gave a set of Burgers vectors that were compatible with the set of DSC vectors for that cell. On this basis, Knowles and Goodhew concluded that the observed interfacial dislocations had these particular Burgers vectors and were accommodating this particular constrained coincident cell.

Figure 7.11 illustrates the two near-coincident cells M_m and M_f corresponding to the fitting case. In figure 7.11(*a*) the lattice sites in the $(010)_f$ and $(1\bar{1}0)_m$ planes are shown superimposed, with the matrix sites represented by the symbol ○ and the fibre sites by the symbol ∗. In the plane of figure

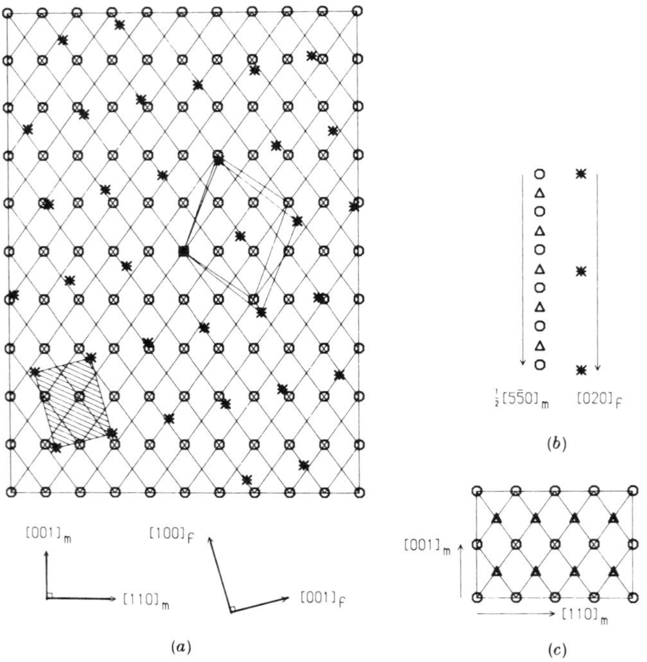

Figure 7.11 (*a*) Representation of fibre (∗) and matrix (○) lattice sites in the $(010)_f \| (1\bar{1}0)_m$ planes at the determined misorientation with a common origin ■; (*b*) stacking of lattice planes in fibre and matrix in the direction normal to $(010)_f \| (1\bar{1}0)_m$ and (*c*) projection of the matrix lattice along $[1\bar{1}0]_m$ where sites marked △ are distant $(1/4)[1\bar{1}0]_m$ above the plane of the ○ sites (Knowles and Goodhew 1983b).

7.11(*a*) the M_m and M_f cells outlined have the common origin marked by the symbol ■ and are specified by the vectors $[11\bar{1}]_m$, $(1/2)[114]_m$ and $[\bar{1}01]_f$, $[101]_f$. The vectors which define the third dimension of the two cells are the out-of-plane vectors $(1/2)[5\bar{5}0]_m$ and $[020]_f$ shown in figure 7.11(*b*). Basis DSC vectors that correspond to the constrained coincident cell arrived at by straining the fibre cell M_f on to the matrix cell M_m by a deformation $\mathbf{S}$ are $\boldsymbol{x} = (1/4)[11\bar{2}]_m$, $\boldsymbol{y} = (1/4)[112]_m$ and $\boldsymbol{z} = (1/4)[1\bar{1}0]_m$. For this constrained coincident cell the deformation $\mathbf{S}$ gives, from equation (7.4) and equations (7.5) for the determined values of $\boldsymbol{n}_1$, $\boldsymbol{n}_2$ and $\boldsymbol{n}_3$, the Burgers vectors

$$\boldsymbol{b}_1 = [0.85 \quad -1.24 \quad 0.64]_m$$

$$\boldsymbol{b}_2 = [0.34 \quad 0.24 \quad -0.56]_m$$

$$\boldsymbol{b}_3 = [0.25 \quad -1.13 \quad 0.01]_m.$$

These determined Burgers vectors were identified with a basis DSC vector and linear combinations of basis DSC vectors as follows:

$$\boldsymbol{b}_1 = 3\boldsymbol{z} - \boldsymbol{x} \quad = (1/2)[1\bar{2}1]_m$$

$$\boldsymbol{b}_2 = \boldsymbol{x} \quad = (1/4)[11\bar{2}]_m$$

$$\boldsymbol{b}_3 = 2\boldsymbol{z} - \boldsymbol{x} - \boldsymbol{y} = [0\bar{1}0]_m.$$

When these combinations of DSC vectors were used to calculate spacings and directions of the dislocations in the arrays *A*, *B* and *C*, the results obtained were not in good agreement with the experimental measurements as can be seen from a comparison of tables 7.1 and 7.2. Knowles and Goodhew attributed the lack of agreement to the uncertainties in the determined orientation relationship between the two phases, and to the use of lattice parameters corresponding to room temperature rather than parameters corresponding to some higher temperature which may have been more appropriate to the state of phase equilibrium being considered. They tested both these possibilities by making small adjustments to the orientation relationship, within the limits of experimental error, and by using lattice

Table 7.1 Observed spacings and directions of dislocations in arrays *A*, *B* and *C*.

Interface	*A*		*B*		*C*	
	Spacing (Å)	Angle ρ°†	Spacing (Å)	Angle ρ°	Spacing (Å)	Angle ρ°
$(104)_f/(332)_m$	260 ± 60	120 ± 14	30 ± 5	6 ± 2	550 ± 100	54 ± 8
$(\bar{1}02)_f/(66\bar{1})_m$	150 ± 20	35 ± 7	160 ± 20	119 ± 9	140 ± 20	171 ± 7
$(504)_f/(6\ 6\ 13)_m$	120 ± 30	152 ± 8	Not observed		140 ± 20	4 ± 3

† ρ° is measured from $[1\bar{1}0]_m$ with the component parallel to $[001]_m$ of the vector describing the line direction taken to be greater than or equal to 0.

Table 7.2 Predicted spacings and directions of dislocations in arrays *A*, *B* and *C*.

Interface	*A*		*B*		*C*	
	Spacing (Å)	Angle ρ°	Spacing (Å)	Angle ρ°	Spacing (Å)	Angle ρ°
$(104)_f/(332)_m$	152	93	26	10	2161	0
$(\bar{1}02)_f/(66\bar{1})_m$	121	53	139	114	133	0
$(504)_f/(6\ 6\ 13)_m$	117	130	15	6	122	0

Table 7.3 Predicted spacings and directions of dislocations in arrays *A*, *B* and *C* for modified orientation relationship and lattice parameters.

Interface	*A*		*B*		*C*	
	Spacing (Å)	Angle ρ°	Spacing (Å)	Angle ρ°	Spacing (Å)	Angle ρ°
$(104)_f/(332)_m$	212	107	28	12	652	85
$(\bar{1}02)_f/(66\bar{1})_m$	158	45	106	127	112	170
$(504)_f/(6\ 6\ 13)_m$	123	146	16	7	110	10

parameters corresponding to 200 °C. In this way they obtained the spacings and directions in table 7.3 which are in much better agreement with the experimental values in table 7.1. As pointed out by Knowles and Goodhew, the determined DSC Burgers vectors were consistent with the observed diffraction contrast in their weak-beam images, although this diffraction contrast could not be used to determine the Burgers vectors.

This work by Knowles and Goodhew was the first quantitative demonstration that the constrained coincidence model had application to real interphase interfaces, and that a geometric analysis could be applied to the determination of the Burgers vectors of interfacial dislocations, when methods involving image contrast could not be used.

7.4.2 Face-centred-cubic/Body-centred-cubic Structures

There have been many electron microscope investigations of the structure of interfaces between FCC and BCC phases in a variety of alloys such as Cu–Zn, Cu–Cr, Ni–Cr, Cu–Fe and duplex stainless steel. In these systems the orientation relationship between the phases is usually close to one of two special orientation relationships, the Kurdjumow–Sachs (KS) orientation relationship, where $(111)_{FCC}\|(110)_{BCC}$ with $[1\bar{1}0]_{FCC}\|[1\bar{1}1]_{BCC}$ and the Nishiyama–Wasserman (NW) orientation relationship where $(111)_{FCC}\|(110)_{BCC}$ with $[0\bar{1}1]_{FCC}\|[001]_{BCC}$.

One of the earliest observations of interfacial dislocations at such interfaces was that of Bäro and Gleiter (1973) for interfaces between the FCC (α) and BCC (β) phases in a Cu – 42 wt% Zn alloy. For the interfaces they studied in which $\{111\}_\alpha$ planes were parallel within 2° to $\{110\}_\beta$ planes, they found a single periodic array of interfacial dislocations with spacings in the range 90–100 Å. They associated these dislocations with the mismatching of the $\{111\}_\alpha$ and $\{110\}_\beta$ planes at the interfaces, suggesting a Burgers vector for each dislocation of $(1/3)\langle 111\rangle_\alpha$ or $(1/2)\langle 110\rangle_\beta$. Later Bollmann (1974) applied his 0-lattice theory to α/β brass interfaces for an NW orientation relationship where $(111)_\alpha \| (110)_\beta$ with $[0\bar{1}1]_\alpha \| [001]_\beta$. He treated the problem in terms of three arrays of interfacial dislocations which accommodated the entire mismatch between the phases for the assumed orientation relationship. This treatment restricted the possible Burgers vectors for the interfacial dislocations to lattice vectors of either the FCC or BCC lattices. For the three independent arrays of interfacial dislocations, he found two with spacings of about 5 Å, which would not have been resolved in the experiments of Bäro and Gleiter, and one with a spacing of 88 Å and a Burgers vector of $(1/2)[110]_\alpha$ or $[010]_\beta$. Bollmann suggested that the array with the dislocation spacing of 88 Å corresponded to that observed by Bäro and Gleiter. However, it should be noted that the 0-lattice model and the simple plane-matching model give a different Burgers vector for the dislocations in this array. This problem has been addressed by Balluffi *et al* (1982) who applied the constrained coincidence model to α/β brass interfaces at the NW orientation. They showed that this model predicted a large DSC vector of $(1/3)[111]_\alpha$ or $(1/2)[110]_\beta$, for the Burgers vector of the dislocations in one array, and two small DSC vectors in the $(111)_\alpha \| (110)_\beta$ plane, $(1/48)[\bar{1}\bar{1}2]_\alpha$ or $(1/22)[\bar{1}10]_\beta$ and $(1/14)[1\bar{1}0]_\alpha$ or $(1/8)[001]_\beta$, for the dislocations in the other two arrays. Balluffi *et al* attribute the large Burgers vector to the coarse array of interfacial dislocations observed by Bäro and Gleiter. Clearly there is a need to positively identify the Burgers vectors of the interfacial dislocations in this alloy in order to determine whether they are lattice vectors or non-lattice vectors. Although recent experimental work by the authors has shown that three arrays of dislocations are present in α/β brass interfaces, detailed image matching to identify the Burgers vectors has yet to be done.

Howell *et al* (1979) found a similar apparent contradiction in the determination of the Burgers vectors of the interfacial dislocations in their work on the structure of interfaces between FCC austenite (γ) and BCC ferrite (α) in a duplex stainless steel (68 wt% Fe, 26 wt% Cr, 6 wt% Ni), for which the orientation relationship was within 2.5° of the KS orientation relationship. In this case, the single array of dislocations observed had a spacing of approximately 80 Å which they showed to be in agreement with a plane matching model involving $(111)_\gamma$ and $(110)_\alpha$ planes. However, they concluded, on the basis of $\boldsymbol{g}\cdot\boldsymbol{b}$ criteria, that the direction of the Burgers vector of the dislocations was consistent with that found by Bollmann (1974) for the coarse array of interfacial dislocations in α/β brass, i.e. parallel to a close-packed $\langle 110 \rangle$ direction in the FCC phase rather than the $\langle 111 \rangle$ direction expected from plane matching.

The structure of interfaces between the γ/α phases in a duplex stainless steel (68.7 wt% Fe, 20.6 wt% Cr, 7.7 wt% Ni, 0.004 wt% C, 0.54 wt% Si) has also been investigated by Penisson and Regheere (1989) using *n*-beam lattice imaging. For a case where the misorientation between the phases appears to be at the exact NW orientation relationship ($(\bar{1}11)_\gamma \| (1\bar{1}0)_\alpha$; $[0\bar{1}1]_\gamma \| [001]_\alpha$), with the plane of the interface parallel to $(\bar{1}11)_\gamma \| (1\bar{1}0)_\alpha$, they found a single array of interfacial dislocations in an *n*-beam image taken with the beam direction along the $[0\bar{1}1]_\gamma \| [001]_\alpha$ line direction of the dislocations. These dislocations were spaced at 22 Å and Penisson and Regheere state that their Burgers vector was either the lattice vector $(1/2)[010]_\gamma$ or the lattice vector $(1/2)[111]_\alpha$. This result is surprising as these vectors have components which would go undetected since they lie along the beam direction. Clearly, more evidence is needed before the Burgers vector can be specified in this case.

Like the interfaces in α/β brass, the interfaces in duplex stainless steel still require quantitative experimental investigation before the types of interfacial dislocation array that occur, and the Burgers vectors of these dislocations can be specified. For general interphase interfaces in duplex stainless steel the authors have observed three independent arrays of dislocations, but again the Burgers vectors have still to be determined.

Rigsbee and Aaronson (1979a,b) have studied the structure of interfaces between FCC austenite (γ) and BCC ferrite (α) in a two-phase Fe–C–Si alloy, for orientation relationships intermediate between NW and KS orientation relationships. They found two intersecting arrays of periodic contrast in weak-beam images and the spacings and directions of these periodic arrays varied with the plane of the interface. One of these arrays, with spacings in the range 15–25 Å, they associated with interfacial dislocations, and the other, with spacings in the range 22–90 Å, they associated with structural ledges. This interpretation of the observations was based on a computer model which maximised good atomic fit over an extensive region of the interface and involved invoking a stepped interface, with discrete regions of good atomic fit on neighbouring parallel $(111)_\gamma$ and $(110)_\alpha$ planes separated by ledges and crossing dislocations. On this model the Burgers vector of the interfacial dislocations is of necessity a lattice vector contained in the $(111)_\gamma/(110)_\alpha$ plane, i.e. vectors of the type $(1/2)\langle \bar{1}10 \rangle_\gamma$. Some support for this type of Burgers vector was obtained using $\boldsymbol{g} \cdot \boldsymbol{b}$ criteria for diffraction conditions where only one phase was diffracting. The possibility that the results of Rigsbee and Aaronson could be interpreted on the constrained coincidence lattice model is discussed by Balluffi *et al* (1982).

Hall and Aaronson (1986), following earlier work by Hall *et al* (1972), have used weak-beam images to study the structure of FCC–BCC interfaces in a Cu – 0.3 wt% Cr alloy, for various precipitate morphologies and orientation relationships varying from near-NW to near-KS orientation relationships. Depending on morphology, either one or two arrays of interfacial dislocations were observed with spacings of about 25 Å or less. Although only one-phase diffracting images were available, they applied $\boldsymbol{g} \cdot \boldsymbol{b}$ criteria and concluded that the Burgers vectors of the interfacial dislocations were consistent with those predicted on their form of the 0-lattice model (Hall *et al* 1986), namely lattice vectors in the $(111)_{Cu}/(110)_{Cr}$ plane.

Luo and Weatherly (1988) report on the structure of interfaces between BCC chromium-rich precipitates and an FCC nickel-rich matrix in an Ni – 45 wt% Cr alloy. The chromium-rich precipitates occurred in the form of coarse laths and the orientation relationship between the laths and the matrix was the KS orientation relationship, $(1\bar{1}1)_{Ni} \| (101)_{Cr}$ with $[\bar{1}01]_{Ni} \| [\bar{1}\bar{1}1]_{Cr}$. The plane of the broad faces of the laths was $(1\bar{2}1)_{Ni} \| (3\bar{1}2)_{Cr}$ and the growth direction, parallel to the long dimension of the laths, was 5.5° away from the $[\bar{1}01]_{Ni}/[\bar{1}\bar{1}1]_{Cr}$ direction towards $[\bar{1}\bar{1}\bar{1}]_{Ni}$. Apart from a low density of growth steps on the faces of the laths, the only structural feature was a single array of interfacial dislocations on the narrow $(313)_{Ni}$ side faces of the laths. Using $\boldsymbol{g} \cdot \boldsymbol{b}$ criteria the contrast behaviour of these dislocations was found to be consistent with a Burgers vector of $(1/3)[1\bar{1}1]_{Ni}$ which is the Burgers vector expected on the plane matching model.

Owing to the lack of positive identification for the Burgers vectors of interfacial dislocations in the electron microscopy of interfaces between FCC and BCC phases described so far, the question remains as to whether the appropriate Burgers vectors are lattice vectors or non-lattice vectors in any given case. In general, lattice vectors would be expected if the observed interfacial dislocations were the counterparts of dislocations in low-angle grain boundaries and accommodated the entire mismatch between the FCC and BCC lattices, whereas non-lattice vectors would be expected if the dislocations were those required by a constrained coincidence model of the interface. These alternatives were addressed by the authors in a study of the interfaces between FCC copper and BCC iron in a two-phase Cu – 25 wt% Fe alloy in which the Burgers vectors of the interfacial dislocation were identified by image matching (Forwood and Clarebrough 1989).

The copper–iron system was chosen to study FCC–BCC interfaces for the following reasons. First, the lattice parameter ratio in this system ($a_{FCC}/a_{BCC} = 1.261$) is very similar to those in the α–β brass, the two-phase steels and the copper–chromium alloy described so far. Secondly, this simple alloy has the advantage over others that copper and iron solid solutions† are the only solid phases that occur. Moreover, these phases can be obtained

† The copper solid solution and the iron solid solution will be referred to henceforth as copper and iron.

in the required volume fractions after solidification without the need for precipitation reactions to produce suitable interfaces. Further, the interfaces between the copper and iron phases are not restricted to a particular habit plane but are general curved surfaces.

The alloy was prepared by argon arc melting from 99.999% Cu and 99.96% Fe. The cast ingots were cold-rolled to produce 75 μm strip and two annealing treatments were used:

(a) four days at 1000 °C followed by four days at 800 °C and
(b) four days at 1000 °C followed by six weeks at 800 °C.

After annealing the alloys were cooled from 800 °C to room temperature at a sufficient rate to prevent precipitation of copper in iron and iron in copper. In the cast condition the alloys consist of dendrites of iron in a matrix of copper and, after rolling to 75 μm strip, the iron is in the form of thin laths in the copper matrix. On annealing at 1000 °C both the iron and the copper recrystallise and spheroidisation of the FCC iron laths occurs. On cooling to 800 °C the iron transforms from FCC to BCC and interfacial structure characteristic of the Cu_{FCC}–Fe_{BCC} interfaces is developed.

After annealing treatments (a) and (b), the grain size of the copper matrix was large (about 1 mm or more) and the distribution of the crystals of iron consisted of both isolated near-spherical single grains and interconnected regions made up of several grains. For specimens given the longer annealing treatment (b), the spheroidisation was more advanced and the proportion of isolated grains of iron to regions of interconnected grains was considerably greater than in specimens given annealing treatment (a).

The ways in which iron particles have been truncated by the specimen surfaces as a result of electropolishing are illustrated in figure 7.12. The electron micrographs in figures 7.12(*a*) and (*b*) are for specimens given heat treatments (a) and (b) respectively and show cases where the iron is in the form of solid caps. In each case the flat surface of the solid cap, defined by the plane of its perimeter, is at the bottom surface of the specimen, with respect to the electron source, and the curved surface of the cap is the interface with the copper matrix. In figure 7.12(*b*) the iron cap is a single crystal, while in figure 7.12(*a*) the cap contains a low-angle boundary along XY. Figure 7.12(*c*) illustrates the other common form of section where the interface between an inner iron crystal and the surrounding copper matrix intersects both surfaces of the specimen.

All interfaces contain two distinct types of dislocation. One type is present as regular arrays of interfacial dislocations and the other type, which gives rise to stronger contrast in electron micrographs, is present as an irregular distribution of lattice dislocations lying in the interface. These lattice dislocations in the interface arise from plastic deformation of the copper matrix, due to differential thermal contraction between the copper and iron during cooling from 800 °C to room temperature. The generation of lattice

transformation matrix **N**, which re-indexes a vector indexed with respect to the BCC iron lattice to indices with respect to the FCC copper lattice, was obtained by multiplying the elements of the rotation matrix by the ratio a_{Fe}/a_{Cu}. For case 1

$$ {}_{Cu}\mathbf{N}_{Fe} = \begin{pmatrix} 0.140\,493 & 0.120\,686 & -0.771\,229 \\ -0.623\,284 & 0.489\,127 & -0.037\,002 \\ 0.469\,975 & 0.612\,607 & 0.181\,478 \end{pmatrix} \quad (7.6) $$

and for case 2

$$ {}_{Cu}\mathbf{N}_{Fe} = \begin{pmatrix} 0.124\,607 & -0.586\,791 & 0.518\,891 \\ 0.069\,141 & 0.531\,604 & 0.584\,564 \\ -0.780\,250 & -0.046\,604 & 0.134\,668 \end{pmatrix}. \quad (7.7) $$

The orientation relationship for case 1 is several degrees away from the KS type, in that the $(\bar{1}0\bar{1})_{Fe}$ and $(11\bar{1})_{Cu}$ planes depart from exact parallelism by 1.10° and the $[11\bar{1}]_{Fe}$ and $[101]_{Cu}$ directions by 5.61°. The orientation relationship for case 2 is closer to the KS orientation relationship in that the $(\bar{1}0\bar{1})_{Fe}$ and $(\bar{1}\bar{1}1)_{Cu}$ planes depart from exact parallelism by only 0.39° and the $[11\bar{1}]_{Fe}$ and $[\bar{1}0\bar{1}]_{Cu}$ directions by only 0.89°. The orientation relationships for cases 1 and 2 cannot be specified by exactly the same indices because the copper crystal for case 2 is twin-related to that of case 1. The orientation relationships for cases 1 and 2 are shown in figures 7.16(*a*) and (*b*) respectively, by using projections along $[\bar{1}0\bar{1}]_{Fe}$ of the copper and iron lattices which have been drawn as interpenetrating. The origin of the lattices is labelled 0 and the lattice sites in the zeroth layer for iron are full circles, which lie in the plane of the page, while those in the zeroth layer for copper are open circles. The key to atom layers below the plane of the page is given in the inset at the top left of figure 7.16(*a*) for case 1, and at the bottom right of figure 7.16(*b*) for case 2. In figures 7.16(*a*) and (*b*) the normal stacking sequence ABAB... for the $(101)_{Fe}$ planes in iron is fully represented because $[101]_{Fe}$ is normal to the page. However, for the copper lattices of figure 7.16, the $[11\bar{1}]_{Cu}$ direction is not normal to the page and, to simplify the diagram, only the zeroth, first and second layers of the ABCABC... stacking sequence of $(11\bar{1})_{Cu}$ planes are plotted.

In some of the work described on the structure of interphase interfaces between FCC and BCC phases, the contrast of the interfacial dislocations in electron microscope images has been interpreted as arising from dislocations with Burgers vectors which are lattice vectors of either the FCC or the BCC lattices. For this reason lattice vectors in the FCC and in the BCC lattices were the first possibilities tested for the Burgers vectors of the interfacial dislocations forming the α and β arrays in cases 1 and 2. This involved comparing experimental images with corresponding theoretical images calculated for all

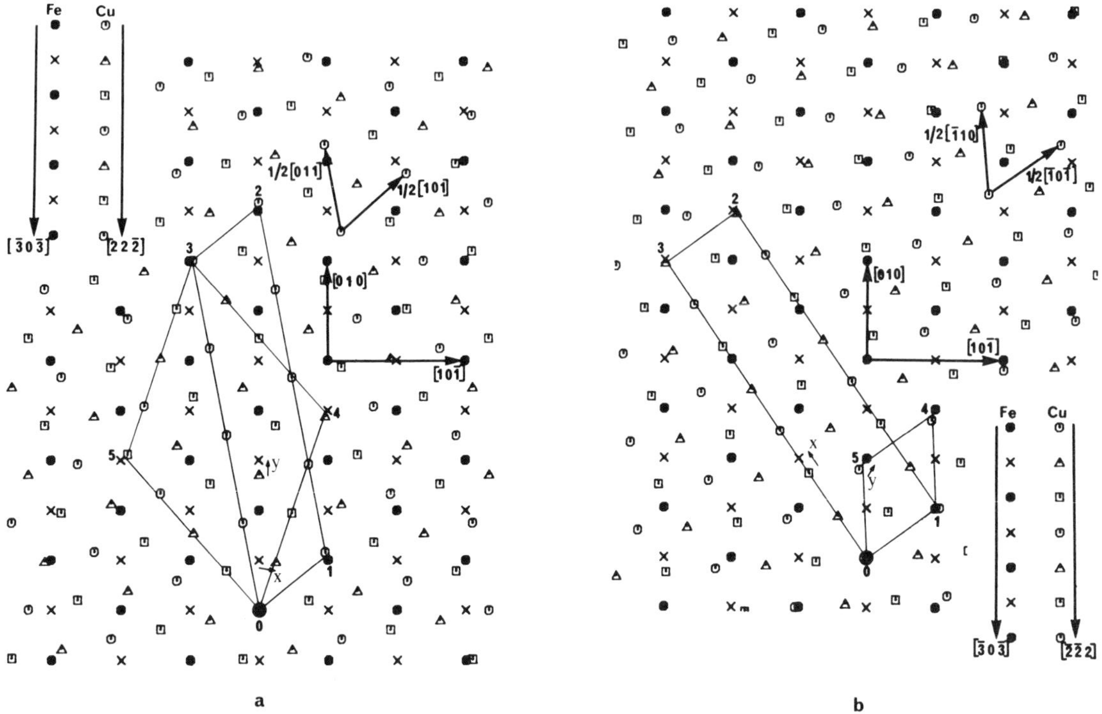

Figure 7.16 Interpenetrating lattices of copper and iron, projected along the $[\bar{1}0\bar{1}]_{Fe}$ direction, (*a*) for the orientation relationship of case 1 and (*b*) the orientation relationship of case 2. (●, ×—iron lattice; ⊙, △, ⊡—copper lattice (see text).) The insets labelled Fe and Cu are schematic illustrations of the stacking sequences below the plane of the page.

possible lattice vectors in copper and in iron up to a magnitude of $\langle 101 \rangle_{Fe}$. However, in no case did any of these lattice vectors give sets of theoretical images which matched the experimental images and examples of these mismatches for lattice vectors will be given later in figures 7.17, 7.19 and 7.20. It was concluded, therefore, that the Burgers vectors of the interfacial dislocations forming the α and β arrays were not lattice vectors. Because of this result a constrained coincidence model was developed for cases 1 and 2 so as to arrive at possible non-lattice Burgers vectors for the interfacial dislocations.

For the interpenetrating copper and iron lattices at the orientation of case 1 (shown in figure 7.16(*a*)) the copper and iron lattice sites have been set in exact coincidence at the origin 0, and there are a number of pairs of lattice sites, comprising one lattice site in copper and the other in iron, such as those labelled 1, 2, 3, 4 and 5, which are close to coincidence. In addition the lattice sites $[\bar{3}0\bar{3}]_{Fe}$ and $[22\bar{2}]_{Cu}$ below the plane of the page are also close to coincidence. When the $[\bar{3}0\bar{3}]_{Fe}$ out-of-plane near-coincident site in iron is chosen with two of the in-plane near-coincident sites in iron a three-dimensional near-coincident cell M_1 is defined in the iron lattice. In the same way a similar three-dimensional near-coincident cell M_2 can be defined in the copper lattice. Two examples of pairs of near-coincident cells M_1 and M_2 are indicated in figure 7.16(*a*) by the near-coincident sites 0, 1, 2, 3 and 0, 4, 3, 5. Consider now the pair of near-coincident cells M_1 and M_2 defined by the pairs of sites 0, 1, 2, 3 and the out-of-plane pair of sites $[\bar{3}0\bar{3}]_{Fe}$ and $[22\bar{2}]_{Cu}$. For the analysis of the present results two ways of forming an exact constrained coincidence lattice (labelled CCSL) will be considered. The constrained coincidence lattice CCSL_1 is formed if the copper lattice is strained so that the corner sites of the copper M_2 cell are brought into coincidence with the corner sites of the iron M_1 cell. Alternatively, a constrained coincidence lattice CCSL_2 is formed if the iron lattice is strained so that the corner sites of the iron M_1 cell are brought into coincidence with the corner sites of the copper M_2 cell. Following the constrained coincidence model outlined in section 7.2, the small transformation **S** required to generate CCSL_1 (or CCSL_2) will require three independent arrays of interfacial dislocations with Burgers vectors of the DSC_1 (or DSC_2) lattice, i.e. Burgers vectors which are differences of lattice vectors in copper and iron when these lattices are constrained to give CCSL_1 (or CCSL_2).

For the pair of near-coincident cells M_1 and M_2 corresponding to 0, 1, 2, 3 in figure 7.16(*a*), the unit cell of CCSL_1 is specified by the vectors $[\bar{3}0\bar{3}]_{Fe}/[22\bar{2}]_{CuM1}$, $\overrightarrow{01} = (1/2)[11\bar{1}]_{Fe}/(1/2)[101]_{CuM1}$ and $\overrightarrow{03} = (1/2)[\bar{1}71]_{Fe}/[022]_{CuM1}$ where the subscript CuM1 has been introduced to represent the deformed copper lattice that is generated in the formation of CCSL_1, i.e. when the corner sites of the M_2 copper cell are brought into coincidence with the corner sites of the M_1 iron cell. Thus, the matrix which re-indexes a vector indexed with respect to the iron lattice into indices with respect to the CuM1 lattice is

$$_{\mathrm{CuM1}}\mathbf{C}_{\mathrm{Fe}} = (1/48)\begin{pmatrix} 5 & 6 & -37 \\ -28 & 24 & -4 \\ 25 & 30 & 7 \end{pmatrix}. \tag{7.8}$$

It follows that basis vectors of the DSC$_1$ lattice for CCSL$_1$ are $\boldsymbol{x}_1 = (1/24)[3\bar{1}\bar{3}]_{\mathrm{Fe}}$, $\boldsymbol{y}_1 = (1/6)[010]_{\mathrm{Fe}}$ and the out-of-plane vector $\boldsymbol{z}_1 = (1/2)[\bar{1}0\bar{1}]_{\mathrm{Fe}}$ where $\boldsymbol{x}_1$ and $\boldsymbol{y}_1$ correspond to the vectors $\boldsymbol{x}$ and $\boldsymbol{y}$ indicated in figure 7.16(a). If CCSL$_1$ is associated with good atomic fit at the interface then the Burgers vectors of the interfacial dislocations will be these basis DSC$_1$ vectors or linear combinations of them. Alternatively, good atomic fit at the interface could be associated with CCSL$_2$ for which basis DSC$_2$ vectors would be $\boldsymbol{x}_2 = (1/48)[5\bar{4}1]_{\mathrm{Cu}}$, $\boldsymbol{y}_2 = (1/48)[145]_{\mathrm{Cu}}$ and the out-of-plane vector $\boldsymbol{z}_2 = (1/3)[11\bar{1}]_{\mathrm{Cu}}$.

Although the interface structure predicted by the constrained coincidence model will be slightly different depending on the choice of CCSL$_1$ or CCSL$_2$, more significant differences will arise for different choices of nearly coincident cells M_1 and M_2. For example, if M_1 and M_2 cells defined by the pairs of sites 0, 4, 3, 5 in figure 7.16(a) are chosen a different interface structure will result. In this case the unit cell of CCSL$_1$ is specified by the vectors $[\bar{3}0\bar{3}]_{\mathrm{Fe}}/[22\bar{2}]_{\mathrm{CuM1}}$, $\overrightarrow{04} = [02\bar{1}]_{\mathrm{Fe}}/[111]_{\mathrm{CuM1}}$ and $\overrightarrow{05} = (1/2)[\bar{1}33]_{\mathrm{Fe}}/[\bar{1}11]_{\mathrm{CuM1}}$. Thus, the matrix which re-indexes a vector indexed with respect to the iron lattice to indices with respect to the CuM1 lattice is

$$_{\mathrm{CuM1}}\mathbf{C}_{\mathrm{Fe}} = (1/33)\begin{pmatrix} 3 & 4 & -25 \\ -21 & 16 & -1 \\ 15 & 20 & 7 \end{pmatrix}. \tag{7.9}$$

Basis DSC$_1$ vectors for CCSL$_1$ in this case are $\boldsymbol{x}_1 = (1/8)[10\bar{1}]_{\mathrm{Fe}}$, $\boldsymbol{y}_1 = (1/8)[010]_{\mathrm{Fe}}$ and the out-of-plane vector $\boldsymbol{z}_1 = (1/2)[\bar{1}0\bar{1}]_{\mathrm{Fe}}$, where again $\boldsymbol{x}_1$ and $\boldsymbol{y}_1$ correspond to the vectors $\boldsymbol{x}$ and $\boldsymbol{y}$ indicated in figure 7.16(a). Similarly, for CCSL$_2$ the basis DSC$_2$ vectors are $\boldsymbol{x}_2 = (1/66)[7\bar{5}2]_{\mathrm{Cu}}$, $\boldsymbol{y}_2 = (1/66)[145]_{\mathrm{Cu}}$ and the out-of-plane vector $\boldsymbol{z}_2 = (1/3)[11\bar{1}]_{\mathrm{Cu}}$.

In determining the Burgers vectors by image matching for the α and β arrays of case 1, the basis DSC$_1$ and DSC$_2$ vectors and their linear combinations for the pairs of near-coincident sites 0, 1, 2, 3 and for the pairs of near-coincident sites 0, 4, 3, 5 were tested. During the identification procedure it became apparent that there were no significant differences between images computed for Burgers vectors based on the DSC$_1$ lattice and those computed

agreement was obtained for the Burgers vectors $\boldsymbol{y}_1 - \boldsymbol{x}_1 = (1/24)[\bar{3}53]_{Fe}$ and $3\boldsymbol{y}_1 - \boldsymbol{x}_1 = (1/24)[\bar{3}\ 13\ 3]_{Fe}$. However, mismatching occurred for other linear combinations. Further, no distinction could be made between the 0, 1, 2, 3 and 0, 4, 3, 5 configurations, because equally good agreement to that in figure 7.18 was obtained between experimental and computed images using Burgers vectors based on the 0, 4, 3, 5 configuration. For the 0, 4, 3, 5 configuration acceptable matching occurred in the range of Burgers vectors from $\boldsymbol{y}_1 - \boldsymbol{x}_1 = (1/8)[\bar{1}11]_{Fe}$ to $3\boldsymbol{y}_1 - \boldsymbol{x}_1 = (1/8)[\bar{1}31]_{Fe}$, but mismatching resulted outside this range. It is concluded from these results that the Burgers vector of the dislocations in the β array is not a lattice vector, but is a vector resulting from a linear combination of the basis DSC_1 vectors in the plane of projection of figure 7.16(a) and is close to $2\boldsymbol{y}_1 - \boldsymbol{x}_1 = (1/8)[\bar{1}31]_{Fe}$.

As for case 1, the orientation of case 2 is illustrated by interpenetrating copper and iron lattices set in exact coincidence at the origin 0 (figure 7.16(b)). Again there are a number of pairs of lattice sites such as 1, 2, 3, 4 and 5 together with the pair of lattice sites $[\bar{3}0\bar{3}]_{Fe}$ and $[\bar{2}\bar{2}2]_{Cu}$ below the plane of the page which are close to coincidence. A pair of near-coincident cells, M_1 and M_2, is associated with the configuration 0, 1, 2, 3 in figure 7.16(b) and a different pair with the configuration 0, 1, 4, 5. For the configuration 0, 1, 2, 3 the corresponding unit cell of CCSL_1 is specified by the vectors $[\bar{3}0\bar{3}]_{Fe}/[\bar{2}\bar{2}2]_{CuM1}$, $\overrightarrow{01} = (1/2)[11\bar{1}]_{Fe}/(1/2)[\bar{1}0\bar{1}]_{CuM1}$ and $\overrightarrow{03} = [\bar{2}31]_{Fe}/(1/2)[\bar{3}43]_{CuM1}$. Thus, the matrix which re-indexes a vector indexed with respect to the iron lattice to indices with respect to the CuM1 lattice is

$$ {}_{\text{CuM1}}\mathbf{C}_{\text{Fe}} = (1/54)\begin{pmatrix} 7 & -32 & 29 \\ 4 & 28 & 32 \\ -43 & -4 & 7 \end{pmatrix}. \tag{7.10} $$

Basis vectors of the DSC_1 lattice are $\boldsymbol{x}_1 = (1/14)[\bar{1}21]_{Fe}$, $\boldsymbol{y}_1 = (1/28)[23\bar{2}]_{Fe}$ corresponding to the vectors $\boldsymbol{x}$ and $\boldsymbol{y}$ in figure 7.16(b), and the out-of-plane vector $\boldsymbol{z}_1 = (1/2)[\bar{1}0\bar{1}]_{Fe}$. Similarly the basis DSC_2 vectors are $\boldsymbol{x}_2 = (1/18)[\bar{1}21]_{Cu}$, $\boldsymbol{y}_2 = (1/54)[\bar{5}1\bar{4}]_{Cu}$ and the out-of-plane vector $\boldsymbol{z}_2 = (1/3)[\bar{1}\bar{1}1]_{Cu}$.

For the configuration 0, 1, 4, 5 the corresponding unit cell of CCSL_1 is specified by the vectors $[\bar{3}0\bar{3}]_{Fe}/[\bar{2}\bar{2}2]_{CuM1}$, $\overrightarrow{01} = (1/2)[11\bar{1}]_{Fe}/(1/2)[\bar{1}0\bar{1}]_{CuM1}$ and $\overrightarrow{05} = [010]_{Fe}/(1/2)[\bar{1}10]_{CuM1}$. Thus, the matrix which re-indexes a vector indexed with respect to the iron lattice to indices with respect to the CuM1 lattice is

$$ {}_{\text{CuM1}}\mathbf{C}_{\text{Fe}} = (1/12)\begin{pmatrix} 1 & -6 & 7 \\ 1 & 6 & 7 \\ -10 & 0 & 2 \end{pmatrix}. \tag{7.11} $$

Basis vectors of the DSC$_1$ lattice are the vectors $(1/6)[10\bar{1}]_{Fe}$, $(1/2)[010]_{Fe}$ and the out-of-plane vector $(1/2)[\bar{1}0\bar{1}]_{Fe}$. Similarly, the basis DSC$_2$ vectors are $(1/2)[\bar{1}\bar{1}2]_{Cu}$, $(1/4)[\bar{1}10]_{Cu}$ and $(1/3)[\bar{1}\bar{1}1]_{Cu}$.

For the α array of case 2, comparison between experimental and theoretical images for the identification of the Burgers vector was made in region 1 of figure 7.13(*b*), where the dislocations in the α array are coarsely spaced and those in the β array are sufficiently finely spaced for their contrast to be weak so that it does not interfere with that of the dislocations in the α array. As for case 1, images computed using the in-plane basis DSC vectors for the Burgers vectors of the dislocations in the α array were always too weak and images computed using lattice vectors in copper and iron did not match the experimental images. Agreement between experimental and theoretical images was only obtained for the Burgers vector $(1/2)[101]_{Fe}$, i.e. the basis out-of-plane DSC vector.

Figure 7.19 illustrates the type of comparison between experimental and computed images which was used to identify the Burgers vector of the dislocations in the α array. Row (i) of figure 7.18 is a series of experimental images, involving three non-coplanar diffracting vectors in copper and in iron, and row (ii) is the set of matching theoretical images computed for the Burgers vector $(1/2)[101]_{Fe}$. Rows (iii)–(viii) are computed images using lattice vectors in both copper and iron as possible Burgers vectors and it can be seen that none of these Burgers vectors results in computed images which match the experimental images. Again these particular mismatching images are presented because the Burgers vectors lie in the $(\bar{1}\bar{1}1)_{Cu}$ and $(\bar{1}0\bar{1})_{Fe}$ planes and are the type found in other work on FCC–BCC interfaces. Higher magnification examples, illustrating more of the detail of matching and mismatching images corresponding to some of the cases in column (*d*) of figure 7.19, are shown in figure 7.20. The experimental image in figure 7.20(*a*i) shows a portion of the region in figure 7.19(*d*i), and figure 7.20(*b*i) shows experimental images for some of the same dislocations in a region near the centre of the specimen (see figure 7.13(*b*)). The character of the experimental images of the dislocations, both near the surface and near the centre of the specimen, is that they are double with stronger and more continuous contrast at the top than at the bottom. This type of contrast is matched very well in figure 7.20(ii) for images computed with the Burgers vector $(1/2)[101]_{Fe}$, but not in figures 7.20, rows (iii), (iv) and (v) for images computed with the Burgers vectors $[101]_{Fe}$, $(1/2)[011]_{Cu}$ and $(1/2)[\bar{1}11]_{Fe}$ respectively. From comparisons of experimental and computed images of the type illustrated in figures 7.19 and 7.20 the Burgers vector of the dislocations in array α for case 2 was identified as $(1/2)[101]_{Fe}$.

For the β array of case 2, comparison of experimental and theoretical images for the identification of the Burgers vector was made in region 2 of figure 7.13(*b*), where the dislocations in the β array are coarsely spaced and where the spacing of the dislocations in the α array is extremely large. Images

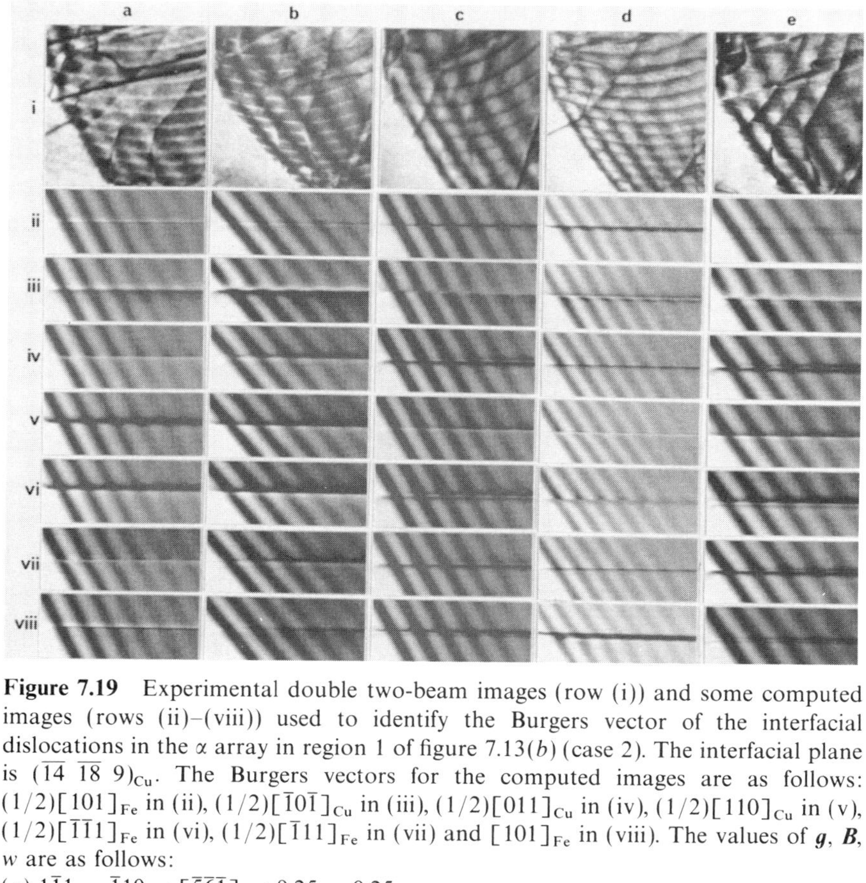

Figure 7.19 Experimental double two-beam images (row (i)) and some computed images (rows (ii)–(viii)) used to identify the Burgers vector of the interfacial dislocations in the α array in region 1 of figure 7.13(*b*) (case 2). The interfacial plane is $(\overline{14}\ \overline{18}\ 9)_{Cu}$. The Burgers vectors for the computed images are as follows: $(1/2)[101]_{Fe}$ in (ii), $(1/2)[\bar{1}0\bar{1}]_{Cu}$ in (iii), $(1/2)[011]_{Cu}$ in (iv), $(1/2)[110]_{Cu}$ in (v), $(1/2)[\bar{1}\bar{1}1]_{Fe}$ in (vi), $(1/2)[\bar{1}11]_{Fe}$ in (vii) and $[101]_{Fe}$ in (viii). The values of $\boldsymbol{g}$, $\boldsymbol{B}$, w are as follows:
(*a*) $1\bar{1}1_{Cu}$, $\bar{1}10_{Fe}$; $[\bar{5}6\bar{1}]_{Cu}$; 0.25_{Cu}, 0.25_{Fe}
(*b*) 200_{Cu}, $\bar{1}01_{Fe}$; $[0\bar{7}4]_{Cu}$; 0.24_{Cu}, 0.17_{Fe}
(*c*) $00\bar{2}_{Cu}$, $0\bar{1}1_{Fe}$; $[1\bar{9}0]_{Cu}$; 0.7_{Cu}, 0.1_{Fe}
(*d*) $00\bar{2}_{Cu}$, $\bar{1}\bar{1}0_{Fe}$; $[\bar{5}\bar{4}0]_{Cu}$; 0.7_{Cu}, 0.7_{Fe}
(*e*) $1\bar{1}\bar{1}_{Cu}$, $\bar{1}01_{Fe}$; $[\bar{2}\bar{7}5]_{Cu}$; 0.3_{Cu}, 0.3_{Fe}.

computed using lattice vectors for the Burgers vectors of the dislocations in the β array did not match the experimental images. Further, images computed using the in-plane basis DSC vectors corresponding to the configuration 0, 1, 2, 3 in figure 7.16(*b*) did not match the experimental images. For this configuration, equally good agreement between experimental and computed images was obtained for two Burgers vectors involving linear combinations of DSC_1 vectors $3\boldsymbol{y}_1 - 2\boldsymbol{x}_1 = (1/28)[10\ 1\ \overline{10}]_{Fe}$ and $2\boldsymbol{y}_1 - 2\boldsymbol{x}_1 = (1/28)[8\bar{2}\bar{8}]_{Fe}$. The type of comparison between experimental and theoretical images used

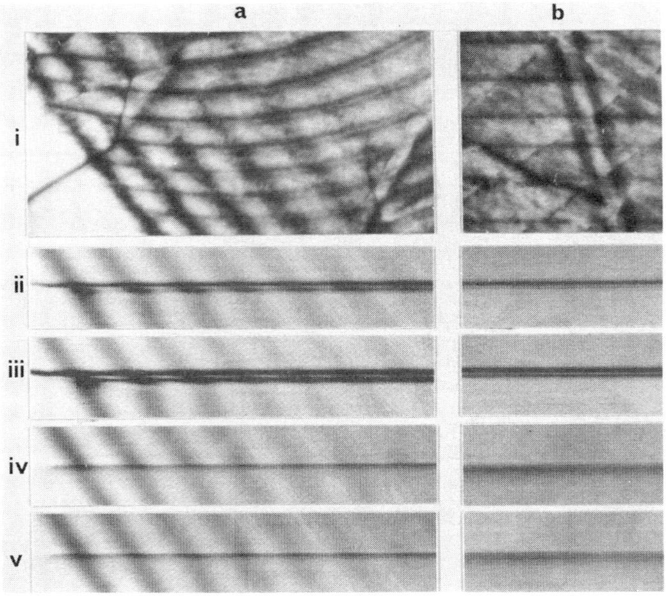

Figure 7.20 Experimental double two-beam images of dislocations in the α array of figure 7.19(*d*) at a higher magnification and at two different positions in the specimen (row (i)) with computed images for different Burgers vectors (rows (ii)–(v)). (*a*) corrresponds to a portion of region 1 of figure 7.13(*b*) near the surface of the specimen where the interface plane is $(\overline{14}\ \overline{18}\ 9)_{Cu}$ and (*b*) corresponds to a region near the centre of the specimen where the interface plane is $(\overline{14}\ \overline{29}\ 12)_{Cu}$. The Burgers vectors for the computed images are as follows: $(1/2)[101]_{Fe}$ in (ii), $[101]_{Fe}$ in (iii), $(1/2)[011]_{Cu}$ in (iv) and $(1/2)[\overline{1}11]_{Fe}$ in (v). The values of $\boldsymbol{g}$, $\boldsymbol{B}$ and w are: $00\overline{2}_{Cu}$, $\overline{1}\overline{1}0_{Fe}$; $[\overline{5}\overline{4}0]_{Cu}$; 0.7_{Cu}, 0.7_{Fe}.

to identify these Burgers vectors is shown in figure 7.21. Row (i) of figure 7.21 shows six experimental images for different combinations of diffracting vectors involving three non-coplanar diffracting vectors in iron and in copper. Row (ii) shows the computed images for the Burgers vector $(1/28)[10\ 1\ \overline{10}]_{Fe}$ and row (iii) the computed images for the Burgers vector $(1/28)[8\overline{2}\overline{8}]_{Fe}$. The agreement between the experimental and theoretical images for these two Burgers vectors is such that it is not possible to choose between them. For example, although in the comparison of experimental and theoretical images in figure 7.21, image (*b*ii) is a slightly better match than (*b*iii), (*a*iii) is a slightly better match than (*a*ii). However, mismatching between experimental and computed images did occur for Burgers vectors involving other linear combinations of DSC vectors. For example, the computed images in row (iv) are for the Burgers vector $\boldsymbol{y}_1 - 2\boldsymbol{x}_1 = (1/28)[6\overline{5}\overline{6}]_{Fe}$ and, for this vector, a

advances have been made in recent work. For example, n-beam lattice imaging is giving detailed information on the atomic structure at particular interphase interfaces, i.e. in those cases where the orientation relationships and interface planes allow n-beam lattice images to be set up simultaneously in both phases with the interface planes parallel to the beam direction. Moreover, recent two-beam electron microscopy, which does not require special orientation relationships or interface planes, has shown, for FCC/orthorhombic and FCC/BCC interfaces, that the Burgers vectors of the interfacial dislocations and the geometry of the dislocation arrays are consistent with a constrained coincidence model for the structure of interphase interfaces. However, more work of this type is required to establish whether this model is in fact generally applicable to interphase interfaces.

References

Amelinckx S and Dekeyser W 1959 *Solid State Phys.* **8** 325–499

Anstis G R and Thompson A L 1989 *Computer Simulation of Electron Microscope Diffraction and Images* ed. W Krakow and M A O'Keefe (Warrendale, PA: The Metallurgical Society)

Ashby M F, Spaepen F and Williams S 1978 *Acta Metall.* **26** 1647–63

Babcock S E and Balluffi R W 1987 *Phil. Mag.* **55** 643–53

Bacmann J-J, Papon A-M, Petit M and Silvestre G 1985 *Phil. Mag.* **51** 697–713

Bacmann J-J, Silvestre G and Petit M 1981 *Phil. Mag.* **43** 189–200

Baillin X, Pelissier J and Bacmann J-J 1987 *Phil. Mag.* A **55** 143–64

Ball C J 1981 *Phil. Mag.* A **44** 1307–17

Balluffi R W, Brokman A and King A H 1982 *Acta Metall.* **30** 1453–70

Balluffi R W, Komem Y and Schober T 1972a *Surf. Sci.* **31** 68–103

Balluffi R W, Sass S L and Schober T 1972b *Phil. Mag.* **26** 585–92

Balluffi R W and Schober T 1972 *Scripta Metall.* **6** 697–706

Balluffi R W, Woolhouse G R and Komem Y 1972c *The Nature and Behaviour of Grain Boundaries* ed. Hsun Hu (New York: Plenum) pp 41–69

Bamford T A, Hardiman B, Shen Z, Clark W A T and Wagoner R H 1986 *Scripta Metall.* **20** 253–8

Bäro G and Gleiter H 1973 *Acta Metall.* **21** 1405–8

Barry D E and Mahajan S 1971 *Phil. Mag.* **23** 727–9

Bilby B A 1955 *Defects in Crystalline Solids* (London: The Physical Society) pp 124–33

Bishop G H and Chalmers B 1968 *Scripta Metall.* **2** 133–9

—— 1971 *Phil. Mag.* **24** 515–26

Bollmann W 1967a *Phil. Mag.* **16** 363–81

—— 1967b *Phil. Mag.* **16** 383–99

—— 1970 *Crystal Defects and Crystalline Interfaces* (Berlin: Springer)

—— 1974 *Phys. Status Solidi* a **21** 543–50

—— 1981 *Phil. Mag.* **44** 991–1003

—— 1982 Crystal Lattices, Matrices (Bollmann)

—— 1984 *Phil. Mag.* **49** 73–9

Bollmann W, Michaut B and Sainfort G 1972 *Phys. Status Solidi* a **13** 637–49

Bonnet R 1981a *Phil. Mag.* **43** 1165–87
—— 1981b *Acta Metall.* **29** 437–45
—— 1982 *Acta Metall.* **30** 311–5
—— 1985 *J. Physique Coll.* C4 **46** 61–8
Bonnet R and Cousineau E 1977 *Acta Crystallogr.* A **33** 850–6
Bonnet R, Cousineau E and Warrington D H 1981 *Acta Crystallogr.* A **37** 184–9
Bonnet R and Durand F 1975a *Phil. Mag.* **32** 997–1006
—— 1975b *Scripta Metall.* **9** 935–9
Boothroyd C B, Crawley A P and Stobbs W M 1986 *Phil. Mag.* A **54** 663–77
Bouchard M, Livak R J and Thomas G 1972 *Surf. Sci.* **31** 275–95
Bragg W L 1940 *Proc. Phys. Soc.* **52** 105–9
Brandon D G 1966 *Acta Metall.* **14** 1479–84
Brandon D G, Ralph B, Ranganathan S and Wald M S 1964 *Acta Metall.* **12** 813–21
Breinan E M, Thompson E R and Tice W K 1972 *Trans. Met. Soc. AIME* **3** 211–9
Bristowe P D and Crocker A G 1975 *Phil. Mag.* **31** 503–17
Brokman A 1981 *Acta Crystallogr.* A **37** 500–6
Brokman A, Bristowe P D and Balluffi R W 1981 *Scripta Metall.* **15** 201–6
Bruggeman G A, Bishop G H and Hartt W H 1972 *The Nature and Behaviour of Grain Boundaries* ed. Hsun Hu (New York: Plenum) pp 83–122
Burgers J M 1940 *Proc. Phys. Soc.* **52** 23–33
Cantor B and Chadwick G A 1974 *J. Cryst. Growth* **23** 12–20
—— 1975 *J. Cryst. Growth* **30** 140–2
Carrington W E, Hale K F and McLean D 1960 *Proc. R. Soc.* A **259** 203–27
Carter C B, Donald A M and Sass S L 1979 *Phil. Mag.* **39** 533–49
Chalmers B and Gleiter H 1971 *Phil. Mag.* **23** 1541–6
Chen F-R and King A H 1987 *Acta Crystallogr.* B **43** 416–22
—— 1988 *Phil. Mag.* A **57** 431–55
Christian J W 1965 *The Theory of Transformations in Metals and Alloys* (Oxford: Pergamon)
—— 1976 *New Aspects of Martensitic Transformations* (Kobe: Japan Institute of Metals)
—— 1985 *Dislocations and Properties of Real Materials* (London: Institute of Metals) pp 94–124
Clarebrough L M 1969 *Aust. J. Phys.* **22** 559–67
—— 1974 *Phil. Mag.* **30** 1295–312
Clarebrough L M and Forwood C T 1978 *9th Int. Congr. on Electron Microscopy* (Toronto) vol III pp 380–90
—— 1980a *Phil. Mag.* A **41** 783–805
—— 1980b *Phys. Status Solidi* a **58** 597–607
—— 1980c *Phys. Status Solidi* a **59** 263–70
—— 1980d *Phys. Status Solidi* a **60** 51–7
—— 1980e *Phys. Status Solidi* a **60** 409–15
—— 1987a *Phil. Mag.* A **55** 217–25
—— 1987b *Phys. Status Solidi* a **104** 51–62
—— 1988 *Phys. Status Solidi* a **105** 131–8
Clark W A T and Smith D A 1978 *Phil. Mag.* A **38** 367–85
Cline H E, Walter J L, Koch E F and Osika L M 1971 *Acta Metall.* **19** 405–14
Cockayne D J H and Gronsky R 1981 *Phil. Mag.* A **44** 159–75

Cockayne D J H, Jenkins M L and Ray I L F 1971a *Phil. Mag.* **24** 1383–92
Cockayne D J H, Parsons J R and Hoelke C W 1971b *Phil. Mag.* **24** 139–53
Cockayne D J H, Ray I L F and Whelan M J 1969 *Phil. Mag.* **20** 1265–70
Cosandey F and Bauer C L 1981 *Phil. Mag.* A **44** 391–403
Cosandey F, Chan S W and Stadelmann P 1988 *Scripta Metall.* **22** 1093–6
Cosandey F, Komem Y, Bauer C L and Carter C B 1978 *Scripta Metall.* **12** 577–82
Cowley J M 1975 *Diffraction Physics* (Amsterdam: North Holland)
Cowley J M and Iijima S 1972 *Z. Naturf.* a **27** 445–51
Cowley J M and Moodie A F 1957 *Acta Crystallogr.* **10** 609–19
Cunningham B, Strunk H P and Ast D G 1982 *Scripta Metall.* **16** 349–52
D'Anterroches C and Bourret A 1984 *Phil. Mag.* A **49** 783–807
Darby T P and Balluffi R W 1977 *Phil. Mag.* **36** 53–66
Darby T P, Schlinder R and Balluffi R W 1978 *Phil. Mag.* A **37** 245–56
Das E S P and Dwarakadasa E S 1974 *J. Appl. Phys.* **45** 574–82
Dash S and Brown N 1963 *Acta Metall.* **11** 1067–75
Davis K G, Teghtsoonian E and Lu A 1966 *Acta Metall.* **14** 1677–84
DeRidder R, Van Landuyt J, Gevers R and Amelinckx S 1968 *Phys. Status Solidi* **30** 797–815
Dingley D J and McLean D 1967 *Acta Metall.* **15** 885–901
Dingley D J and Pond R C 1979 *Acta Metall.* **27** 667–82
Doyle P A and Turner P S 1968 *Acta Crystallogr.* A **24** 390–7
Ecob R C and Ralph B 1984 *Phil. Mag.* A **49** L19–24
Elkajbaji M and Thibault-Desseaux J 1988 *Phil. Mag.* A **58** 325–45
Erlings J G 1979 *A Study of Crystalline Interfaces by Means of Electron Diffraction and Transmission Electron Microscopy* (Delft University) pp 47–53
Erlings J G and Schapink F W 1977 *Scripta Metall.* **11** 427–9
—— 1978 *Phys. Status Solidi* a **46** 653–7
—— 1979 *Phys. Status Solidi* a **52** 529–39
Eshelby J D, Read W T and Shockley W 1953 *Acta Metall.* **1** 251–9
Fionova L K, Andreeva A V and Zhukova T I 1981 *Phys. Status Solidi* a **67** K15–9
Forwood C T and Clarebrough L M 1977 *Phil. Mag.* **36** 1131–45
—— 1978 *Phil. Mag.* A **37** 837–9
—— 1981 *Phil. Mag.* A **44** 31–41
—— 1982 *Acta Metall.* **30** 1443–51
—— 1983 *Phil. Mag.* A **47** L35–8
—— 1984 *Acta Metall.* **32** 757–71
—— 1985a *Aust. J. Phys.* **38** 449–69
—— 1985b *Phil. Mag.* A **51** 589–606
—— 1986a *Phil. Mag.* A **53** L31–4
—— 1986b *Phil. Mag.* A **53** 863–86
—— 1988 *Phys. Status Solidi* a **105** 365–75
—— 1989 *Phil. Mag.* B **59** 637–65
Forwood C T and Humble P 1975 *Phil. Mag.* **31** 1025–48
Frank F C 1950 *Plastic Deformation of Crystalline Solids* (Carnegie Institute of Technology) pp 150–1
Garg A, Clark W A T and Hirth J P 1989 *Phil. Mag.* A **59** 479–99
Garmong G and Rhodes C G 1974 *Acta Metall.* **22** 1373–82
—— 1975 *Trans. Met. Soc. AIME* **6** A 2209–16

Subject Index

Author Index